Methods in Molecular Biology

Series Editor
John M. Walker
School of Life and Medical Sciences
University of Hertfordshire
Hatfield, Hertfordshire, AL10 9AB, UK

For further volumes:
http://www.springer.com/series/7651

Th9 Cells

Methods and Protocols

Edited by

Ritobrata Goswami

School of Bio Science, Sir JC Bose Laboratory Complex,
Indian Institute of Technology, Kharagpur, West Bengal, India

Humana Press

Editor
Ritobrata Goswami
School of Bio Science
Sir JC Bose Laboratory Complex
Indian Institute of Technology
Kharagpur, West Bengal, India

ISSN 1064-3745 ISSN 1940-6029 (electronic)
Methods in Molecular Biology
ISBN 978-1-4939-6876-3 ISBN 978-1-4939-6877-0 (eBook)
DOI 10.1007/978-1-4939-6877-0

Library of Congress Control Number: 2017937939

Printed on acid-free paper

This Humana Press imprint is published by Springer Nature
The registered company is Springer Science+Business Media LLC
The registered company address is: 233 Spring Street, New York, NY 10013, U.S.A.

Preface

T helper cells that play an important role in adaptive immune response have a new member, Th9 cells. Th9 cells secrete IL-9, a pleiotropic cytokine having biological effects on distinct cell types. It has been more than 25 years since the cloning of IL-9. IL-9 can be produced by various cell types including long-term T cell lines and mast cells; however, T helper cells produce copious amounts of IL-9 in the presence of IL-4 and TGF-β. This discovery has propelled the study of Th9 cells with enthusiasm.

Over the last eight years, several studies have tried to optimize the conditions for the production of Th9 cells, transcriptional regulation of Th9 cells, and the *in vivo* function of Th9 cells. One of the goals of this book is to present comprehensive laboratory protocols that have been used to generate Th9 cells both *in vitro* and *in vivo*. Th9 cells have been ascribed to be involved in several diseases having both beneficial and detrimental roles. In this book techniques used to study the role of Th9 cells in various inflammatory diseases models have been described. This includes allergic inflammation model, parasite model, tumor model, and EAE and IBD model. This book will use the knowledge of expert scientists in the field to provide the reader with the laboratory techniques used to generate Th9 cells for specific downstream events.

West Bengal, India ***Ritobrata Goswami***

Contents

Preface . . . . *v*
Contributors . . . . *ix*

1 Th9 Cells: New Member of T Helper Cell Family 1
Ritobrata Goswami

2 IL-9: Function, Sources, and Detection 21
Wilmer Gerardo Rojas-Zuleta and Elizabeth Sanchez

3 IL-9 Signaling Pathway: An Update 37
Dijendra Nath Roy and Ritobrata Goswami

4 A Method to In Vitro Differentiate Th9 Cells from Mouse Naïve CD4+ T Cells 51
Duy Pham

5 T Cell Receptor and Co-Stimulatory Signals for Th9 Generation 59
Françoise Meylan and Julio Gomez-Rodriguez

6 Polarizing Cytokines for Human Th9 Cell Differentiation 73
Prabhakar Putheti

7 Determining the Frequencies of Th9 Cells from Whole Blood 83
Anuradha Rajamanickam and Subash Babu

8 IL-9 Production by Nonconventional T helper Cells 93
Silvia C.P. Almeida and Luis Graca

9 Prediction and Validation of Transcription Factors Binding Sites in the *Il9* Locus 111
William Orent and Wassim Elyaman

10 Flow Cytometric Assessment of STAT Molecules in Th9 Cells 127
Lucien P. Garo, Vanessa Beynon, and Gopal Murugaiyan

11 Transcription Factors Downstream of IL-4 and TGF-β Signals: Analysis by Quantitative PCR, Western Blot, and Flow Cytometry 141
Atsushi Sugimoto, Ryoji Kawakami, and Norihisa Mikami

12 Retroviral Transduction and Reporter Assay: Transcription Factor Cooperation in Th9 Cell Development 155
Rukhsana Jabeen

13 Transcription Factor Binding Studies in CD4+ T Cells: siRNA Transfection, Chromatin Immunoprecipitation, and Liquid Luminescent DNA Precipitation Assay 167
Etienne Humblin, François Ghiringhelli, and Frédérique Végran

14 Defining Epigenetic Regulation of the Interleukin-9 Gene by Chromatin Immunoprecipitation 179
Alla Skapenko and Hendrik Schulze-Koops

15 Allergic Inflammation and Atopic Disease: Role of Th9 Cells 189
Pornpimon Angkasekwinai

16 Characterization of Th9 Cells in the Development of EAE and IBD 201
Sakshi Malik, Valerie Dardalhon, and Amit Awasthi

17 B16 Lung Melanoma Model to Study the Role of Th9 Cells in Cancer 217
Alka Dwivedi, Sushant Kumar, and Rahul Purwar

18 Th9 Cells and Parasitic Inflammation: Use of *Nippostrongylus* and *Schistosoma* Models. 223
Miguel Enrique Serrano Pinto and Paula Licona-Limón

19 Isolation and Purification of Th9 Cells for the Study of Inflammatory Diseases in Research and Clinical Settings 247
Patricia Keating and James X. Hartmann

Index *257*

Contributors

SILVIA C.P. ALMEIDA • *Faculdade de Medicina, Instituto de Medicina Molecular, Universidade de Lisboa, Lisbon, Portugal; Instituto Gulbenkian de Ciencia, Oeiras, Portugal*

PORNPIMON ANGKASEKWINAI • *Department of Medical Technology, Faculty of Allied Health Sciences, Thammasat University, Pathumthani, Thailand; Graduate Program, Faculty of Allied Health Sciences, Thammasat University, Pathumthani, Thailand*

AMIT AWASTHI • *Center for Human Microbial Ecology (CHME), Translational Health Science & Technology Institute (THTI), Faridabad, Haryana, India*

SUBASH BABU • *National Institutes of Health - International Center for Excellence in Research, National Institute of Research in Tuberculosis (Formerly Tuberculosis Research Center), Chennai, India*

VANESSA BEYNON • *Ann Romney Center for Neurologic Diseases, Brigham and Women's Hospital and Harvard Medical School, Boston, MA, USA*

VALERIE DARDALHON • *Institut de Génétique Moléculaire de Montpellier, Centre National de la Recherche Scientifique UMR5535, Université de Montpellier, Montpellier, France*

ALKA DWIVEDI • *Department of Bioscience and Bioengineering, Indian Institute of Technology Bombay (IIT Bombay), Mumbai, Maharashtra, India*

WASSIM ELYAMAN • *Ann Romney Center for Neurologic Diseases, Brigham and Women's Hospital and Harvard Medical School, Boston, MA, USA; Program in Translational Neurogenomics and Neuroimmunology, Department of Neurology, Brigham and Women's Hospital, Harvard Medical School, Broad Institute at Harvard University and MIT, Boston, MA, USA*

LUCIEN P. GARO • *Ann Romney Center for Neurologic Diseases, Brigham and Women's Hospital and Harvard Medical School, Boston, MA, USA*

FRANÇOIS GHIRINGHELLI • *Université Bourgogne Franche-Comté, Dijon, France; Centre de Recherche INSERM LNC-UMR1231, Dijon, France; Plateforme de Transfert en Biologie Cancérologique, Centre GF Leclerc, Dijon, France*

JULIO GOMEZ-RODRIGUEZ • *Genetic Disease Research Branch, National Human Genome Research Institute, National Institutes of Health, Bethesda, MD, USA*

RITOBRATA GOSWAMI • *School of Bio Science, Sir JC Bose Laboratory Complex, Indian Institute of Technology, Kharagpur, West Bengal, India*

LUIS GRACA • *Faculdade de Medicina, Instituto de Medicina Molecular, Universidade de Lisboa, Lisbon, Portugal; Instituto Gulbenkian de Ciencia, Oeiras, Portugal*

JAMES X. HARTMANN • *Florida Atlantic University, Boca Raton, FL, USA*

ETIENNE HUMBLIN • *Université Bourgogne Franche-Comté, Dijon, France; Centre de Recherche INSERM LNC-UMR1231, Dijon, France*

RUKHSANA JABEEN • *HB Wells Center for Pediatric Research, Indiana School of Medicine, Indianapolis, IN, USA*

RYOJI KAWAKAMI • *Department of Experimental Immunology, Immunology Frontier Research Center, Osaka University, Osaka, Japan*

PATRICIA KEATING • *Florida Atlantic University, Boca Raton, FL, USA*

SUSHANT KUMAR • *Department of Bioscience and Bioengineering, Indian Institute of Technology Bombay (IIT Bombay), Mumbai, Maharashtra, India*
PAULA LICONA-LIMÓN • *Departamento de Biología Celular y del Desarrollo, Instituto de Fisiología Celular, Universidad Nacional Autónoma de México, Mexico City, Mexico*
SAKSHI MALIK • *Center for Human Microbial Ecology (CHME), Translational Health Science & Technology Institute (THTI), Faridabad, Haryana, India*
FRANÇOISE MEYLAN • *Immunoregulation Section, Autoimmunity Branch, National Institute of Arthritis and Musculoskeletal and Skin Diseases, National Institutes of Health, Bethesda, MD, USA*
NORIHISA MIKAMI • *Department of Experimental Immunology, Immunology Frontier Research Center, Osaka University, Osaka, Japan*
GOPAL MURUGAIYAN • *Ann Romney Center for Neurologic Diseases, Brigham and Women's Hospital and Harvard Medical School, Boston, MA, USA*
WILLIAM ORENT • *Ann Romney Center for Neurologic Diseases, Brigham and Women's Hospital and Harvard Medical School, Boston, MA, USA; Program in Translational Neurogenomics and Neuroimmunology, Department of Neurology, Brigham and Women's Hospital, Harvard Medical School, Broad Institute at Harvard University and MIT, Boston, MA, USA*
DUY PHAM • *Department of Pathology, University of Alabama at Birmingham, Birmingham, AL, USA*
MIGUEL ENRIQUE SERRANO PINTO • *Departamento de Biología Celular y del Desarrollo, Instituto de Fisiología Celular, Universidad Nacional Autónoma de México, Mexico City, Mexico*
RAHUL PURWAR • *Department of Bioscience and Bioengineering, Indian Institute of Technology Bombay (IIT Bombay), Mumbai, Maharashtra, India*
PRABHAKAR PUTHETI • *Department of Medicine, Weill-Cornell Medical College, New York, NY, USA*
ANURADHA RAJAMANICKAM • *National Institutes of Health - International Center for Excellence in Research, National Institute of Research in Tuberculosis (Formerly Tuberculosis Research Center), Chennai, India*
WILMER GERARDO ROJAS-ZULETA • *Department of Rheumatology, Universidad de Antioquia, Medellín, Colombia*
DIJENDRA NATH ROY • *Department of Bioengineering, National Institute of Technology, Jirania, NIT-Agartala, Tripura, India*
ELIZABETH SANCHEZ • *Department of Physiology, Universidad Nacional de Colombia, Bogotá, Colombia*
HENDRIK SCHULZE-KOOPS • *Division of Rheumatology and Clinical Immunology, Department of Medicine IV, University of Munich, Munich, Germany*
ALLA SKAPENKO • *Division of Rheumatology and Clinical Immunology, Department of Medicine IV, University of Munich, Munich, Germany*
ATSUSHI SUGIMOTO • *Department of Experimental Immunology, Immunology Frontier Research Center, Osaka University, Osaka, Japan*
FRÉDÉRIQUE VÉGRAN • *Université Bourgogne Franche-Comté, Dijon, France; Centre de Recherche INSERM LNC-UMR1231, Dijon, France; Plateforme de Transfert en Biologie Cancérologique, Centre GF Leclerc, Dijon, France*

Chapter 1

Th9 Cells: New Member of T Helper Cell Family

Ritobrata Goswami

Abstract

T Helper cells (CD4+ T cells) constitute one of the key arms of adaptive immune responses. Differentiation of naïve CD4+ T cells into multiple subsets ensure a proper protection against multitude of pathogens in immunosufficient individual. After differentiation, T helper cells secrete specific cytokines that are critical to provide immunity against various pathogens. The recently discovered Th9 cells secrete the pleiotropic cytokine, IL-9. Although IL-9 was cloned more than 25 years ago and characterized as a Th2 cell-specific cytokine, not many studies were carried out to define the function of IL-9. This cytokine has been demonstrated to act on multiple cell types as a growth factor. After the discovery of Th9 cells as an abundant source of IL-9, renewed focus has been generated. In this chapter, I discuss the biology and development of IL-9-secreting Th9 cells. Furthermore, I highlight the role of Th9 cells and IL-9 in health and diseases.

Key words Th9 cells, IL-9, Transcription factors, Epigenetic modification, Allergic inflammation, Autoimmune disorder, Tumor immunity, Helminthic infection

1 Introduction

An adaptive immune response begins when a naïve CD4+ T cell interacts with an antigen presenting cell with a nonself peptide in the context of class II MHC molecule. Following this interaction, the naïve CD4+ T cell differentiates into distinct T helper subsets. Differentiation into distinct T helper subset would depend on cytokines present in the microenvironment and each of these subsets would express their signature cytokines. The newest member of the ever growing T helper subset is the interleukin-9 (henceforth to be known as IL-9) secreting T helper cells, also known as Th9 cells. T helper cells are characterized by their distinct functions. Th1 cells are responsible to fight against intracellular pathogens, Th2 cells provide immunity against extracellular parasites, while Th17 cells mediate immunity against fungal infections and extracellular bacteria. Even though IL-9 was cloned almost three decades back, we have started unraveling the factors that control the expression and function of the gene recently. The cytokine microenvironment leading to the production of IL-9 by mouse CD4+ T cells was first

Ritobrata Goswami (ed.), *Th9 Cells: Methods and Protocols*, Methods in Molecular Biology, vol. 1585,
DOI 10.1007/978-1-4939-6877-0_1, © Springer Science+Business Media LLC 2017

described by Schmitt et al. in 1994 [1]. Naïve CD4+ T cells are primed into IL-9-secreting Th9 cells in the presence of the cytokines TGF-β and IL-4 that is augmented by IL-2. The cytokine IL-4 is required for the development of Th2 cells [2]. IL-9-secreting T cells were initially thought to be a part of Th2 responses in vivo. Seminal studies by Veldhoen et al. and Dardalhon et al. demonstrated that in the presence of TGF-β, Th2 cells can be polarized into IL-9-producing T cells [3, 4]. Other cytokines including IL-25, IL-1β, IL-6, IL-10, IL-12, and IL-21 can aid in IL-9 production by human Th9 cells [5, 6]. Type I interferons can induce the expression of IL-9 via IL-21 [5]. However, type II interferon, IFN-γ inhibits the production of IL-9 [1]. Several studies have given insight on the transcriptional regulation of Th9 cells [7]. It is argued that TGF-β might be responsible for secretion of IL-9 by opposite T helper cells including Treg and Th17 cells. Further studies are going on to understand how different factors regulate the beneficial and detrimental functions of Th9 cells.

We have started appreciating the role of IL-9 and Th9 cells very recently. This chapter reviews the biology of IL-9, the development and in vivo function of Th9 cells. Furthermore, I incorporate the current view of the role of Th9 cells in different inflammatory diseases.

2 Biology of IL-9

Characterized as mast cell and T cell growth factor, IL-9 was cloned in 1989 [8]. Initially known as P40, the growth factor was later renamed as IL-9 based on its effect on myeloid and lymphoid cells [9, 10]. Human *IL9* gene maps to chromosome 5, while mouse *Il9* gene maps to chromosome 13 [11]. T lymphocytes including antigen-specific T cells, naïve mouse T cells and long term T cells are the key source of IL-9 [12]. IL-9 signals via IL-9 receptor (IL-9R) that has two subunits: the IL-9Rα chain and common γ chain [13]. Signal transduction of IL-9 leads to activation of multiple STAT molecules including STAT1, STAT3, and STAT5 and MAPK and IRS-PI3K pathway [14–16]. T Cell lines and effector T cells including Th2 and Th17 cells express IL-9R [17, 18].

IL-9, a pleiotropic cytokine, can impart its effect on multiple cell types. Acting as a growth factor on T cells, IL-9 can act on CD4+ T cells including Th2 and Th17 cells [19]. IL-9 specifically acts on B1 B cells and enhances IL-4-mediated IgE and IgG production from human B cells [19]. Growth of mast cells is evident in the presence of IL-9 and stem cell factor [20]. IL-9 also regulates hematopoiesis [21]. IL-9 can act on both airway smooth muscle cells and airway epithelial cells leading to enhanced production of cytokine and goblet cell metaplasia, respectively [22, 23].

3 Transcriptional Regulation of Th9 Cells

Transcription factors play a crucial role in the development of T helper cells. Master regulators and signal transducers and activators of transcription (STAT) molecules are the essential players in the development of Th cells [24]. Master regulators were thought to be necessary and sufficient for the development of each specific T helper cell. However, these transcription factors are presently referred to as lineage-specific transcription factors due to plasticity of T helper cells [25]. When the phenotype of IL-9-secreting Th cells were first reported, there were no studies stating the transcription factors required for the development of Th9 cells. The transcription factors downstream of IL-4 and TGF-β signals are now well described. However, the question remained about the transcription factors that would work downstream of both these cytokines and lead to the transcription and optimal expression of *Il9* gene. Studies thus far have not been able to identify any 'master regulator' for Th9 cells. This section of the chapter reviews the present understanding of transcriptional and epigenetic regulation of *Il9* gene in Th9 cells.

3.1 In search of "Master Regulator" of Th9 Cells

Development of T helper cells is associated with a "master regulator." T-bet acts as master regulator of Th1 cells, while GATA3 acts master regulator of Th2 cells. Activin A, a member of TGF-β superfamily has also been shown to drive Th9 cell differentiation [26]. In 2008 when the reports of IL-9-secreting T helper cells corroborated with the findings of Schmitt et al. in 1994, investigators began to identify the transcription factor network that govern Th9 cell differentiation. TGF-β can suppress the production of Th2-specific cytokines including IL-4. The transcription factor PU.1 (a member of ETS family of transcription factors) is shown to regulate Th2 heterogeneity by interfering with GATA3-DNA binding [27]. It was therefore thought that PU.1 might positively regulate the development of Th9 cells. Indeed, T cell specific deletion of PU.1 lead to impaired IL-9 production both in vitro and in vivo [28]. The expression of *Sfpi1* (PU.1 encoding gene) was higher in Treg cells compared to Th9 cells with some expression in Th2 and Th17 cells [29]. Since TGF-β signal is required for the development of both Treg and Th9 cells, it was hypothesized that TGF-β regulates the expression of PU.1. There was TGF-β dose dependent induction of PU.1, while altered IL-4 dose did not change the expression of PU.1. The study by Chang et al. was one of the first studies to describe that the transcription factor PU.1 is required for Th9 cell development [28]. In the absence of PU.1 in murine CD4+ T cells, there is impaired production of IL-9, while overexpression of PU.1 in developing Th9 cells enhances IL-9 production [28]. PU.1 binds to the *Il9* promoter [28]. The importance of PU.1 was also demonstrated using CD4+ T cells from PBMCs and skewing under Th9-cell conditions [28]. PU.1 acts downstream of TGF-β signal in Th9 cells [29]. A recent study has reported that Etv5, another transcription factor and member of ETS family, functions along with PU.1 for Th9 cell development. Similar to PU.1, Etv5 recruits histone acetyltransferase to bind to the *Il9* gene locus to promote IL-9 production [30]. The Smad family of transcription factors has been shown to regulate T helper cell differentiation [31]. Smad2 and Smad4 acting downstream of TGF-β have been demonstrated to be important for Th9 cell development, even though neither of the factors regulate the expression of PU.1 [32]. Both Smad2 and Smad3 bind to the *Il9* promoter [33]. Notch protein plays an important role in Smad3 binding to *Il9* gene [34].

It is therefore a fine tuning between transcription factors downstream of both TGF-β and IL-4 signals that is required for the optimum expression of IL-9 in Th9 cells. A slew of transcription factors work downstream of the IL-4/STAT6 signal that regulate Th9 cell development. Naïve T cells from Gata3-null mice polarized under Th9 differentiating conditions did not secrete IL-9 [4]. However, the role of GATA3 is not understood completely as over expression of this factor in Th9 cells attenuate the production of IL-9 [29]. It is suggested that GATA3 might reduce the expression of the transcription factor Foxp3 that would otherwise inhibit the production of IL-9.

The transcription factor IRF4, a target of IL-4 in Th2 cells, is also required for Th9 cell development [35]. Naïve CD4+ T cells from IRF4-deficient mice have impaired production of IL-9 during Th9 polarization, while siRNA-mediated IRF4 silencing attenuates the production of IL-9 [35]. IRF4 binds to the *Il9* gene directly and overexpression of IRF4 in developing Th9 cells leads to enhanced production of IL-9 [35, 36]. Ectopic expression of IRF4 in the absence of IL-4/STAT6 signaling failed to rescue IL-9 production suggesting that other transcription factors downstream of IL-4 is required for Th9 cell development. Not only IRF4 induces the expression of *Il9* gene, it also blocks the expression of transcription factors that negatively regulate IL-9 production from Th9 cells. IRF4 also plays a crucial role in the development of Th17 cells suggesting that IRF4 is important for the development of multiple T helper cells. IRF4 also forms a complex with Smad2 and Smad3, while Smad and IRF4 requirement is reciprocal with respect to binding to the *Il9* promoter [33]. The Tec family of cytosolic tyrosine kinsase, Itk is required for Th9 cell differentiation [37]. Itk induces the expression of IRF4 in Th9 cells [37]. In human Th9 cells the positive role of Itk has been demonstrated [37].

The transcription factor BATF, an AP1 family transcription factor and downstream of IL-4, is also required for Th9 cell development [36]. BATF regulated the expression of IL-9 and Th9-associated genes in both human and murine systems [36]. Both BATF and IRF4 work together to promote Th9 cell development as co-transduction of BATF and IRF4 resulted in enhanced IL-9 production in either BATF-deficient or IRF4-deficient Th9 cells greater than produced in wild-type control transduced cells [36]. In the absence of BATF, the expression of *Irf4*, *Gata3*, and *Maf* is attenuated [36]. BATF, like IRF4 not only is required for the development multiple T helper cells but also binds directly to the *Il9* gene [36]. BATF-transgenic mice or ectopic expression of BATF in Th9 cells leads to the enhanced production of IL-9 [36]. In contrast to IRF4, BATF regulates the expression of majority of Th9-associated genes [36]. It is thought that BATF and IRF4 could form a transcriptional module in Th9 cells akin to Th17 cells [38].

Ubiquitously expressed transcription factors also play an important role in the development of Th9 cells. OX40, a receptor expressed on T cells, induces Th9 cell development via noncanonical NF-κB pathway even though canonical NF-kB pathway is also activated by OX40 [39]. The absence of PU.1 does not alter the expression of OX40 [39]. Nuclear factor of activated T cells 1 (NFAT1), regulated by calcium acts in tandem with other transcription factors to regulate gene expression in various T helper cells [40]. In the absence of NFAT1, IL-9 production is significantly impaired [41]. Reconstituting NFAT1 in NFAT1-deficient Th9 cells rescues IL-9 production [41]. Glucocorticoid-induced tumor necrosis factor receptor (TNFR)-related protein (GITR), which shows antitumor activity, enhances Th9 cell differentiation in a NF-κB-dependent fashion by shifting from inducible Treg cells toward IL-9-producing cells [42]. The transcription factor Id3 is downregulated in the presence of TGF-β and IL-4 [43]. The kinase TAK1 regulates the expression of Id3 [43]. In the absence of Id3, the binding of Gata3 to the *Il9* promoter region is enhanced, leading to augmented *Il9* transcription [43]. In human CD4+ T cells, Id3 regulates IL-9 production [43]. Thus, the development of Th9 cells require a coordinated interaction of transcription factors downstream of both IL-4 and TGF-β signal.

3.2 STAT Molecules in Th9 Cell Development

STAT6 is a critical molecule in the differentiation of Th2 cells. Since IL-4 is a common cytokine required for the development of both Th2 and Th9 cells, the role of STAT6 in Th9 cells development has been determined. In the absence of STAT6 signal, IL-9 production was severely impaired [29]. However, the level of activated STAT6 was unaltered in the absence of Itk [37]. The transcription factor and master regulator of Th2 cells, GATA3, though expressed at a lower level in Th9 cells, is a STAT6 target gene. However, in the absence of STAT6 there was unaltered expression of PU.1 in Th9 cells [29].

In addition to STAT6, the role of other STAT molecules has been investigated in Th9 cell development. Though STAT3 is activated and required for Th2 cells, STAT3 is dispensable for Th9 cell development [29]. Cytokine signals including IL-6 that activate STAT3 negatively regulate IL-9 production from Th9 cells in STAT3-dependent manner [44]. However, other STAT molecules could play an important role in Th9 cell development. IL-2 activated STAT5 is required for Th9 cell development [45]. In IL-2-deficient T helper cells differentiated under Th9 polarizing conditions, external IL-2 can recover IL-9 production [1, 45]. STAT5 binds to and activates the *Il9* promoter [46]. STAT5 binds to the *Irf4* promoter, while IL-2 restores IRF4 expression [37]. In addition to IL-2, the cytokine TSLP also activates STAT5 molecule. It has been demonstrated that TSLP induces the expression of activated STAT5 during Th9 cell differentiation leading to increased IL-9 production from Th9 cells [47]. TL1A, a TNF superfamily member, increases the production of IL-9 from Th9 cells presumably via IL-2/STAT5 pathway independently of STAT6 and PU.1 [48].

The transcription factor Bcl6, which is the master regulator of T follicular helper cells, is expressed at lower level in Th9 cells and binds to the *Il9* promoter competing with STAT5 [45, 46]. Ectopic expression of Bcl6 attenuates the production of IL-9 in Th9 cells, while IL-2/STAT5 signaling suppresses Bcl6 expression by binding to the *Il9* gene [45, 46]. Blocking IL-2/STAT5 signaling leads to attenuated expression of Bcl6. STAT3 acts as a negative regulator of STAT5 activation during Th9 cell development [44]. Ectopic expression of constitutively active STAT5 in developing Th9 cells excludes the ability of IL-6 to reduce IL-9 production [44].

STAT1 also regulates the development of Th9 cells. Activation of STAT1 plays an important role in Th1 cell differentiation by IFN-γ and in autoimmune disease such as inflammatory bowel disease [49, 50]. Activation of STAT1 occurs via different cytokines including type I and II IFNs, IL-6, IL-21, and IL-27 [51]. When Th9 cells are treated in the presence of IFN-γ, the production of IL-9 is reduced associated with activation of STAT1 molecule [1]. In contrast, in the absence of TYK2 (molecule required for activating STAT1) there is increased IL-9 production [52]. IFN-γ inhibits the development of Th9 cells via IL-27 (produced by dendritic cells) that is partially dependent on STAT1 and T-bet [53]. Other studies point toward a positive role of STAT1 in the development of Th9 cells. The transcription factor IRF1 augments Th9 effector cell function with IL-1β inducing the phosphorylation of STAT1 molecules [54]. STAT1 activation in Th9 cells is mediated by the tyrosine kinase Fyn [54]. A similar observation has also been found out in human Th9 cells. Human Th9 cells cultured in the presence of STAT1-activating cytokines lead to enhanced IL-9 production [5]. These results suggest a complex role of activated STAT1 molecule in Th9 cell development. STAT4, the key STAT molecule involved in Th1 cell development also inhibit IL-9 production by Th9 cells [29].

3.3 Epigenetics and Th9 Cell Differentiation

Transcription factors are affected by various epigenetic regulation that govern the differentiation of T helper cells [55]. In Th9 cell development, multiple transcription factors control the *Il9* gene locus epigenetically. Multiple conserved noncoding sequences (CNS) have been identified in *Il9* gene. CNS1 is located at the *Il9* promoter, while CNS2, conserved between mouse and human is located 5kb downstream of the *Il9* promoter [56]. CNS0, a third regulatory region has been identified ~6kb upstream of the *Il9* promoter [56]. Acetylation of total H3 and H4 and specific histone modification of H3K9 and H3K18 were demonstrated to be highest in Th9 cells at both CNS1 and CNS2 [28]. In contrast, Th9 cells had the lowest amount of trimethylated H3K27, a negative chromatin modification among other T helper cells [28]. Total H3 acetylation was attenuated in the absence of PU.1 in Th9 cells, while the level of H3K27 trimethylation remained unchanged [28]. PU.1 deficiency leads to reduced amount of specific histone acetylation marks at CNS0 and CNS1 of *Il9* gene in Th9 cells associated with diminished binding of histone acetyl transferases Gcn5 and PCAF [57]. PU.1 associates with Gcn5 and inhibition of Gcn5 leads to impaired IL-9 production [57]. In contrast, an enhanced binding of histone deacetylases was observed at the *Il9* gene in the absence of PU.1 [57]. Th9 cell differentiation is also regulated by chromatin modifications of PU.1 [58]. In naïve T cells, PU.1 promoter has restrictive chromatin marks compared to memory T cells that limits Th9 cell differentiation [58]. There is increased accessibility of PU.1 promoter during the transition of naïve T cells to memory T cells mirrored by less intense stimulation required for Th9 cell differentiation [58]. Downstream of TGF-β, Smad molecules regulate histone marks of *Il9* gene in Th9 cells. Smad2-deficient Th9 cells have significantly reduced total acetylation of H3 and H4 as well as trimethylation of H3K4 [32, 33]. Smad2 and Smad3 interact and transactivate the *Il9* promoter with IRF4 [33]. Smad3 also cooperates with Notch and RBP-Jκ (molecule downstream of TGF-β) to transactivate the *Il9* promoter. There is increased acetylation of H3 and H4 at Smad3 and RBP-Jκ sites [34]. Concomitantly, there is also enhanced permissive H3K4 monomethylation and reduced restrictive H3K27 trimethylation at Smad3 and RBP-Jκ sites in the *Il9* promoter [34].

Transcription factors downstream of IL-4 signaling and required for Th9 cell differentiation also modify *Il9* gene epigenetically. IRF4 binds to the *Il9* promoter to maintain IL-9 production in Th9 cells [35]. Both BATF and IRF4 bind in abundance to the *Il9* gene in Th9 cells compared to Th2 cells [36]. Binding of either of the transcription factors to the *Il9* promoter is attenuated in the reciprocal gene-deficient cells. However, BATF is not required for PU.1 binding to the *Il9* gene in Th9 cells [36]. IRF4 is required for BATF binding to its target genes in Th9 cells [36].

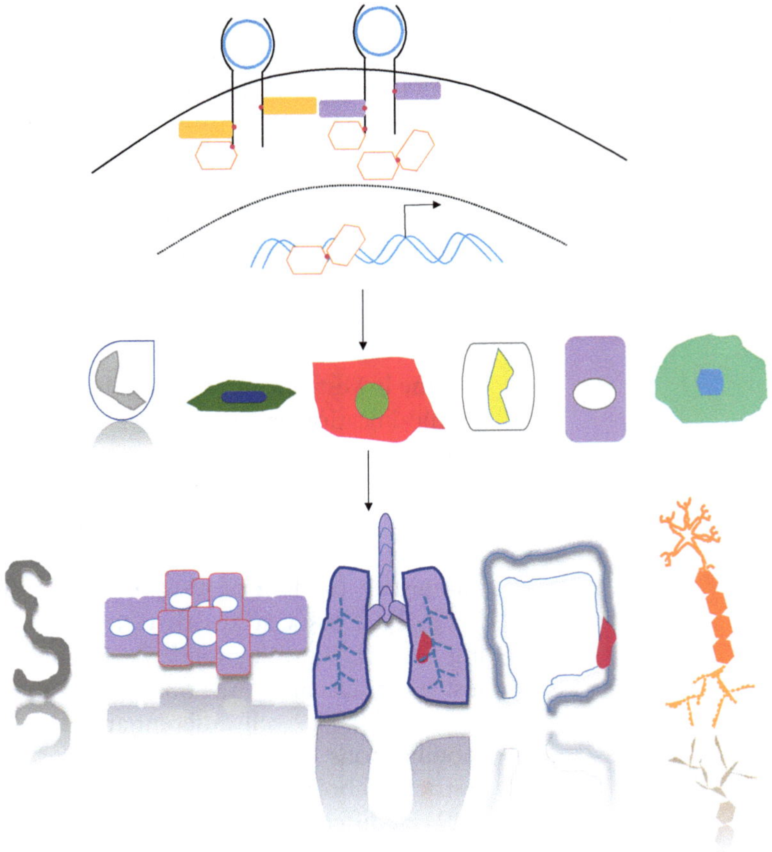

Fig. 1 Naïve CD4+ T cells in the presence of IL-4 and TGF-β secrete copious amount of IL-9 that is aided by transcription factors downstream of polarizing cytokines. IL-9 can act on multiple cells including mast cells, B cells, eosinophils, smooth muscle cells, and airway epithelial cells. The effect of Th9 cells on these cells has a key role in having either protective or detrimental effects on multiple diseases including allergic inflammation, helminthic infection, tumor immunity, IBD, and EAE

3.4 Th9 Cells in Health and Diseases

As is evident in other T helper cells, Th9 cells have been demonstrated to have roles in vivo, thereby having ramifications in health and diseases both directly and indirectly (Fig. 1). Though a major source of IL-9, Th9 cells are not the only source of IL-9, making it difficult to ascertain the in vivo function of Th9 cells. Given below are the roles of Th9 cells in human health.

3.4.1 Tumor Immunity

Several studies have highlighted the antitumor property of Th9 cells. The ability of Th9 cells to limit tumor growth has been attributed to cytokines including IL-9. Melanoma patients have decreased number of Th9 cells in blood and skin compared to healthy volunteers [59]. There is enhanced risk of cutaneous malignant melanoma associated with IL-9 SNP [60]. In a lung adenocarcinoma model, IL-9 produced by Th9 cells play a protective role [59]. In a B16 melanoma model, blocking IL-9 induced tumor growth [61]. Adoptive transfer of tumor-specific Th9 cells led to production of CCL20 resulting in the recruitment of DCs to promote tumor cell destruction [61]. These Th9 cells elicit a strong tumor-specific CD8+ cytotoxic T cell response that is Ccr6-dependent [61]. The chemokine CCL20 could induce the recruitment of Th9 cells into the lung [62]. When tumor-reactive CD8+ T cells are cultured under Th9 polarizing conditions and adoptively transferred in recipient mice, they provided better antitumor responses against large tumors that depend on the generation of IL-9 in vivo [62]. Mast cells play an important role in IL-9-mediated antitumor activity [59]. There is increased survival of DCs when they are co-cultured with Th9 cells [64]. When DCs are primed by Th9 cells they promote robust antitumor responses in B16 lung melanoma model of mice due to secretion of IL-3, but not IL-9 by Th9 cells [64]. Furthermore, IL-1β induces cytokine production from Th9 cells, which when adoptively transferred into mice having lung melanoma or adenocarcinoma leads to strong antitumor activity [54]. Mechanistically this antitumor response depends on Th9-specific expression of IL-21 and the transcription factor IRF1 [54]. IL-9 also prevents cell growth of HTB-72 melanoma cell line via upregulation of p21 and TRAIL [65]. In CT26-colon cancer cells, tumor-specific Th2 cells convert Tregs into Th9 cells providing potent growth inhibition [66]. The co-stimulatory GITR expressed on T cells enhances Th9 cell differentiation while promoting tumor-specific cytotoxic responses that is IL-9-dependent [42]. GITR also triggers Th9 differentiation under iTreg conditions by chromatin remodeling at both *Foxp3* and *Il9* gene loci [67]. The transcription factor Id3 plays a role in the polarization of Th9 cells. In the absence of Id3 there is increased Th9 cell differentiation, *Il9* expression and increased antitumor responses in mice [43]. However, IL-9 and Th9 cells can aid tumor cell proliferation in human studies. In malignant pleural effusion Th9 cells are increased and IL-9 promotes lung tumor cell proliferation and migration [68]. IL-9 and Th9 cells may initiate antitumor responses against solid tumors but not against non-solid tumors. Patients with Hodgkin lymphoma and anaplastic lymphoma have a strong expression of *IL9* [69]. That IL-9 aids in the proliferation of tumor cells and inhibition of apoptosis of tumor cells has been observed in in vitro studies [70]. Hepatocellular carcinoma patients (HCC) have increased frequencies of IL-9-producing Th9 cells compared to healthy volunteers. However, HCC patients with enhanced Th9 cells also had signifi-canlty reduced disease-free survival period [71] IL-9 can promote tumor formation in a T lymphoblastic lymphoma mouse model [72]. Whether IL-9 could be used as a therapeutic to treat cancer would depend on the type of cancer and expression of molecules in the milieu.

3.4.2 Helminthic Infection

Parasitic or helminthic infection affect a lot of people worldwide, especially in the developing countries. T helper cells play an important role in the disease pathophysiology, and recent studies have indicated that Th9 cells may play a critical function in fighting the infection. T-Cell specific TGF-β receptor type 2 deficient mice display reduced IL-9 and diminished mast cell numbers associated with significantly higher worm burden in *Trichuris muris* infection model [3]. However, patients with lymphatic filariasis demonstrate increased antigen-specific Th9 cells [73]. Th9 cells limited *Nippostrongylus brasiliensis* worm burden when adoptively transferred to RAG-deficient mice that is associated with increased numbers of eosinophils, basophils, and mast cells [74]. IL-9 promotes ILC2 survival which could lead to Th9 cell-mediated increased ILC2 numbers and activity [74, 75]. The relative contribution of Th9 cells and ILC2 to IL-9 production in helminthic infection is not well understood. Using a transcriptional reporter mice, one group has demonstrated that mainly CD4+ T cells are producer of IL-9 in GI tract during *N. brasiliensis* infection [74]. When adoptively transferred into *Il9*-deficient mice, Th9 cells attenuated the adverse effects associated with worm expulsion [74]. However, in pulmonary *N. brasiliensis* model using a IL-9 fate reporter mice, ILC2s were the dominant IL-9 producers [76]. IL-9 is indispensable to clear worm during *N. brasiliensis* and *T. muris* infection [77, 78]. Transgenic infection of IL-9 provides resistance against *T. muris* infections [79]. However, IL-9 may be dispensable for expulsion of *Trichenella spiralis* [80, 81]. Therefore, IL-9 may not be critical to protect from all parasitic infections. Reports indicate that parasite antigen-specific Th9 cells positively correlates with disease severity [73]. Even though the role of IL9/Th9 cells has been investigated using animal models not much is known about the role of Th9 cells in human helminth infections. One study has reported that *Strongyloidis stercoralis* infected individuals have increased frequencies of Th9 cells which are induced by IL-10 and TGF-β [82]. Further studies are required to have a better understanding of the role of Th9 cells during worm infection.

3.4.3 EAE

Th1 and Th17 cells play a critical role in the development of EAE, the animal model of multiple sclerosis [83]. Th9 cells have also been shown to be important for the development of EAE that depends on IL-9 [53]. IL-9 enhances chemokine expression in astrocytes [84]. The phenotype of neuroinflammation is different when Th9 cells are adoptively transferred compared to recipients of either Th1 or Th17 cells [85]. Mice given proteolipid protein peptide ($PLP_{180\text{-}199}$) to induce EAE demonstrate the presence of Th9 cells in draining lymph nodes and CNS [86]. Increased expression of IL-9R is observed in astrocytes, microglia, and oligodendrocytes in mice during EAE [84]. RAG-deficient mice developed EAE when MOG-specific T cells polarized under Th9 cell differentiation conditions were adoptively transferred [53, 85]. However, IL-9 may play a protective role against EAE by enhancing the function of Tregs [87]. Sometimes Th9 cells could also secrete IL-17 and IFN-γ, thereby promoting EAE severity [85]. Use of non-adoptive transfer model could unravel the function of Th9 cells and IL-9 in multiple sclerosis.

3.4.4 Inflammatory Bowel Disease

Chronic inflammation of the GI tract characterize IBD that has two distinct forms: ulcerative colitis (UC) and Crohn's disease (CD). IBD patients have increased number of Th17 cells with UC patients having predominantly Th2 cytokines while CD patients have Th1 cytokines [88]. Th9 cell specific transcription factors are expressed in patients with IBD. Both PU.1 and IRF4+ T cells are observed in GI tract of IBD patients [89, 90]. Psoriatic arthritis patients have significantly enhanced expression of IL-9 and differentiated Th9 cells in inflamed gut [91]. In vitro derived Th9 cells when adoptively transferred, resulted in increased development of colitis in RAG-deficient mice [4]. This effect was IL-9-dependent [89]. IL-9-deficient and IRF4-deficient mice display reduced colitis score in oxazolone-induced colitis model [89, 92]. Th9 cells and IL-9 modulate epithelial cells by either modifying composition of tight junction protein or by altering epithelial cell proliferation, thereby contributing to IBD [89]. In a TNBS-colitis model, IL-9 modulates intestinal epithelial cells by altering the expression of tight junction proteins [93]. Interestingly, IL-9R expression is enhanced in GI epithelial cells of UC patients and in paneth cells of psoriatic arthritis patients [89, 91, 94]. IL-9 effect on epithelial cells could be indirect as mast cells play an important role of antigen-induced anaphylaxis model [95]. In DSS-induced colitis mice model, neutralization of IL-9 ameliorated the disease severity [96]. However, the role of Th9 cells and IL-9 is not properly understood as disease development is limited by IL-9 in Th1 cell-mediated colitis model that resembles some signatures of CD [97, 98]. Therefore, Th9 cells and IL-9 could play a protective role in IBD depending on the inflammatory microenvironment.

3.4.5 Allergic Disorder

Th2 cytokines and IgE-mediated immediate hypersensitivity characterize allergic disorders including atopic dermatitis and asthma [99, 100]. In humans Th9 cell signature molecules have been associated with the development of asthma and other allergies [7]. Atopic patients expressed the Th9 cell-related proteins including IL-17RB, IRF4, and PU.1 by IL-9-secreting T cells [101, 102]. Itk is required for IRF4 expression in Th9 cells, while mice deficient in Itk is protected from papain-induced lung damage [37]. Circulating T cells have the propensity to secrete IL-9 when activated with pollen, dander in allergic patients [7]. Allergic asthma patients have activated STAT6 and PU.1 in Th9 cells in late phase airway inflammation compared to healthy controls [103]. Th9 cell number and IL-9 production are enhanced significantly in atopic children and patients suffering from atopic dermatitis and psoriasis [104, 105]. In atopic patients, serum IL-9 and Th9 cell numbers positively correlate with allergen-specific IgE titers [26, 105, 106]. Via IL-9 production, Th9 cells mediate atopic disease in mice model. Th9 cells are present in the draining lymph nodes and respiratory tract in mouse model of asthma [26, 86]. Blocking Th9 cell polarization by neutralizing either TGF-β or activin A (a TGF-β family member) attenuates the progress of allergic disorder [26]. Using IL-9 fate reporter mouse, it has been demonstrated that the major source of IL-9 in vivo is Th9 cells during allergic airway inflammation [76]. Expression of IL-9 and level of Th9 cells are enhanced when the cytokines IL-25 and TSLP are expressed during mouse model of asthma [6, 47]. Transcription factors required for the development of Th9 cells play a crucial role in the development of allergic inflammation. Mice having CD4+ T cell-specific deletion of either PU.1 or IRF4 have reduced allergic inflammation [28, 35]. This phenotype is also observed in BATF-deficient mice [36]. The roles of Th2 and Th9 cells are different as PU.1-deficient mice have diminished Th9 cell differentiation and attenuated allergic inflammation in the OVA sensitization model but normal Th2 cell development [28].

When Th9 cells are adoptively transferred accumulation of mast cells, eosinophils and mucus production is observed [7]. However, these effects are dependent of IL-9 as adoptive transfer of Th9 cells in IL-9-deficient mice or in mice where IL-9 was neutralized fail to promote the accumulation of mast cells and eosinophils as well as bronchial hyperresponsiveness [35, 36, 47]. Mast cell proliferation and in vivo activation is IL-9-dependent [74, 107]. IL-9 neutralization or PU.1 deficiency within the T cell compartment attenuates mucus hyperplasia, mast cell accumulation, and lung remodeling and airway hyperreactivity in a house dust mite-induced asthma model [108, 109]. Transgenic expression of IL-9 is sufficient in itself to cause bronchial hyperresponsiveness via its effects on the respiratory epithelium and the enhancement of Th2-type cytokine release. Th9 cells are therefore capable of triggering allergic inflammation in allergic transfer models.

4 Conclusion

What was initially thought to be a cytokine cloned more than 25 years ago and produced by Th2 cells, IL-9 has indeed come a long way. Despite being characterized as growth factor, IL-9 was not paid enough attention to ascertain its other biological functions. After it has been demonstrated that T helper cells in the presence of TGF-β and IL-4 produce abundant level of IL-9, scientific community has renewed focus on IL-9 biology. Several studies have identified the transcription network that govern the development of Th9 cells, yet no "master regulator" has been identified with PU.1 being the closest one. Other studies have revealed that apart from Th9 cells other T helper and non-T helper cells can produce IL-9. The function of Th9 cells has been a work in progress in various animal and human studies. Th9 cells impart both protective and harmful effects in our body. Th9 cells play an important role against tumors and mount protective immune responses against helminths. In contrast, Th9 cells may be responsible for allergic inflammation and distinct phenotype of autoimmune disorders. Whether IL-9 produced by Th9 cells can be targeted as therapeutic has been studied. Humanized neutralizing IL-9 antibody MEDI-528 has been used as medical intervention and has shown hope to treat mild to moderate asthma. However, further clinical trials are required to underscore the potential of targeting IL-9 as a therapeutic. Th9 cells have been suggested to be involved in the pathogenesis of Takayasu's arteritis and patients with immune thrombocytopenia. Future studies will reveal the role of IL-9 and Th9 cells in other diseases.

References

1. Schmitt E, Germann T, Goedert S, Hoehn P, Huels C, Koelsch S, Kuhn R, Muller W, Palm N, Rude E (1994) IL-9 production of naive CD4+ T cells depends on IL-2, is synergistically enhanced by a combination of TGF-beta and IL-4, and is inhibited by IFN-gamma. J Immunol 153(9):3989–3996
2. Mosmann TR, Cherwinski H, Bond MW, Giedlin MA, Coffman RL (1986) Two types of murine helper T cell clone. I Definition according to profiles of lymphokine activities and secreted proteins. J Immunol 136(7):2348–2357
3. Veldhoen M, Uyttenhove C, van Snick J, Helmby H, Westendorf A, Buer J, Martin B, Wilhelm C, Stockinger B (2008) Transforming growth factor-beta 'reprograms' the differentiation of T helper 2 cells and promotes an interleukin 9-producing subset. Nat Immunol 9(12):1341–1346. doi:10.1038/ni.1659
4. Dardalhon V, Awasthi A, Kwon H, Galileos G, Gao W, Sobel RA, Mitsdoerffer M, Strom TB, Elyaman W, Ho IC, Khoury S, Oukka M, Kuchroo VK (2008) IL-4 inhibits TGF-beta-induced Foxp3+ T cells and, together with TGF-beta, generates IL-9+ IL-10+ Foxp3(−) effector T cells. Nat Immunol (12):1347–1355. doi:10.1038/ni.1677
5. Wong MT, Ye JJ, Alonso MN, Landrigan A, Cheung RK, Engleman E, Utz PJ (2010) Regulation of human Th9 differentiation by type I interferons and IL-21. Immunol Cell Biol 88(6):624–631. doi:10.1038/icb.2010.53
6. Angkasekwinai P, Chang SH, Thapa M, Watarai H, Dong C (2010) Regulation of IL-9 expression by IL-25 signaling. Nat Immunol 11(3):250–256. doi:10.1038/ni.1846
7. Kaplan MH, Hufford MM, Olson MR (2015) The development and in vivo function of T helper 9 cells. Nat Rev Immunol 15(5):295–307. doi:10.1038/nri3824
8. Van Snick J, Goethals A, Renauld JC, Van Roost E, Uyttenhove C, Rubira MR, Moritz RL, Simpson RJ (1989) Cloning and characterization of a cDNA for a new mouse T cell growth factor (P40). J Exp Med 169(1):363–368
9. Uyttenhove C, Simpson RJ, Van Snick J (1988) Functional and structural characterization of P40, a mouse glycoprotein with T-cell growth factor activity. Proc Natl Acad Sci U S A 85(18):6934–6938
10. Yang YC, Ricciardi S, Ciarletta A, Calvetti J, Kelleher K, Clark SC (1989) Expression cloning of cDNA encoding a novel human hematopoietic growth factor: human homologue of murine T-cell growth factor P40. Blood 74(6):1880–1884
11. Mock BA, Krall M, Kozak CA, Nesbitt MN, McBride OW, Renauld JC, Van Snick J (1990) IL9 maps to mouse chromosome 13 and human chromosome 5. Immunogenetics 31(4):265–270

12. Schmitt E, Van Brandwijk R, Van Snick J, Siebold B, Rude E (1989) TCGF III/P40 is produced by naive murine CD4+ T cells but is not a general T cell growth factor. Eur J Immunol 19(11):2167–2170. doi:10.1002/eji.1830191130
13. Renauld JC, Druez C, Kermouni A, Houssiau F, Uyttenhove C, Van Roost E, Van Snick J (1992) Expression cloning of the murine and human interleukin 9 receptor cDNAs. Proc Natl Acad Sci U S A 89(12):5690–5694
14. Bauer JH, Liu KD, You Y, Lai SY, Goldsmith MA (1998) Heteromerization of the gammac chain with the interleukin-9 receptor alpha subunit leads to STAT activation and prevention of apoptosis. J Biol Chem 273(15):9255–9260
15. Demoulin JB, Uyttenhove C, Van Roost E, DeLestre B, Donckers D, Van Snick J, Renauld JC (1996) A single tyrosine of the interleukin-9 (IL-9) receptor is required for STAT activation, antiapoptotic activity, and growth regulation by IL-9. Mol Cell Biol 16(9):4710–4716
16. Demoulin JB, Louahed J, Dumoutier L, Stevens M, Renauld JC (2003) MAP kinase activation by interleukin-9 in lymphoid and mast cell lines. Oncogene 22(12):1763–1770. doi:10.1038/sj.onc.1206253
17. Cosmi L, Liotta F, Angeli R, Mazzinghi B, Santarlasci V, Manetti R, Lasagni L, Vanini V, Romagnani P, Maggi E, Annunziato F, Romagnani S (2004) Th2 cells are less susceptible than Th1 cells to the suppressive activity of CD25+ regulatory thymocytes because of their responsiveness to different cytokines. Blood 103(8):3117–3121. doi:10.1182/blood-2003-09-3302
18. Nowak EC, Weaver CT, Turner H, Begum-Haque S, Becher B, Schreiner B, Coyle AJ, Kasper LH, Noelle RJ (2009) IL-9 as a mediator of Th17-driven inflammatory disease. J Exp Med 206(8):1653–1660. doi:10.1084/jem.20090246
19. Goswami R, Kaplan MH (2011) A brief history of IL-9. J Immunol 186(6):3283–3288. doi:10.4049/jimmunol.1003049
20. Matsuzawa S, Sakashita K, Kinoshita T, Ito S, Yamashita T, Koike K (2003) IL-9 enhances the growth of human mast cell progenitors under stimulation with stem cell factor. J Immunol 170(7):3461–3467
21. Holbrook ST, Ohls RK, Schibler KR, Yang YC, Christensen RD (1991) Effect of interleukin-9 on clonogenic maturation and cell-cycle status of fetal and adult hematopoietic progenitors. Blood 77(10):2129–2134
22. Gounni AS, Hamid Q, Rahman SM, Hoeck J, Yang J, Shan L (2004) IL-9-mediated induction of eotaxin1/CCL11 in human airway smooth muscle cells. J Immunol 173(4):2771–2779
23. Temann UA, Geba GP, Rankin JA, Flavell RA (1998) Expression of interleukin 9 in the lungs of transgenic mice causes airway inflammation, mast cell hyperplasia, and bronchial hyperresponsiveness. J Exp Med 188(7):1307–1320
24. O'Shea JJ, Lahesmaa R, Vahedi G, Laurence A, Kanno Y (2011) Genomic views of STAT function in CD4+ T helper cell differentiation. Nat Rev Immunol 11(4):239–250. doi:10.1038/nri2958
25. Oestreich KJ, Weinmann AS (2012) Master regulators or lineage-specifying? Changing views on CD4+ T cell transcription factors. Nat Rev Immunol 12(11):799–804. doi:10.1038/nri3321
26. Jones CP, Gregory LG, Causton B, Campbell GA, Lloyd CM (2012) Activin A and TGF-beta promote T(H)9 cell-mediated pulmonary allergic pathology. J Allergy Clin Immunol 129(4):1000–1010 . doi:10.1016/j.jaci.2011.12.965e1003
27. Chang HC, Zhang S, Thieu VT, Slee RB, Bruns HA, Laribee RN, Klemsz MJ, Kaplan MH (2005) PU.1 expression delineates heterogeneity in primary Th2 cells. Immunity 22(6):693–703. doi:10.1016/j.immuni.2005.03.016
28. Chang HC, Sehra S, Goswami R, Yao W, Yu Q, Stritesky GL, Jabeen R, McKinley C, Ahyi AN, Han L, Nguyen ET, Robertson MJ, Perumal NB, Tepper RS, Nutt SL, Kaplan MH (2010) The transcription factor PU.1 is required for the development of IL-9-producing T cells and allergic inflammation. Nat Immunol 11(6):527–534. doi:10.1038/ni.1867
29. Goswami R, Jabeen R, Yagi R, Pham D, Zhu J, Goenka S, Kaplan MH (2012) STAT6-dependent regulation of Th9 development. J Immunol 188(3):968–975. doi:10.4049/jimmunol.1102840
30. Koh B, Hufford MM, Pham D, Olson MR, Wu T, Jabeen R, Sun X, Kaplan MH (2016) The ETS Family Transcription Factors Etv5 and PU.1 Function in Parallel To Promote Th9 Cell Development. J Immunol 197(6):2465–2472
31. Lu L, Wang J, Zhang F, Chai Y, Brand D, Wang X, Horwitz DA, Shi W, Zheng SG (2010) Role of SMAD and non-SMAD signals in the development of Th17 and regulatory T cells. J Immunol 184(8):4295–4306. doi:10.4049/jimmunol.0903418
32. Wang A, Pan D, Lee YH, Martinez GJ, Feng XH, Dong C (2013) Cutting edge: Smad2 and Smad4 regulate TGF-beta-mediated Il9

gene expression via EZH2 displacement. J Immunol 191(10):4908–4912. doi:10.4049/jimmunol.1300433
33. Tamiya T, Ichiyama K, Kotani H, Fukaya T, Sekiya T, Shichita T, Honma K, Yui K, Matsuyama T, Nakao T, Fukuyama S, Inoue H, Nomura M, Yoshimura A (2013) Smad2/3 and IRF4 play a cooperative role in IL-9-producing T cell induction. J Immunol 191(5):2360–2371. doi:10.4049/jimmunol.1301276
34. Elyaman W, Bassil R, Bradshaw EM, Orent W, Lahoud Y, Zhu B, Radtke F, Yagita H, Khoury SJ (2012) Notch receptors and Smad3 signaling cooperate in the induction of interleukin-9-producing T cells. Immunity 36(4):623–634. doi:10.1016/j.immuni.2012.01.020
35. Staudt V, Bothur E, Klein M, Lingnau K, Reuter S, Grebe N, Gerlitzki B, Hoffmann M, Ulges A, Taube C, Dehzad N, Becker M, Stassen M, Steinborn A, Lohoff M, Schild H, Schmitt E, Bopp T (2010) Interferon-regulatory factor 4 is essential for the developmental program of T helper 9 cells. Immunity 33(2):192–202. doi:10.1016/j.immuni.2010.07.014
36. Jabeen R, Goswami R, Awe O, Kulkarni A, Nguyen ET, Attenasio A, Walsh D, Olson MR, Kim MH, Tepper RS, Sun J, Kim CH, Taparowsky EJ, Zhou B, Kaplan MH (2013) Th9 cell development requires a BATF-regulated transcriptional network. J Clin Invest 123(11):4641–4653. doi:10.1172/JCI69489
37. Gomez-Rodriguez J, Meylan F, Handon R, Hayes ET, Anderson SM, Kirby MR, Siegel RM, Schwartzberg PL (2016) Itk is required for Th9 differentiation via TCR-mediated induction of IL-2 and IRF4. Nat Commun 7:10857. doi:10.1038/ncomms10857
38. Schraml BU, Hildner K, Ise W, Lee WL, Smith WA, Solomon B, Sahota G, Sim J, Mukasa R, Cemerski S, Hatton RD, Stormo GD, Weaver CT, Russell JH, Murphy TL, Murphy KM (2009) The AP-1 transcription factor Batf controls T(H)17 differentiation. Nature 460(7253):405–409. doi:10.1038/nature08114
39. Xiao X, Balasubramanian S, Liu W, Chu X, Wang H, Taparowsky EJ, Fu YX, Choi Y, Walsh MC, Li XC (2012) OX40 signaling favors the induction of T(H)9 cells and airway inflammation. Nat Immunol 13(10):981–990. doi:10.1038/ni.2390
40. Hermann-Kleiter N, Baier G (2010) NFAT pulls the strings during CD4+ T helper cell effector functions. Blood 115(15):2989–2997. doi:10.1182/blood-2009-10-233585
41. Jash A, Sahoo A, Kim GC, Chae CS, Hwang JS, Kim JE, Im SH (2012) Nuclear factor of activated T cells 1 (NFAT1)-induced permissive chromatin modification facilitates nuclear factor-kappaB (NF-kappaB)-mediated interleukin-9 (IL-9) transactivation. J Biol Chem 287(19):15445–15457. doi:10.1074/jbc.M112.340356
42. Kim IK, Kim BS, Koh CH, Seok JW, Park JS, Shin KS, Bae EA, Lee GE, Jeon H, Cho J, Jung Y, Han D, Kwon BS, Lee HY, Chung Y, Kang CY (2015) Glucocorticoid-induced tumor necrosis factor receptor-related protein co-stimulation facilitates tumor regression by inducing IL-9-producing helper T cells. Nat Med 21(9):1010–1017. doi:10.1038/nm.3922
43. Nakatsukasa H, Zhang D, Maruyama T, Chen H, Cui K, Ishikawa M, Deng L, Zanvit P, Tu E, Jin W, Abbatiello B, Goldberg N, Chen Q, Sun L, Zhao K, Chen W (2015) The DNA-binding inhibitor Id3 regulates IL-9 production in CD4(+) T cells. Nat Immunol 16(10):1077–1084. doi:10.1038/ni.3252
44. Olson MR, Verdan FF, Hufford MM, Dent AL, Kaplan MH (2016) STAT3 impairs STAT5 activation in the development of IL-9-secreting T cells. J Immunol. doi:10.4049/jimmunol.1501801
45. Liao W, Spolski R, Li P, Du N, West EE, Ren M, Mitra S, Leonard WJ (2014) Opposing actions of IL-2 and IL-21 on Th9 differentiation correlate with their differential regulation of BCL6 expression. Proc Natl Acad Sci U S A 111(9):3508–3513. doi:10.1073/pnas.1301138111
46. Bassil R, Orent W, Olah M, Kurdi AT, Frangieh M, Buttrick T, Khoury SJ, Elyaman W (2014) BCL6 controls Th9 cell development by repressing Il9 transcription. J Immunol 193(1):198–207. doi:10.4049/jimmunol.1303184
47. Yao W, Zhang Y, Jabeen R, Nguyen ET, Wilkes DS, Tepper RS, Kaplan MH, Zhou B (2013) Interleukin-9 is required for allergic airway inflammation mediated by the cytokine TSLP. Immunity 38(2):360–372. doi:10.1016/j.immuni.2013.01.007
48. Richard AC, Tan C, Hawley ET, Gomez-Rodriguez J, Goswami R, Yang XP, Cruz AC, Penumetcha P, Hayes ET, Pelletier M, Gabay O, Walsh M, Ferdinand JR, Keane-Myers A, Choi Y, O'Shea JJ, Al-Shamkhani A, Kaplan MH, Gery I, Siegel RM, Meylan F (2015) The TNF-family ligand TL1A and its receptor DR3 promote T cell-mediated allergic immunopathology by enhancing differentiation and pathogenicity of IL-9-producing T cells. J Immunol 194(8):3567–3582. doi:10.4049/jimmunol.1401220

49. Szabo SJ, Sullivan BM, Peng SL, Glimcher LH (2003) Molecular mechanisms regulating Th1 immune responses. Annu Rev Immunol 21:713–758. doi:10.1146/annurev.immunol.21.120601.140942
50. Schreiber S, Rosenstiel P, Hampe J, Nikolaus S, Groessner B, Schottelius A, Kuhbacher T, Hamling J, Folsch UR, Seegert D (2002) Activation of signal transducer and activator of transcription (STAT) 1 in human chronic inflammatory bowel disease. Gut 51(3):379–385
51. Khodarev NN, Roizman B, Weichselbaum RR (2012) Molecular pathways: interferon/stat1 pathway: role in the tumor resistance to genotoxic stress and aggressive growth. Clin Cancer Res 18(11):3015–3021. doi:10.1158/1078-0432.CCR-11-3225
52. Ubel C, Graser A, Koch S, Rieker RJ, Lehr HA, Muller M, Finotto S (2014) Role of Tyk-2 in Th9 and Th17 cells in allergic asthma. Sci Rep 4:5865. doi:10.1038/srep05865
53. Murugaiyan G, Beynon V, Pires Da Cunha A, Joller N, Weiner HL (2012) IFN-gamma limits Th9-mediated autoimmune inflammation through dendritic cell modulation of IL-27. J Immunol 189(11):5277–5283. doi:10.4049/jimmunol.1200808
54. Vegran F, Berger H, Boidot R, Mignot G, Bruchard M, Dosset M, Chalmin F, Rebe C, Derangere V, Ryffel B, Kato M, Prevost-Blondel A, Ghiringhelli F, Apetoh L (2014) The transcription factor IRF1 dictates the IL-21-dependent anticancer functions of TH9 cells. Nat Immunol 15(8):758–766. doi:10.1038/ni.2925
55. Wilson CB, Rowell E, Sekimata M (2009) Epigenetic control of T-helper-cell differentiation. Nat Rev Immunol 9(2):91–105. doi:10.1038/nri2487
56. Perumal NB, Kaplan MH (2011) Regulating Il9 transcription in T helper cells. Trends Immunol 32(4):146–150. doi:10.1016/j.it.2011.01.006
57. Goswami R, Kaplan MH (2012) Gcn5 is required for PU.1-dependent IL-9 induction in Th9 cells. J Immunol 189(6):3026–3033. doi:10.4049/jimmunol.1201496
58. Ramming A, Druzd D, Leipe J, Schulze-Koops H, Skapenko A (2012) Maturation-related histone modifications in the PU.1 promoter regulate Th9-cell development. Blood 119(20):4665–4674. doi:10.1182/blood-2011-11-392589
59. Purwar R, Schlapbach C, Xiao S, Kang HS, Elyaman W, Jiang X, Jetten AM, Khoury SJ, Fuhlbrigge RC, Kuchroo VK, Clark RA, Kupper TS (2012) Robust tumor immunity to melanoma mediated by interleukin-9-producing T cells. Nat Med 18(8):1248–1253. doi:10.1038/nm.2856
60. Yang XR, Pfeiffer RM, Wheeler W, Yeager M, Chanock S, Tucker MA, Goldstein AM (2009) Identification of modifier genes for cutaneous malignant melanoma in melanoma-prone families with and without CDKN2A mutations. Int J Cancer 125(12):2912–2917. doi:10.1002/ijc.24622
61. Lu Y, Hong S, Li H, Park J, Hong B, Wang L, Zheng Y, Liu Z, Xu J, He J, Yang J, Qian J, Yi Q (2012) Th9 cells promote antitumor immune responses in vivo. J Clin Invest 122(11):4160–4171. doi:10.1172/JCI65459
62. Bu XN, Zhou Q, Zhang JC, Ye ZJ, Tong ZH, Shi HZ (2013) Recruitment and phenotypic characteristics of interleukin 9-producing CD4+ T cells in malignant pleural effusion. Lung 191(4):385–389. doi:10.1007/s00408-013-9474-4
63. Lu Y, Hong B, Li H, Zheng Y, Zhang M, Wang S, Qian J, Yi Q (2014) Tumor-specific IL-9-producing CD8+ Tc9 cells are superior effector than type-I cytotoxic Tc1 cells for adoptive immunotherapy of cancers. Proc Natl Acad Sci U S A 111(6):2265–2270. doi:10.1073/pnas.1317431111
64. Park J, Li H, Zhang M, Lu Y, Hong B, Zheng Y, He J, Yang J, Qian J, Yi Q (2014) Murine Th9 cells promote the survival of myeloid dendritic cells in cancer immunotherapy. Cancer Immunol Immunother 63(8):835–845. doi:10.1007/s00262-014-1557-4
65. Fang Y, Chen X, Bai Q, Qin C, Mohamud AO, Zhu Z, Ball TW, Ruth CM, Newcomer DR, Herrick EJ, Nicholl MB (2015) IL-9 inhibits HTB-72 melanoma cell growth through upregulation of p21 and TRAIL. J Surg Oncol 111(8):969–974. doi:10.1002/jso.23930
66. Liu JQ, Li XY, Yu HQ, Yang G, Liu ZQ, Geng XR, Wang S, Mo LH, Zeng L, Zhao M, Fu YT, Sun HZ, Liu ZG, Yang PC (2015) Tumor-specific Th2 responses inhibit growth of CT26 colon-cancer cells in mice via converting intratumor regulatory T cells to Th9 cells. Sci Rep 5:10665. doi:10.1038/srep10665
67. Xiao X, Shi X, Fan Y, Zhang X, Wu M, Lan P, Minze L, Fu YX, Ghobrial RM, Liu W, Li XC (2015) GITR subverts Foxp3(+) Tregs to boost Th9 immunity through regulation of histone acetylation. Nat Commun 6:8266. doi:10.1038/ncomms9266
68. Ye ZJ, Zhou Q, Yin W, Yuan ML, Yang WB, Xiong XZ, Zhang JC, Shi HZ (2012) Differentiation and immune regulation of

IL-9-producing CD4+ T cells in malignant pleural effusion. Am J Respir Crit Care Med 186(11):1168–1179. doi:10.1164/rccm.201207-1307OC

69. Merz H, Houssiau FA, Orscheschek K, Renauld JC, Fliedner A, Herin M, Noel H, Kadin M, Mueller-Hermelink HK, Van Snick J et al (1991) Interleukin-9 expression in human malignant lymphomas: unique association with Hodgkin's disease and large cell anaplastic lymphoma. Blood 78(5):1311–1317
70. Chen N, Lu K, Li P, Lv X, Wang X (2014) Overexpression of IL-9 induced by STAT6 activation promotes the pathogenesis of chronic lymphocytic leukemia. Int J Clin Exp Pathol 7(5):2319–2323
71. Tan H, Wang S, Zhao L (2016) A tumor-promoting role of Th9 cells in hepatocellular carcinoma through CCL20 and STAT3 pathways. Clin Exp Pharmacol Physiol Oct 31. doi:10.1111/1440 1681.12689
72. Lange K, Uckert W, Blankenstein T, Nadrowitz R, Bittner C, Renauld JC, van Snick J, Feller AC, Merz H (2003) Overexpression of NPM-ALK induces different types of malignant lymphomas in IL-9 transgenic mice. Oncogene 22(4):517–527. doi:10.1038/sj.onc.1206076
73. Anuradha R, George PJ, Hanna LE, Chandrasekaran V, Kumaran P, Nutman TB, Babu S (2013) IL-4-, TGF-beta-, and IL-1-dependent expansion of parasite antigen-specific Th9 cells is associated with clinical pathology in human lymphatic filariasis. J Immunol 191(5):2466–2473. doi:10.4049/jimmunol.1300911
74. Licona-Limon P, Henao-Mejia J, Temann AU, Gagliani N, Licona-Limon I, Ishigame H, Hao L, Herbert DR, Flavell RA (2013) Th9 Cells Drive Host Immunity against Gastrointestinal Worm Infection. Immunity 39(4):744–757. doi:10.1016/j.immuni.2013.07.020
75. Turner JE, Morrison PJ, Wilhelm C, Wilson M, Ahlfors H, Renauld JC, Panzer U, Helmby H, Stockinger B (2013) IL-9-mediated survival of type 2 innate lymphoid cells promotes damage control in helminth-induced lung inflammation. J Exp Med 210(13):2951–2965. doi:10.1084/jem.20130071
76. Wilhelm C, Hirota K, Stieglitz B, Van Snick J, Tolaini M, Lahl K, Sparwasser T, Helmby H, Stockinger B (2011) An IL-9 fate reporter demonstrates the induction of an innate IL-9 response in lung inflammation. Nat Immunol 12(11):1071–1077. doi:10.1038/ni.2133
77. Khan WI, Richard M, Akiho H, Blennerhasset PA, Humphreys NE, Grencis RK, Van Snick J, Collins SM (2003) Modulation of intestinal muscle contraction by interleukin-9 (IL-9) or IL-9 neutralization: correlation with worm expulsion in murine nematode infections. Infect Immun 71(5):2430–2438
78. Richard M, Grencis RK, Humphreys NE, Renauld JC, Van Snick J (2000) Anti-IL-9 vaccination prevents worm expulsion and blood eosinophilia in Trichuris muris-infected mice. Proc Natl Acad Sci U S A 97(2):767–772
79. Faulkner H, Renauld JC, Van Snick J, Grencis RK (1998) Interleukin-9 enhances resistance to the intestinal nematode Trichuris muris. Infect Immun 66(8):3832–3840
80. Angkasekwinai P, Srimanote P, Wang YH, Pootong A, Sakolvaree Y, Pattanapanyasat K, Chaicumpa W, Chaiyaroj S, Dong C (2013) Interleukin-25 (IL-25) promotes efficient protective immunity against Trichinella spiralis infection by enhancing the antigen-specific IL-9 response. Infect Immun 81(10):3731–3741. doi:10.1128/IAI.00646-13
81. Faulkner H, Humphreys N, Renauld JC, Van Snick J, Grencis R (1997) Interleukin-9 is involved in host protective immunity to intestinal nematode infection. Eur J Immunol 27(10):2536–2540. doi:10.1002/eji.1830271011
82. Anuradha R, Munisankar S, Bhootra Y, Jagannathan J, Dolla C, Kumaran P, Nutman TB, Babu S (2016) IL-10- and TGFbeta-mediated Th9 Responses in a Human Helminth Infection. PLoS Negl Trop Dis 10(1):e0004317. doi:10.1371/journal.pntd.0004317
83. Kebir H, Kreymborg K, Ifergan I, Dodelet-Devillers A, Cayrol R, Bernard M, Giuliani F, Arbour N, Becher B, Prat A (2007) Human TH17 lymphocytes promote blood-brain barrier disruption and central nervous system inflammation. Nat Med 13(10):1173–1175. doi:10.1038/nm1651
84. Ding X, Cao F, Cui L, Ciric B, Zhang GX, Rostami A (2015) IL-9 signaling affects central nervous system resident cells during inflammatory stimuli. Exp Mol Pathol 99(3):570–574. doi:10.1016/j.yexmp.2015.07.010
85. Jager A, Dardalhon V, Sobel RA, Bettelli E, Kuchroo VK (2009) Th1, Th17, and Th9 effector cells induce experimental autoimmune encephalomyelitis with different pathological phenotypes. J Immunol 183(11):7169–7177. doi:10.4049/jimmunol.0901906
86. Kara EE, Comerford I, Bastow CR, Fenix KA, Litchfield W, Handel TM, McColl SR (2013) Distinct chemokine receptor axes reg-

ulate Th9 cell trafficking to allergic and autoimmune inflammatory sites. J Immunol 191(3):1110–1117. doi:10.4049/jimmunol.1203089
87. Elyaman W, Bradshaw EM, Uyttenhove C, Dardalhon V, Awasthi A, Imitola J, Bettelli E, Oukka M, van Snick J, Renauld JC, Kuchroo VK, Khoury SJ (2009) IL-9 induces differentiation of TH17 cells and enhances function of FoxP3+ natural regulatory T cells. Proc Natl Acad Sci U S A 106(31):12885–12890. doi:10.1073/pnas.0812530106
88. Strober W, Fuss IJ (2011) Proinflammatory cytokines in the pathogenesis of inflammatory bowel diseases. Gastroenterology 140(6):1756–1767. doi:10.1053/j.gastro.2011.02.016
89. Gerlach K, Hwang Y, Nikolaev A, Atreya R, Dornhoff H, Steiner S, Lehr HA, Wirtz S, Vieth M, Waisman A, Rosenbauer F, McKenzie AN, Weigmann B, Neurath MF (2014) TH9 cells that express the transcription factor PU.1 drive T cell-mediated colitis via IL-9 receptor signaling in intestinal epithelial cells. Nat Immunol 15(7):676–686. doi:10.1038/ni.2920
90. Mudter J, Yu J, Zufferey C, Brustle A, Wirtz S, Weigmann B, Hoffman A, Schenk M, Galle PR, Lehr HA, Mueller C, Lohoff M, Neurath MF (2011) IRF4 regulates IL-17A promoter activity and controls RORgammat-dependent Th17 colitis in vivo. Inflamm Bowel Dis 17(6):1343–1358. doi:10.1002/ibd.21476
91. Ciccia F, Guggino G, Ferrante A, Raimondo S, Bignone R, Rodolico V, Peralta S, Van Tok M, Cannizzaro A, Schinocca C, Ruscitti P, Cipriani P, Giacomelli R, Alessandro R, Dieli F, Rizzo A, Baeten D, Triolo G (2016) IL-9 over-expression and Th9 polarization characterize the inflamed gut, the synovial tissue and the peripheral blood of patients with psoriatic arthritis. Arthritis Rheumatol. doi:10.1002/art.39649
92. Mudter J, Amoussina L, Schenk M, Yu J, Brustle A, Weigmann B, Atreya R, Wirtz S, Becker C, Hoffman A, Atreya I, Biesterfeld S, Galle PR, Lehr HA, Rose-John S, Mueller C, Lohoff M, Neurath MF (2008) The transcription factor IFN regulatory factor-4 controls experimental colitis in mice via T cell-derived IL-6. J Clin Invest 118(7):2415–2426. doi:10.1172/JCI33227
93. Gerlach K, McKenzie AN, Neurath MF, Weigmann B (2015) IL-9 regulates intestinal barrier function in experimental T cell-mediated colitis. Tissue Barriers 3(1–2):e983777. doi:10.4161/21688370.2014.983777
94. Nalleweg N, Chiriac MT, Podstawa E, Lehmann C, Rau TT, Atreya R, Krauss E, Hundorfean G, Fichtner-Feigl S, Hartmann A, Becker C, Mudter J (2015) IL-9 and its receptor are predominantly involved in the pathogenesis of UC. Gut 64(5):743–755. doi:10.1136/gutjnl-2013-305947
95. Forbes EE, Groschwitz K, Abonia JP, Brandt EB, Cohen E, Blanchard C, Ahrens R, Seidu L, McKenzie A, Strait R, Finkelman FD, Foster PS, Matthaei KI, Rothenberg ME, Hogan SP (2008) IL-9- and mast cell-mediated intestinal permeability predisposes to oral antigen hypersensitivity. J Exp Med 205(4):897–913. doi:10.1084/jem.20071046
96. Yuan A, Yang H, Qi H, Cui J, Hua W, Li C, Pang Z, Zheng W, Cui G (2015) IL-9 antibody injection suppresses the inflammation in colitis mice. Biochem Biophys Res Commun 468(4):921–926. doi:10.1016/j.bbrc.2015.11.057
97. Kim HS, Chung DH (2013) IL-9-producing invariant NKT cells protect against DSS-induced colitis in an IL-4-dependent manner. Mucosal Immunol 6(2):347–357. doi:10.1038/mi.2012.77
98. Neurath MF, Weigmann B, Finotto S, Glickman J, Nieuwenhuis E, Iijima H, Mizoguchi A, Mizoguchi E, Mudter J, Galle PR, Bhan A, Autschbach F, Sullivan BM, Szabo SJ, Glimcher LH, Blumberg RS (2002) The transcription factor T-bet regulates mucosal T cell activation in experimental colitis and Crohn's disease. J Exp Med 195(9):1129–1143
99. Brandt EB, Sivaprasad U (2011) Th2 cytokines and atopic dermatitis. J Clin Cell Immunol 2(3). doi:10.4172/2155-9899.1000110
100. Barnes PJ (2001) Th2 cytokines and asthma: an introduction. Respir Res 2(2):64–65
101. Brough HA, Cousins DJ, Munteanu A, Wong YF, Sudra A, Makinson K, Stephens AC, Arno M, Ciortuz L, Lack G, Turcanu V (2014) IL-9 is a key component of memory TH cell peanut-specific responses from children with peanut allergy. J Allergy Clin Immunol 134(6):1329–1338 . doi:10.1016/j.jaci.2014.06.032e1310
102. Froidure A, Shen C, Gras D, Van Snick J, Chanez P, Pilette C (2014) Myeloid dendritic cells are primed in allergic asthma for thymic stromal lymphopoietin-mediated induction of Th2 and Th9 responses. Allergy 69(8):1068–1076. doi:10.1111/all.12435
103. Hoppenot D, Malakauskas K, Lavinskiene S, Sakalauskas R (2015) p-STAT6, PU.1, and NF-kappaB are involved in allergen-induced late-phase airway inflammation in asthma patients. BMC Pulm Med 15:122. doi:10.1186/s12890-015-0119-7
104. Schlapbach C, Gehad A, Yang C, Watanabe R, Guenova E, Teague JE, Campbell L,

Yawalkar N, Kupper TS, Clark RA (2014) Human TH9 cells are skin-tropic and have autocrine and paracrine proinflammatory capacity. Sci Transl Med 6(219):219218. doi:10.1126/scitranslmed.3007828

105. Yao W, Tepper RS, Kaplan MH (2011) Predisposition to the development of IL-9-secreting T cells in atopic infants. J Allergy Clin Immunol 128(6):1357–1360. doi:10.1016/j.jaci.2011.06.019e1355

106. Devos S, Cormont F, Vrtala S, Hooghe-Peters E, Pirson F, Snick J (2006) Allergen-induced interleukin-9 production in vitro: correlation with atopy in human adults and comparison with interleukin-5 and interleukin-13. Clin Exp Allergy 36(2):174–182. doi:10.1111/j.1365-2222.2006.02422.x

107. Townsend JM, Fallon GP, Matthews JD, Smith P, Jolin EH, McKenzie NA (2000) IL-9-deficient mice establish fundamental roles for IL-9 in pulmonary mastocytosis and goblet cell hyperplasia but not T cell development. Immunity 13(4):573–583

108. Kearley J, Erjefalt JS, Andersson C, Benjamin E, Jones CP, Robichaud A, Pegorier S, Brewah Y, Burwell TJ, Bjermer L, Kiener PA, Kolbeck R, Lloyd CM, Coyle AJ, Humbles AA (2011) IL-9 governs allergen-induced mast cell numbers in the lung and chronic remodeling of the airways. Am J Respir Crit Care Med 183(7):865–875. doi:10.1164/rccm.200909-1462OC

109. Sehra S, Yao W, Nguyen ET, Glosson-Byers NL, Akhtar N, Zhou B, Kaplan MH (2015) TH9 cells are required for tissue mast cell accumulation during allergic inflammation. J Allergy Clin Immunol 136(2):433–440. doi:10.1016/j.jaci.2015.01.021e431

Chapter 2

IL-9: Function, Sources, and Detection

Wilmer Gerardo Rojas-Zuleta and Elizabeth Sanchez

Abstract

IL-9 is a pleiotropic cytokine produced in different amounts by a wide variety of cells including mast cells, NKT cells, Th2, Th17, Treg, ILC2, and Th9 cells. Th9 cells are considered to be the main CD4+ T cells that produce IL-9. IL-9 exerts its effects on multiple types of cells and different tissues. To date, its main role has been found in the immune responses against parasites and pathogenesis of allergic diseases such as asthma and bronchial hyperreactivity. Additionally, it induces the proliferation of hematologic neoplasias, including Hodgkin's lymphoma in humans. However, IL-9 also has antitumor properties in solid tumors such as melanoma. The objective of this review is to describe IL-9, its function, sources, and methods of detection.

Key words Interleukin-9, P 40, T Lymphocyte growth factor, Mast cell growth-enhancing activity, Th9 cells

1 Introduction

Interleukin-9 (IL-9) is a pleiotropic cytokine produced in varying amounts in different immune cells. Among these are mast cells, NKT cells, Th2, Th17, Treg cells, and Th9 cells, with the greatest amount being produced by the latter [1]. It is a 14-kD glycoprotein, which is composed of 144 amino acids and one signal peptide of 18 amino acids [2]. The human IL-9 gene is located on the long arm of chromosome 5. It is found in the Th2 cytokine gene cluster [3]. In murine, it is localized in the chromosome 13 and is not linked to the same gene cluster [4]. The human *IL9* gene has promoter sequences to transcription factors such as PU.1 and IRF4, among others that regulate their expression [5, 6].

IL-9 receptor is a part of the cytokine family that share the common γ chain receptor, including IL-2, IL-4, IL-7, IL-15, and IL-21 [7]. IL-9 receptor is made up of IL-9Rα apart from the common γ chain. When IL-9 binds to its cognate receptor, it recruits Janus Kinase 1 (JAK1) and JAK 3 to α and γ chains, respectively [8]. These are cross-phosphorylated and activate STAT family of transcription

Ritobrata Goswami (ed.), *Th9 Cells: Methods and Protocols*, Methods in Molecular Biology, vol. 1585,
DOI 10.1007/978-1-4939-6877-0_2,

factors STAT1, STAT3, and STAT5, which form either homodimers or heterodimers. Afterwards, they move to the nucleus to activate the expression of IL-9-induced genes [1].

1.1 Discovery

IL-9 was first described in 1989 [1, 9]. Initially discovered in cell clones (TUC2 and TUC7) derived from C57BL/6 mice immunized with concanavalin A, the supernatant of these cell lines was observed to support the growth of specific T helper cell clones in the absence of other stimulus [10]. This supernatant was purified and this T cell growth factor was originally called P 40, T lymphocyte growth factor (TCGFIII) or MEA (mast cell growth-enhancing activity). At first, it was thought to be part of a cytokine repertoire produced by Th2 cells. Nevertheless, it was not known if it was produced by IL-4-producing Th2 cells [11] or if it was secreted by another other immune cells [12].

In 1994, Schmitt et al. [13] for the first time described the IL-9 production by naïve CD4+ murine effector lymphocytes in presence of IL-2 along with TGF-β. Likewise, they found that adding IL-4 enhanced this effect. These cytokines acted synergistically and had a dose-dependent effect to induce IL-9. In contrast, when IFN-γ was added to the medium with TGF-β, IL-9 secretion was inhibited. Schmitt et al. also proposed that the synergistic effect of IL-4 in IL-9 production is secondary to its capacity to inhibit IFN-γ production. Thus, the inhibiting effect it has on the IL-9 expression is neutralized [13].

Later on, studies by Veldhoen et al. showed that IL-4 producing Th2 cells cultured in the presence of TGF-β would stimulate production of IL-9 [14]. Dardalhon et al. also demonstrated that with the stimulus of IL-4 and TGF-β on a specific T cell subset would subsequently promote the differentiation to polarized cells that would produce high quantities of IL-9 as well as IL-10 in mice [15]. This study found that IL-4 was able to suppress TGF-β-induced Foxp3 expression and prevent the generation of $Foxp3^+$ Treg cells, and promoted predominantly $Foxp3^-IL9^+IL\text{-}10^+$ $CD4^+$ T cells that did not suppress T cell responses [14, 15]. These cells were termed as Th9 cells.

These findings sparked a renewed interest in IL-9 and its role in immune regulation as well as in IL-9-secreting other cells. Several studies have revealed other cellular sources of IL-9 [1]. Human Th17 cells have been demonstrated to secrete concomitantly IL-9 and IL-17 under TGF-β stimulation [16]. In addition to being the target of IL-9, activated mast cells have also been identified as IL-9 producers [17]. Natural killer T cells are also capable of producing IL-9, initially observed in $DX5^+CD3^+$ T-NK cells derived from murine splenocytes [18], but later was observed in human nasal NKT lymphoma cells [19]. And while Treg cells have been shown to produce IL-9, to date, debate persists regarding the in vivo conditions this occurs [16]. New research is currently focusing on discovering different factors involved in influencing IL-9 production in different cells and the multiple effects that this cytokine exerts.

1.2 Function

IL-9 exerts its effect on multiple types of cells and different tissues (Fig. 1), and initially was considered as a growth factor of activated T cells [20]. Later, its potent proliferative effects were demonstrated in other cell types mainly mast cells [21], hematopoietic erythroid precursors and on myeloid leukemia cell lines [22].

1.2.1 Allergic Inflammatory Processes

IL-9 plays an important role in the regulation of airway inflammation and airway hyperresponsiveness. IL-9 transgenic mice develop an asthma like phenotype and lymphocytic and eosinophilic lung inflammation [23]. It has been demonstrated that IL-9 exerts proliferative effects on goblet cells and cells that produce mucin in the airways [24], which is reflected with an increased production of mucus, favoring allergic inflammation in the respiratory tracts [23]. However, increased Th9 cell numbers in peripheral blood of allergic patients correlated with IgE titers [25]. In B lymphocytes, IL-9 in the presence of IL-4 increases secretion of IgG1 and IgE and it also promotes an isotype switch [23], contributing to the pathogenesis of allergic diseases of the respiratory tract, specifically in asthma and bronchial hyperreactivity [26]. The pathogenic role

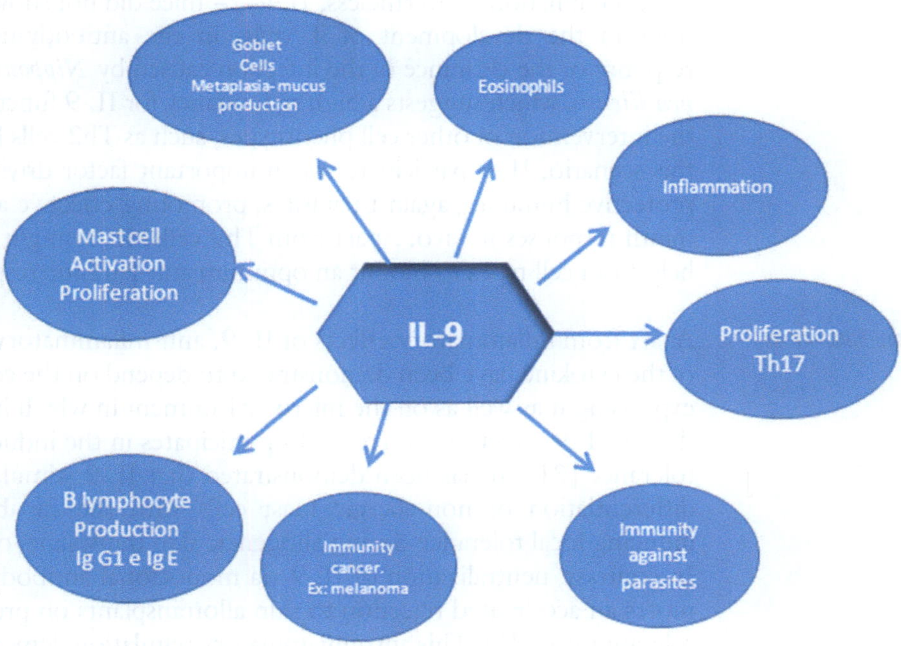

Fig. 1 Functions of interleukin-9. IL-9 participates in a great variety of physiological processes ranging from promoting inflammatory mechanisms up to immunity against parasites and tumors. Nevertheless, because of its effects on the activation of mast cells, eosinophils and the induction of IgE production, just like the production of mucus by goblet cells, it participates in the pathogenesis of allergic diseases like asthma

of IL-9 in allergic diseases was demonstrated in a *Rag2−/−* murine model via the induction of allergic inflammation in airways mediated by ovalbumin and the adoptive transfer of Th9 cells. An induction of goblet cell metaplasia and an increase in bronchial reactivity was observed. These effects were reduced when an anti-IL-9 monoclonal antibody was administered [27].

1.2.2 Effect on Neoplasia

IL-9 has been demonstrated to play an important role immune regulation in neoplasia. Among these, one important aspect is related to hematologic neoplasms. In vitro studies have demonstrated that ectopic expression of IL-9 induces the proliferation of mouse thymic lymphomas. In humans, in vitro studies have observed increase of IL-9 production in cells of Hodgkin's lymphoma by promoting the growth of these cultured cells [8]. The effect of IL-9 in neoplasia may depend whether the tumor is solid or not. In solid tumors, specifically in melanoma, it has been demonstrated that Th9 and IL-9 have an important antitumor effect favoring the recruitment of both innate adaptive immune cells, reducing tumor burden [28].

1.2.3 Immunity Against Parasites

In addition, it has also been shown that IL-9 participates in immunity against parasites. IL-9 transgenic mice overexpressing IL-9 eradicate *Trichinella spiralis* infections faster than wild-type mice [29]. This nematode requires a great amount of intestinal mast cells for its elimination. Nevertheless, *IL-9−/−* mice did not show alterations in the development of T cells, in the antibody-mediated response or the clearance of the infection caused by *Nippostrongylus brasiliensis*, which suggests a high redundancy for IL-9 function and the intervention of other cell phenotypes, such as Th2 cells [30]. In this scenario, IL-9 participates as an important factor driving host protective immunity against parasites, promoting effective anti-helminth responses in vivo. Apart from Th9 cells, IL-9 might require help from cell types to mount an optimum anti-parasitic response.

1.2.4 Anti-inflammatory Effects

Apart from inflammatory effects of IL-9, anti-inflammatory effects of the cytokine have been demonstrated to depend on the cell types expressing it as well as on the microenvironment in which it is produced. IL-9 secretion by Treg cells participates in the induction of tolerance [31]. It has been demonstrated that IL-9 stimulates the differentiation of non-allergic mast cells with the capability of inducing local tolerance during allogeneic skin transplants on mice. In contrast, neutralization of IL-9 via monoclonal antibodies promotes an accelerated rejection to skin allotransplants on previously tolerant mice [31]. This anti-inflammatory regulation demonstrates the important role IL-9 plays in immune tolerance.

1.2.5 Autoimmune Diseases

IL-9 has been implicated in numerous pathogenic processes of diseases, mainly allergic diseases as asthma and atopic dermatitis [32]. IL-9 serum levels are elevated in patients with systemic lupus erythematosus [33], rheumatoid arthritis [34, 35] and systemic sclerosis [36] but their clinical significance are still not completely understood. It has been debated if IL-9 has a role in the pathogenesis of these diseases, or if its presence is due to an epiphenomenon caused by a broad activation of inflammatory mechanisms, and this has made it difficult to define its function in the development of the disease; thus further studies are required to elucidate the pathogenic role in those rheumatic disease [37, 38].

1.3 Sources

As previously mentioned a variety of cells produce IL-9, in vivo studies are rare to assess the amount of IL-9 being produced. During the discovery of this cytokine, multiple studies have been able to discover the cells that express IL-9 and the stimulation that triggers this production.

1.3.1 Th2 Cells

As discussed earlier, when IL-9 was first discovered, Th2 cells were one of the first cells studied in association to this cytokine. As mentioned above, they were studied in murine models infected in vivo with *Leishmania major* and these cells were found to co-express other cytokines as well including IL-4, IL-5, and IL-13. Initially believed to be the main producers of IL-9, a correlation was found between Th2 cell expansion and IL-9 levels. Additionally, IL-4 stimulation was found to play a key role in Th2 cell differentiation and necessary for IL-9 production in these cells [39]. Further studies have demonstrated the co-expression of IL-4 and IL-9 in differentiated Th2 cells; however, these quantities are very low [40]. The discovery of other cell sources has revealed larger quantities of this cytokine originating from other cells, suggesting that Th2 cells are not the main producers of IL-9.

1.3.2 NKT Cells

It has been demonstrated that under certain conditions, natural killer T (NKT) cells can produce IL-9. Studies using NKT cells from naive mice have shown that after stimulation with IL-2, these cells can produce IL-9 [18]. IL-2 stimulation also triggers the expression of IL-4, IL-5, and IL-13 in NKT cells, but not IFN-γ, suggesting that these cells assist in the humoral immune response [18]. Naïve NKT cells in the presence of TGF-β and IL-4 polarize and secrete IL-9 in murine and human thymic iNKT cells [41]. Jones et al. observed that in the absence of CD1d, pulmonary NKT cells decrease IL-9 expression accompanied by decrease in mast cell recruitment to the lungs in allergic airway inflammation [42]. Additionally, peripheral iNKT cells under the influence of TGF-β and IL-4 adopt an IL-9-producing NKT cell phenotype able to mediate pro-inflammatory effects observed in vivo, namely granulocyte and mast cell recruitment to the lungs [41]. Other studies have also discovered that NKT cells that have been involved in nasal NKT cell lymphoma can also produce IL-9 that acts as an autocrine growth factor and promotes disease progression [19].

1.3.3 Mast Cells

It has also been discovered that activated mast cells can secrete IL-9. Several cytokines have been found to stimulate IL-9 production by mast cells, while IL-9 acts as a growth factor and promotes mast cell expansion [17]. Mast cells are stimulated in an autocrine manner in response to IL-9-induced signals and the cross-linking of IgE molecules on the surface of mast cells triggers release of numerous other cytokines [17, 43]. Histamine and IL-1β, two cytokines released after mast cell degranulation, have been found to further IL-9 production, and along with IL-9 itself, seem to behave in a positive feedback loop inducing IL-9 production [43]. Mast cells have been observed to produce IL-9 in response in a p38 MAPK-dependent manner, and the activation of GATA1 has been found to increase *Il9* promoter activation [44]. Wiener et al. found that IL-9 production in conjunction with ionomycin can trigger the expression at mRNA level of *Il4*, *Il5*, *Il9*, *Il10*, *Il1β*, *Il1ra*, *Il6*, and *MIF* [43]. In addition, IL-9 can induce mast cell production of pro-inflammatory factors, such as IL-1α, IL-1β, IL-1Rα, IL-3, IL-4, IL-5, and IL-6, which is believed to contribute to airway hyperresponsiveness [43]. Chen et al. has recently discovered multifunctional IL-9-producing mucosal mast cells in mice in intestinal mucosa [45]. Intestinal mucosal mast cell production of IL-9 and IL-13 could also play a role in the development of food allergies in mice [45]. Further studies are needed to determine the importance of mast cell IL-9 production in development of food allergies.

1.3.4 Th9 Cells

In 2008, Veldhoen et al. [14] discovered a distinct $CD4^+$ T subpopulation based on the cultivating $CD4^+$ murine lymphocytes under different groups of inductor cytokines which polarized the differentiation toward Th1, Th2, Th17, Treg, and $CD4^+IL\text{-}9^+$ cells. There was evidence that these cells, which acquired the IL-9 phenotype lost expression of other characteristic cytokine of T effector lymphocytes including IL-4, IL-5, IL-13 (Th2), IL-17-α (Th17), or IFN-γ (Th1), and they expressed very low level of transcription factors T-bet for Th1 [46], GATA3 for Th2 [47], Foxp3 for Treg [48], and RORγt for Th17 cells [49]. This suggested that this subset of lymphocytes is a different T-lymphocyte subpopulation characterized by the expression and secretion high amounts IL-9 and IL-10 and hence they were named Th9 lymphocytes. To date, a wide variety of stimuli has been described which contribute to Th9 cell differentiation, as IL-2 [13], IL-25 [50], the peptide related to the calcitonin gene [51], and thymic stromal lymphopoietin [52], among others. This redundancy suggests cell function diversity and heterogeneity.

1.3.5 Th17 Cells

Studies have demonstrated that polarized mouse Th17 cells can produce IL-9 while co-expressing IL-17 as well [16]. However, IL-23 has been observed to suppress IL-9 production and given its importance in the maintenance of Th17 cells it remains unclear whether this IL-9 production by Th17 cells is transient [53]. In vitro studies have shown that human Th17 cells also produce IL-9. Differentiated naïve cells need repeated stimulation by Th17 inducing conditions to co-express IL-17 and IL-9. Memory CD4$^+$ T cells subjected to Th17 inducing cytokines such as IL-1β and IL-21 result in the co-expression of IL-9 and IL-17 [16].

1.3.6 Treg Cells

Few studies have suggested the production of IL-9 by Treg cells. While a couple of studies have confirmed that IL-9 is produced by Treg cells, there are conflicting reports under the circumstances which this occurs [1, 12, 31]. One study reported co-expression of forkhead box P3 (Foxp3) and IL-9 in Treg cells in tolerant murine allografts [31]. This has not been reported in other studies, which studied the function of Treg cells in vitro [54]. Additionally, in human donors the co-expression of Foxp3 and IL-9 has not been reported either [16]. Further studies are needed to explore the role of IL-9 production in Treg cells and their role in immune regulation in humans.

1.3.7 ILC2 Cells

Research has discovered a novel subset of innate lymphoid cells (ILC) that release type 2 cytokines named group 2 ILC (ILC2) cells [55]. Studies with IL-9 reporter mice in vivo have demonstrated that in certain inflammatory milieu, ILC2 cells have been found to express IL-9 cells to variety of stimuli [56–58]. In a papain-induced lung inflammation model in mice, Wilhelm et al. discovered that IL-9 was largely produced by ILC2 cells. This production was demonstrated to be dependent on IL-2, but rapidly diminished as other the production of other cytokines, such as IL-13 and IL-5 increased [56]. When IL-9 production was neutralized in ILC2 cells, a lower expression in IL-13 and IL-5 was observed, suggesting that the production of IL-9 by ILC2 cells may play a role in regulation of Th2 cells [56]. Another study in IL-9R-deficient mice demonstrated that ILC2 expression of IL-9R was important in the production of IL-5 and IL-13 in infection with *Nippostrongylus brasiliensis* in the lung [59]. The absence of IL-9 signaling in these mice resulted in reduced lung ILC2 recruitment and suggests that the production of IL-9 works as an autocrine amplifier in the function and survival of ILC2 cells [59]. IL-9 production in ILC2 cells seems to be dependent on other cytokine stimulation and play a role in immune response in the lung.

1.4 Detection

Finally, depending on the environment in which one wants to quantify either fluid or tissue there are several methods for detecting IL-9. Current methods for quantitative IL-9 detection in serum, plasma and cell culture include ELISA and proliferation assays and tissue or cell expression can be detected through quantitative real-time PCR for *Il9* mRNA expression.

1.4.1 ELISA

Presently, ELISA kits for IL-9 are widely commercially available and have been used in various studies for quantification of IL-9 levels in serum, plasma and supernatant samples. Various manufacturers have developed ELISA kits that use sandwich ELISA technique measuring IL-9 between capture and detection antibody. These ELISA kits utilize both human and mouse antibodies. The methodology to carry out the ELISA depends on the manufacturer's instructions, which vary between commercial houses. To date there have been no studies comparing performance between these different ELISA kits (*see* **Note 1**).

1.4.2 Quantitative Real-Time PCR

Quantitative real-time reverse transcription followed by polymerase chain reaction (RT-PCR) is the most appropriate method for the detection and quantification of cytokine mRNA quantification [60]. It enables quantification of nucleic acids through amplification in a cyclic process (high temperature–low temperature) that generates a large amount of identical copies of the sequence to be analyzed. For analysis of *Il9* expression, isolating mRNA from the plasma or supernatant sample that is initially required and cDNA is transcribed using a reverse transcriptase. In this case, the reverse transcriptase utilized is a specific mRNA transcript *Il9* hybridizing polyA tail to the mRNA (3′-end). The amount of amplified product is monitored during the course of the reaction by measuring the fluorescence (through labeled primers), which is proportional to the amount of product formed [61]. For quantitative analysis, it is necessary to perform a normalization process to compensate for differences between the amounts of biological material samples each. This will be done by comparing the result with the relative expression of a gene known constant expression (e.g., β-catenin) by using the 2-ΔΔCt method [62]. This method is highly sensitive and specific, has good reproducibility and wide range quantification, which makes the method of choice for quantifying the expression of multiple cytokines.

1.4.3 Proliferation Assays

Proliferation assays act as alternative method to detect IL-9 through the capacity of IL-9 to induce proliferation certain cellular lines. Several cell lines that can be used for this purpose. Here we describe two cellular lines that are predominantly used (*see* **Note 2**).

TS1h9RA3 (TS1) is a murine cell line transfected with a human IL-9 receptor commonly used in proliferation assays [10]. These cells are receptive to stimulation murine IL-4, human IL-9 or murine IL-9. These cells are cultured in a serial dilution of known IL-9 concentrations. Then their growth or proliferation can be measured through hexosaminidase method or [3H] thymidine incorporation [63]. In the latter, the determination of cell proliferation is assessed by measuring the incorporation of [3H] thymidine into cellular DNA [64]. This radioactive label is added during the last 4–24 h of the culture. The harvest of the cultures is carried out with a semi-automated cell harvesting apparatus. This will lyse the cells with water and precipitate the labeled DNA on glass fiber filters. After these filters are dried, they are counted by standard liquid scintillation counting techniques in a β counter [64].

Furthermore, this method has the advantage that only human IL-9 is able to promote proliferation, which makes it ideal in a medium with several cytokines present simultaneously. Nonetheless, one of the major disadvantages of this method is that in a medium with high cell density and low availability of IL-9, this effect is no longer observed in cell growth.

The M-O7e cells are human megakaryoblastic leukemia cellular line dependent on IL-3 or granulocyte macrophage colony-stimulating factor for proliferation. Other factors such as human IL-2, IL-4, IL-6 and murine or human IL-9 also stimulate the growth of these cells [65]. These cells are grown in different known dilutions of IL-9 and these known concentrations are subsequently compared to the growth that results using the sample with the unknown concentration of IL-9 and the amount of [3H] thymidine that is incorporated into the cells. They have the disadvantage of requiring cytokine free medium so that there is no interference in the measurement of IL-9, since these cells can proliferate in the presence of other cytokines (*see* **Note 3**). Therefore, this is not a useful tool when measuring IL-9 in a medium that has other cytokines. In the following section we describe the method of IL-9 proliferation assay.

2 Materials

The use of these materials has been described by Jean-Christophe Renauld and Jacques Van Snick and included below with authors' permission [63].

1. Complete DMEM-10 medium: Dulbecco's modified Eagle's medium, 10% fetal bovine serum (v/v), 2 mM L-glutamine, 100 U/mL penicillin G, 100 μg/mL streptomycin.
2. 1000 U/mL human or mouse IL-9 reference standard.
3. Unknown samples containing IL-9.
4. TS1 cell culture (3- to 4-day-old).
5. PBS, pH 7.4.
6. 50 mM citrate buffer, pH 5.0.
7. 0.25% Triton X-100.
8. 3.75 mM *p*-nitrophenyl *N*-acetyl-β-D-glucosaminide.
9. Glycine buffer: 10 mM EDTA, 0.1 M glycine, pH 10.4, store at room temperature.
10. Multichannel pipette and tips

11. 96-well flat-bottom microtiter plates with lids.
12. 15 mL conical centrifuge tube
13. Jouan GR422 centrifuge (or equivalent).

3 Methods

3.1 Measurement of IL-9 Activity Using the TS1h9RA3 Cell Proliferation Assay

A basic protocol has been described by Jean-Christophe Renauld and Jacques Van Snick and included below with authors' permission [63].

1. Using a multichannel pipette, add 100 μL of complete DMEM-10 medium to each well of a 96-well flat-bottom microtiter plate.
2. Thaw the 1000 U/mL IL-9 standard and dilute with complete DMEM-10 medium to 600 U/mL. Prepare serial dilutions of the IL-9 standard as follows:
   (a) Add 50 μL of 600 U/mL IL-9 to row A, columns 1–3.
   (b) Using a multichannel pipette, transfer 50 μL from row A, columns 1–3, to the wells containing complete medium (from **step 1**) in row B, columns 1–3.
   (c) Mix the standard by pipetting up and down several times.
   (d) Transfer 50 μL from row B, columns 1–3, to row C, columns 1–3; mix thoroughly and continue this stepwise transfer of 50 μL through row G.
   (e) When the last dilution is made in columns 1–3 of row G, mix thoroughly and discard 50 μL from these wells so that each well in columns 1–3, rows A to G, contains 100 μL of medium plus standard.
3. Add 50 μL/well of three unknown samples containing IL-9 in triplicate starting from row A, columns 4–6, 7–9, and 10–12. Repeat with another set of three samples in row E, columns 4–6, 7–9, and 10–12. Prepare three serial dilutions in rows A–D and then in rows E–H as described above in **step 2** (to span a eightfold dilution range for each unknown sample).
4. Harvest TS1 cells from the tissue-culture flask in active log-phase growth. Transfer the cells to a 15 mL conical centrifuge tube and centrifuge for 5 min at 300 × *g* (1500 rpm in Jouan GR422 centrifuge), room temperature. Discard the supernatant and wash the cells at least two times in 14 mL of complete DMEM-10 medium to remove residual cytokine. Resuspend the cells in a small volume (1 mL) of complete DMEM-10 medium.

5. Count viable cells by the trypan blue exclusion method. Resuspend the cells at a concentration of 3 × 10^4 cells/mL using complete DMEM-10 medium. Using a multichannel pipette, add 100 μL of cell suspension to each well of the 96-well plate containing IL-9 standards or samples.
6. Cover the plate with a lid and incubate the plates for 3 days in a humidified 37 °C, 5% CO_2 incubator.
7. After 3 days, wash cells two times with PBS, pH 7.4. To wash cells, centrifuge plates 5 min at 300 × *g* at room temperature; discard supernatants and add 200 μL of PBS per well. After the second wash, resuspend the cells in 60 μL of 50 mM citrate buffer containing 0.25% Triton X-100 and 3.75 mM *p*-nitrophenyl *N*-acetyl-β-D-glucosaminide. Incubate for 1.5 h at 37 °C.
8. Stop the colorimetric reaction by adding 90 μL of glycine buffer.
9. Quantify the IL-9 activity either directly or by comparison to a standard IL-9 sample. The concentration at which half-maximal effect is observed is defined as 1 U/mL of IL-9. With purified IL-9, 1 U/mL should correspond to 25 pg/mL for the mouse protein and 50 pg/mL for the human protein. Thus, the dilution factor required to obtain half-maximal proliferation with one particular sample corresponds to the number of U/mL in this sample.

3.2 Conclusion

IL-9 is a multifaceted cytokine involved in multiple biological processes such as immune responses against parasites, pathogenesis of allergic diseases (bronchial asthma and hyperreactivity), and immunity against solid neoplasias. It is produced by several cells including mast cells, NKT cells, Th2, Th17, Treg, and ILC2 cells, but Th9 cells are the predominant producers of IL-9. Methods of detection include proliferation assays, ELISA for the detection in serum plasma and supernatant samples, and quantitative real-time PCR assay for relative expression in cells and tissues. Of these, ELISA and qPCR are established as the most suitable and widely used tools for IL-9 detection.

4 Notes

1. Due to the fact that there are no studies comparing performance between different ELISA kits, the decision on which one to use depends on the preference of the individual researcher.
2. From a practical standpoint, we would recommend the use of ELISA kits over the use proliferation assays to measure levels of IL-9. As previously mentioned, there are currently no studies comparing the performance of ELISA kits on the market between

the commercial houses or comparison studies to proliferation assays. Despite the lack of evidence, there are various advantages of using the ELISA method. Cell cultures require more materials and frequent changing of culture medium. Despite these measures, the cells can still die or fail to grow properly.

3. In the event the medium has a variety of other interleukins, there could be interference with measurement of IL-9. These competing interleukins would have to be blocked or removed from the culture medium before assessing the level of IL-9 in order to prevent false readings. In this scenario, the addition of antibodies against IL-9 is required to verify the modification of cellular proliferation. There could be proliferation despite the blocking of IL-9, and the addition of antibodies directed against other possible cytokines present would be required.

References

1. Noelle RJ, Nowak EC (2010) Cellular sources and immune functions of interleukin-9. Nat Rev Immunol 10(10):683–687. doi:10.1038/nri2848
2. Renauld JC, Goethals A, Houssiau F, Van Roost E, Van Snick J (1990) Cloning and expression of a cDNA for the human homolog of mouse T cell and mast cell growth factor P40. Cytokine 2(1):9–12
3. Nicolaides NC, Holroyd KJ, Ewart SL, Eleff SM, Kiser MB, Dragwa CR, Sullivan CD, Grasso L, Zhang LY, Messler CJ, Zhou T, Kleeberger SR, Buetow KH, Levitt RC (1997) Interleukin 9: a candidate gene for asthma. Proc Natl Acad Sci U S A 94(24):13175–13180
4. Mock BA, Krall M, Kozak CA, Nesbitt MN, McBride OW, Renauld JC, Van Snick J (1990) IL9 maps to mouse chromosome 13 and human chromosome 5. Immunogenetics 31(4):265–270
5. Kaplan MH (2013) Th9 cells: differentiation and disease. Immunol Rev 252(1):104–115. doi:10.1111/imr.12028
6. Perumal NB, Kaplan MH (2011) Regulating IL-9 transcription in T helper cells. Trends Immunol 32(4):146–150. doi:10.1016/j.it.2011.01.006
7. Demoulin JB, Renauld JC (1998) Signalling by cytokines interacting with the interleukin-2 receptor gamma chain. Cytokines Cell Mol Ther 4(4):243–256
8. Knoops L, Renauld JC (2004) IL-9 and its receptor: from signal transduction to tumorigenesis. Growth Factors 22(4):207–215. doi:10.1080/08977190410001720879
9. Van Snick J, Goethals A, Renauld JC, Van Roost E, Uyttenhove C, Rubira MR, Moritz RL, Simpson RJ (1989) Cloning and characterization of a cDNA for a new mouse T cell growth factor (P40). J Exp Med 169(1):363–368
10. Uyttenhove C, Simpson RJ, Van Snick J (1988) Functional and structural characterization of P40, a mouse glycoprotein with T-cell growth factor activity. Proc Natl Acad Sci U S A 85(18):6934–6938
11. Schmitt E, Van Brandwijk R, Van Snick J, Siebold B, Rude E (1989) TCGF III/P40 is produced by naive murine CD4+ T cells but is not a general T cell growth factor. Eur J Immunol 19(11):2167–2170. doi:10.1002/eji.1830191130
12. Stassen M, Schmitt E, Bopp T (2012) From interleukin-9 to T helper 9 cells. Ann N Y Acad Sci 1247:56–68. doi:10.1111/j.1749-6632.2011.06351.x
13. Schmitt E, Germann T, Goedert S, Hoehn P, Huels C, Koelsch S, Kuhn R, Muller W, Palm N, Rude E (1994) IL-9 production of naive CD4+ T cells depends on IL-2, is synergistically enhanced by a combination of TGF-beta and IL-4, and is inhibited by IFN-gamma. J Immunol 153(9):3989–3996
14. Veldhoen M, Uyttenhove C, van Snick J, Helmby H, Westendorf A, Buer J, Martin B, Wilhelm C, Stockinger B (2008) Transforming growth factor-beta 'reprograms' the differentiation of T helper 2 cells and promotes an interleukin 9-producing subset. Nat Immunol 9(12):1341–1346. doi:10.1038/ni.1659

15. Dardalhon V, Awasthi A, Kwon H, Galileos G, Gao W, Sobel RA, Mitsdoerffer M, Strom TB, Elyaman W, Ho IC, Khoury S, Oukka M, Kuchroo VK (2008) IL-4 inhibits TGF-beta-induced Foxp3+ T cells and, together with TGF-beta, generates IL-9+ IL-10+ Foxp3(−) effector T cells. Nat Immunol (12):1347–1355. doi:10.1038/ni.1677

16. Beriou G, Bradshaw EM, Lozano E, Costantino CM, Hastings WD, Orban T, Elyaman W, Khoury SJ, Kuchroo VK, Baecher-Allan C, Hafler DA (2010) TGF-beta induces IL-9 production from human Th17 cells. J Immunol 185(1):46–54. doi:10.4049/jimmunol.1000356

17. Stassen M, Arnold M, Hultner L, Muller C, Neudorfl C, Reineke T, Schmitt E (2000) Murine bone marrow-derived mast cells as potent producers of IL-9: costimulatory function of IL-10 and kit ligand in the presence of IL-1. J Immunol 164(11):5549–5555

18. Lauwerys BR, Garot N, Renauld JC, Houssiau FA (2000) Cytokine production and killer activity of NK/T-NK cells derived with IL-2, IL-15, or the combination of IL-12 and IL-18. J Immunol 165(4):1847–1853

19. Nagato T, Kobayashi H, Kishibe K, Takahara M, Ogino T, Ishii H, Oikawa K, Aoki N, Sato K, Kimura S, Shimizu N, Tateno M, Harabuchi Y (2005) Expression of interleukin-9 in nasal natural killer/T-cell lymphoma cell lines and patients. Clin Cancer Res 11(23):8250–8257. doi:10.1158/1078-0432.ccr-05-1426

20. Houssiau FA, Renauld JC, Stevens M, Lehmann F, Lethe B, Coulie PG, Van Snick J (1993) Human T cell lines and clones respond to IL-9. J Immunol 150(7):2634–2640

21. Hultner L, Druez C, Moeller J, Uyttenhove C, Schmitt E, Rude E, Dormer P, Van Snick J (1990) Mast cell growth-enhancing activity (MEA) is structurally related and functionally identical to the novel mouse T cell growth factor P40/TCGFIII (interleukin 9). Eur J Immunol 20(6):1413–1416. doi:10.1002/eji.1830200632

22. Yang YC, Ricciardi S, Ciarletta A, Calvetti J, Kelleher K, Clark SC (1989) Expression cloning of cDNA encoding a novel human hematopoietic growth factor: human homologue of murine T-cell growth factor P40. Blood 74(6):1880–1884

23. Temann UA, Laouar Y, Eynon EE, Homer R, Flavell RA (2007) IL9 leads to airway inflammation by inducing IL13 expression in airway epithelial cells. Int Immunol 19(1):1–10. doi:10.1093/intimm/dxl117

24. Zhao P, Xiao X, Ghobrial RM, Li XC (2013) IL-9 and Th9 cells: progress and challenges. Int Immunol 25(10):547–551. doi:10.1093/intimm/dxt039

25. Jones CP, Gregory LG, Causton B, Campbell GA, Lloyd CM (2012) Activin A and TGF-beta promote T(H)9 cell-mediated pulmonary allergic pathology. J Allergy Clin Immunol 129(4):1000–1010.e1003. doi:10.1016/j.jaci.2011.12.965

26. Xing J, Wu Y, Ni B (2011) Th9: a new player in asthma pathogenesis? J Asthma 48(2):115–125. doi:10.3109/02770903.2011.554944

27. Staudt V, Bothur E, Klein M, Lingnau K, Reuter S, Grebe N, Gerlitzki B, Hoffmann M, Ulges A, Taube C, Dehzad N, Becker M, Stassen M, Steinborn A, Lohoff M, Schild H, Schmitt E, Bopp T (2010) Interferon-regulatory factor 4 is essential for the developmental program of T helper 9 cells. Immunity 33(2):192–202. doi:10.1016/j.immuni.2010.07.014

28. Purwar R, Schlapbach C, Xiao S, Kang HS, Elyaman W, Jiang X, Jetten AM, Khoury SJ, Fuhlbrigge RC, Kuchroo VK, Clark RA, Kupper TS (2012) Robust tumor immunity to melanoma mediated by interleukin-9-producing T cells. Nat Med 18(8):1248–1253. doi:10.1038/nm.2856

29. Faulkner H, Humphreys N, Renauld JC, Van Snick J, Grencis R (1997) Interleukin-9 is involved in host protective immunity to intestinal nematode infection. Eur J Immunol 27(10):2536–2540. doi:10.1002/eji.1830271011

30. Townsend JM, Fallon GP, Matthews JD, Smith P, Jolin EH, McKenzie NA (2000) IL-9-deficient mice establish fundamental roles for IL-9 in pulmonary mastocytosis and goblet cell hyperplasia but not T cell development. Immunity 13(4):573–583

31. Lu LF, Lind EF, Gondek DC, Bennett KA, Gleeson MW, Pino-Lagos K, Scott ZA, Coyle AJ, Reed JL, Van Snick J, Strom TB, Zheng XX, Noelle RJ (2006) Mast cells are essential intermediaries in regulatory T-cell tolerance. Nature 442(7106):997–1002. doi:10.1038/nature05010

32. Yao W, Tepper RS, Kaplan MH (2011) Predisposition to the development of IL-9-secreting T cells in atopic infants. J Allergy Clin Immunol 128(6):1357–1360.e1355. doi:10.1016/j.jaci.2011.06.019

33. Ouyang H, Shi Y, Liu Z, Feng S, Li L, Su N, Lu Y, Kong S (2013) Increased interleukin9 and CD4+IL-9+ T cells in patients with systemic lupus erythematosus. Mol Med Rep 7(3):1031–1037. doi:10.3892/mmr.2013.1258

34. Dantas AT, Marques CD, da Rocha Junior LF, Cavalcanti MB, Goncalves SM, Cardoso PR, Mariz Hde A, Rego MJ, Duarte AL, Pitta Ida R, Pitta MG (2015) Increased serum interleukin-9 levels in rheumatoid arthritis and systemic lupus erythematosus: pathogenic role or just an epiphenomenon? Dis Markers 2015:519638. doi:10.1155/2015/519638
35. Ciccia F, Guggino G, Rizzo A, Manzo A, Vitolo B, La Manna MP, Giardina G, Sireci G, Dieli F, Montecucco CM, Alessandro R, Triolo G (2015) Potential involvement of IL-9 and Th9 cells in the pathogenesis of rheumatoid arthritis. Rheumatology (Oxford, UK). doi:10.1093/rheumatology/kev252
36. Yanaba K, Yoshizaki A, Asano Y, Kadono T, Sato S (2011) Serum interleukin 9 levels are increased in patients with systemic sclerosis: association with lower frequency and severity of pulmonary fibrosis. J Rheumatol 38(10):2193–2197. doi:10.3899/jrheum.110268
37. Pan HF, Leng RX, Li XP, Zheng SG, Ye DQ (2013) Targeting T-helper 9 cells and interleukin-9 in autoimmune diseases. Cytokine Growth Factor Rev 24(6):515–522
38. Rojas-Zuleta WG, Vásquez G (2016) Th9 lymphocytes: a recent history from IL-9 to its potential role in rheumatic diseases. Autoimmun Rev 15(7):649–655. doi:10.1016/j.autrev.2016.02.020
39. Gessner A, Blum H, Rollinghoff M (1993) Differential regulation of IL-9-expression after infection with leishmania major in susceptible and resistant mice. Immunobiology 189(5):419–435. doi:10.1016/s0171-2985(11)80414-6
40. Chang HC, Sehra S, Goswami R, Yao W, Yu Q, Stritesky GL, Jabeen R, McKinley C, Ahyi AN, Han L, Nguyen ET, Robertson MJ, Perumal NB, Tepper RS, Nutt SL, Kaplan MH (2010) The transcription factor PU.1 is required for the development of IL-9-producing T cells and allergic inflammation. Nat Immunol 11(6):527–534. doi:10.1038/ni.1867
41. Monteiro M, Agua-Doce A, Almeida CF, Fonseca-Pereira D, Veiga-Fernandes H, Graca L (2015) IL-9 expression by invariant NKT cells is not imprinted during thymic development. J Immunol 195(7):3463–3471. doi:10.4049/jimmunol.1403170
42. Jones TG, Hallgren J, Humbles A, Burwell T, Finkelman FD, Alcaide P, Austen KF, Gurish MF (2009) Antigen-induced increases in pulmonary mast cell progenitor numbers depend on IL-9 and CD1d-restricted NKT cells. J Immunol 183(8):5251–5260. doi:10.4049/jimmunol.0901471
43. Wiener Z, Falus A, Toth S (2004) IL-9 increases the expression of several cytokines in activated mast cells, while the IL-9-induced IL-9 production is inhibited in mast cells of histamine-free transgenic mice. Cytokine 26(3):122–130. doi:10.1016/j.cyto.2004.01.006
44. Stassen M, Klein M, Becker M, Bopp T, Neudorfl C, Richter C, Heib V, Klein-Hessling S, Serfling E, Schild H, Schmitt E (2007) p38 MAP kinase drives the expression of mast cell-derived IL-9 via activation of the transcription factor GATA-1. Mol Immunol 44(5):926–933. doi:10.1016/j.molimm.2006.03.019
45. Chen C-Y, Lee J-B, Liu B, Ohta S, Wang P-Y, Kartashov Andrey V, Mugge L, Abonia JP, Barski A, Izuhara K, Rothenberg Marc E, Finkelman Fred D, Hogan Simon P, Wang Y-H (2015) Induction of interleukin-9-producing mucosal mast cells promotes susceptibility to IgE-mediated experimental food allergy. Immunity 43(4):788–802. doi:10.1016/j.immuni.2015.08.020
46. Szabo SJ, Kim ST, Costa GL, Zhang X, Fathman CG, Glimcher LH (2000) A novel transcription factor, T-bet, directs Th1 lineage commitment. Cell 100(6):655–669
47. Zheng W, Flavell RA (1997) The transcription factor GATA-3 is necessary and sufficient for Th2 cytokine gene expression in CD4 T cells. Cell 89(4):587–596
48. Hori S, Nomura T, Sakaguchi S (2003) Control of regulatory T cell development by the transcription factor Foxp3. Science (New York, NY) 299(5609):1057–1061. doi:10.1126/science.1079490
49. Ivanov II, McKenzie BS, Zhou L, Tadokoro CE, Lepelley A, Lafaille JJ, Cua DJ, Littman DR (2006) The orphan nuclear receptor RORgammat directs the differentiation program of proinflammatory IL-17+ T helper cells. Cell 126(6):1121–1133. doi:10.1016/j.cell.2006.07.035
50. Angkasekwinai P, Chang SH, Thapa M, Watarai H, Dong C (2010) Regulation of IL-9 expression by IL-25 signaling. Nat Immunol 11(3):250–256. doi:10.1038/ni.1846
51. Mikami N, Miyagi Y, Sueda K, Takatsuji M, Fukada S, Yamamoto H, Tsujikawa K (2013) Calcitonin gene-related peptide and cyclic adenosine 5′-monophosphate/protein kinase A pathway promote IL-9 production in Th9 differentiation process. J Immunol 190(8):4046–4055. doi:10.4049/jimmunol.1203102
52. Yao W, Zhang Y, Jabeen R, Nguyen ET, Wilkes DS, Tepper RS, Kaplan MH, Zhou B (2013) Interleukin-9 is required for allergic airway inflammation mediated by the cytokine TSLP. Immunity 38(2):360–372. doi:10.1016/j.immuni.2013.01.007

53. Elyaman W, Bradshaw EM, Uyttenhove C, Dardalhon V, Awasthi A, Imitola J, Bettelli E, Oukka M, van Snick J, Renauld JC, Kuchroo VK, Khoury SJ (2009) IL-9 induces differentiation of TH17 cells and enhances function of FoxP3+ natural regulatory T cells. Proc Natl Acad Sci U S A 106(31):12885–12890. doi:10.1073/pnas.0812530106
54. Liu Y, Teige I, Birnir B, Issazadeh-Navikas S (2006) Neuron-mediated generation of regulatory T cells from encephalitogenic T cells suppresses EAE. Nat Med 12(5):518–525. doi:10.1038/nm1402
55. Roediger B, Weninger W (2015) Group 2 innate lymphoid cells in the regulation of immune responses. Adv Immunol 125:111–154. doi:10.1016/bs.ai.2014.09.004
56. Wilhelm C, Hirota K, Stieglitz B, Van Snick J, Tolaini M, Lahl K, Sparwasser T, Helmby H, Stockinger B (2011) An IL-9 fate reporter demonstrates the induction of an innate IL-9 response in lung inflammation. Nat Immunol 12(11):1071–1077. doi:10.1038/ni.2133
57. Wilhelm C, Turner JE, Van Snick J, Stockinger B (2012) The many lives of IL-9: a question of survival? Nat Immunol 13(7):637–641. doi:10.1038/ni.2303
58. Spits H, Di Santo JP (2011) The expanding family of innate lymphoid cells: regulators and effectors of immunity and tissue remodeling. Nat Immunol 12(1):21–27. doi:10.1038/ni.1962
59. Turner JE, Morrison PJ, Wilhelm C, Wilson M, Ahlfors H, Renauld JC, Panzer U, Helmby H, Stockinger B (2013) IL-9-mediated survival of type 2 innate lymphoid cells promotes damage control in helminth-induced lung inflammation. J Exp Med 210(13):2951–2965. doi:10.1084/jem.20130071
60. Stordeur P, Poulin LF, Craciun L, Zhou L, Schandene L, de Lavareille A, Goriely S, Goldman M (2002) Cytokine mRNA quantification by real-time PCR. J Immunol Methods 259(1–2):55–64
61. Kubista M, Andrade JM, Bengtsson M, Forootan A, Jonak J, Lind K, Sindelka R, Sjoback R, Sjogreen B, Strombom L, Stahlberg A, Zoric N (2006) The real-time polymerase chain reaction. Mol Aspects Med 27(2–3):95–125. doi:10.1016/j.mam.2005.12.007
62. Livak KJ, Schmittgen TD (2001) Analysis of relative gene expression data using real-time quantitative PCR and the 2(−Delta Delta C(T)) Method. Methods (San Diego, CA) 25(4):402–408. doi:10.1006/meth.2001.1262
63. Renauld JC, Van Snick J (2002) Measurement of mouse and human interleukin 9. Curr Protoc Immunol Chapter 6:Unit 6.13. doi:10.1002/0471142735.im0613s51
64. Shevach EM (2001) Labeling cells in microtiter plates for determination of [3H]thymidine uptake. Curr Protoc Immunol Appendix 3:Appendix 3D. doi:10.1002/0471142735.ima03ds21
65. Avanzi GC, Brizzi MF, Giannotti J, Ciarletta A, Yang YC, Pegoraro L, Clark SC (1990) M-07e human leukemic factor-dependent cell line provides a rapid and sensitive bioassay for the human cytokines GM-CSF and IL-3. J Cell Physiol 145(3):458–464. doi:10.1002/jcp.1041450310

Chapter 3

IL-9 Signaling Pathway: An Update

Dijendra Nath Roy and Ritobrata Goswami

Abstract

Since the discovery of IL-9 almost three decades back as a growth factor, we have come a long way to understand its pleiotropic functions in the immune system. Despite its many functions, IL-9 still remains as an understudied cytokine. In the last decade, renewed emphasis has been provided to understand the biology of IL-9. Any growth factor or cytokine signals via its cognate receptor to mediate biological functions. In this chapter, we discuss the IL-9 signal transduction in different cell types, which would then exert its distinct functions.

Key words IL-9, Mast cells, B cells, Epithelial cells, Eosinophils, Innate lymphoid cells

1 Introduction

IL-9, a T cell and mast cell growth factor, can have pleiotropic function on multiple immune and nonimmune cells. IL-9 belongs to the four-helix cytokine family [1]. IL-9 signal transduction requires the two chain IL-9 receptors: the IL-9Rα and the common γ chain, which is also involved in IL-2, IL-4, IL-7, and IL-15 signaling. IL-9 receptor is found in membrane-bound and soluble forms. As evident in other growth factors and cytokine signaling, activation of Jak-STAT pathway is observed in during IL-9-mediated signaling [2]. STAT1, 3, and 5 become activated when one tyrosine residue of the IL-9 receptor is induced after IL-9 binding [3]. This activation leads to the transport of the specific STAT molecules to the nucleus to drive transcription of STAT-dependent genes. There is also activation and recruitment of ERK/MAPK pathway and IRS1 molecules after the induction of IL-9 [3–5]. In T cells and mast cells, IL-9 enhances the expression of Bcl3 thereby regulating NF-κB activity [6]. Recruitment of IRS1 and PI3K pathway after IL-9 stimulation leads to activation of Akt pathway [7]. The expression of SOCS family of genes is also induced by IL-9 that would inhibit Jak-STAT pathway [8].

Ritobrata Goswami (ed.), *Th9 Cells: Methods and Protocols*, Methods in Molecular Biology, vol. 1585,
DOI 10.1007/978-1-4939-6877-0_3, © Springer Science+Business Media LLC 2017

IL-9 can have different effects when it acts on various cell types. When activated by IL-4, IL-9 augments IgG, IgM, and IgE production by normal human B cells [9]. IL-9 potentiates the proliferation of mast cells in the presence of IL-3 [10]. While acting on mast cells, IL-9 is responsible for IL-6 induction by mast cell lines leading to functional changes [10]. In hematopoietic cells there is increased BFU-E formation as a result of synergistic effects of IL-9 with stem cell factor [11, 12]. IL-9 also affects airway epithelial cells and airway smooth muscle cells [2]. In this chapter, we will discuss the pathways induced by IL-9 signaling in different immune and nonimmune cells thereby exerting its functions. The role of IL-9 expressed by Th9 cells is described in this book elsewhere.

2 IL-9 Signaling in B Cells

As is common with cytokine signaling, IL-9/IL-9 receptor signaling activates the evolutionary conserved Jak-STAT pathway. Jak-STAT pathway plays central role in the innate and adaptive immune system. The first evidence established that IL-9 signaling could affect B cells is because of the expression of the IL-9Rα chain on primary human tonsillar B lymphocytes; with higher expression on germinal center (GC) B cells and B cells with a memory phenotype [13]. IL-9 receptor is also expressed by peritoneal B-1 cells [14]. However, administration of IL-9 in *xid* (mice bearing X-linked immunodeficiency mutation) mice fails to restore the function of IL-9 [15]. The presence of IL-9Rα on GC cells was specifically shown to be functional, because stimulation with IL-9 resulted in increased phosphorylation of STAT3 and STAT5 transcription factors and the capability of IL-9 to potentiate IgE production [16]. Besides, the appearance of CD27 on IL-9Rα low-density B cells describes the phenotype of IgE producers in the context of CD40 stimulation of B cells as memory B cells [16]. This study suggested that the role of IL-9 is in the late phase of B cell differentiation in secondary lymphoid follicles, which is associated with regulation of peripheral B cell differentiation and immunoglobulin production in human beings [16]. Another study demonstrated that there was an elevated serological level of IL-9 in some B cell Non-Hodgkin lymphoma patients including several diffuse large B cell lymphoma (DLBCL) cases that correlated with the effect of IL-9 on the biological behavior of DLBCL cell lines in vitro [17]. Another study demonstrated the activities of IL-9 and IL-9R in diffuse DLBCL using pathological tissues from DLBCL patients, where there was a correlation of higher expression of IL-9R within DLBCL tissues compared with hyperplasic lymph nodes [18]. Immunohistochemical analysis suggested the membrane localization of IL-9R in approximately 60% of DLBCL cases [18]. The upregulated expression of IL-9R correlated to the serum levels of β2-microglobulin and albumin in which low level of albumin and β2-microglobulin might act as a biomarker of aggressive progression of disease. The authors further tried to find out the association between IL-9R and Ki-67 expression, which could be correlated to the pro-proliferative activity of IL-9 [18]. The expression intensity of IL-9R correlated with adverse prognostic indicators of DLBCL and Ki-67 expression in pathological sections of tumor [18]. The participation of IL-9 in DLBCL indicates that specific silencing of the IL-9R gene may serve as potentially therapeutic in the clinical management of DLBCL [17].

IL-9 provides anti-tumor immunity. Neutralizing IL-9 with antibodies resulted in attenuated inhibition of tumor growth in murine models of B cell lymphoma [19]. The overexpression of IL-9 controls lung fibrosis induced by silica particles in mice [20], which is associated with an expansion of lung B lymphocytes [21]. To explore the influence of these cells while assessing the protective effect of IL-9, IL-9 transgenic mice and B-cell-deficient mice were crossed to check whether IL-9 acted as anti-fibrotic factor [20, 21]. Interestingly, anti-fibrotic effect was abolished in mice deficient in B lymphocytes and was restored by reconstituting these mice with B lymphocytes. The expression of the anti-fibrotic mediator prostaglandin E2 was markedly increased in the lung of IL-9 transgenic mice compared to baseline, which was similar in both wild-type and transgenic strains upon silica treatment [20, 21]. This prostaglandin E2 expression was completely abolished in B cell deficient mice, both at baseline and upon silica administration [20, 21]. Overall, these results suggest that IL-9 signaling impacts B cell leading to regulation of immune system.

3 IL-9 Signaling in Mast Cells

IL-9 imparts its effects on mast cells along with other cytokines and growth factors. In the presence of stem cell factor (SCF) and IL-3, IL-9 regulates mast cell growth and differentiation [22]. IL-9 augments the expression of mast cell proteases [10, 23]. Apart from being a growth factor for mast cells, IL-9 signaling is required for mast cell functional responses. Along with the SCF or FcεRl, IL-9 augments pro-inflammatory cytokines secreted by mast cells [2]. By enhancing mast cell numbers and function in the airways, IL-9 is responsible for airway hyperresponsiveness. Mast cells and granulocytes-macrophages (GM) are important regulatory cellular module of the host defense system against worm infections, including schistosomiasis. In 1996, one report demonstrated the kinetics of changes in the number of colony-forming cells (CFC) for mast cells and GM during the course of a primary experimental infection of mice with *Schistosoma mansoni cercariae* over a period of 24-week post-infection [24]. In that particular study, increased GM-CFC numbers correlated with elevated level of IL-3, while increased mast cell-CFC numbers associated with increased IL-9 concentration in spleen cell-conditioned medium [24]. The result demonstrated that during the course of infection of mice with *S. mansoni*, the synchronization requirement for mast cells and GM is at least partly regulated at the stage of progenitor cell commitment in the bone marrow and spleen. The generation of mature effector cells is stimulated by IL-3 and IL-9 in mice [24]. Another study evaluated the activity of IL-9 on human alveolar macrophages (AM) from healthy volunteers [25]. That particular study suggested that IL-9 and IL-4 act through different mechanisms [25] including IFN-γ that antagonise the effect of IL-4, but not the IL-9-mediated inhibition of AM oxidative burst. There is down-regulation of CD14 expression by IL-4 but not by IL-9; production of TGF-β by activated AM is stimulated by IL-9 and not by IL-4, which is also required for the IL-9-mediated inhibition of AM oxidative burst [25]. These results provide evidence related to the activity of IL-9 in the lung, which may have an inflammatory role and/or fibrotic function in lung disease. *Il9* mRNA has been shown to be significantly increased in patients with chronic bronchitis [26]. Co-localization studies demonstrated more than 80% of all IL-9 positive cells are CD3+ T cells. This result concludes that an increased presence of T lymphocytes and up-regulation of IL-9 and IFN-γ-inducible protein-10 mRNA expressions in the bronchial tissue may contribute to the airway obstruction in chronic bronchitis patients [27].

4 IL-9 Signaling in Eosinophils

IL-9 imparts its biological activities in multiple cell types of hematopoietic lineage including eosinophils. As eosinophils play important role in the pathophysiology of asthma, one study investigated the role of IL-9 in asthma using IL-9 transgenic mice in which expression of the murine *Il9* cDNA was regulated by the rat clara cell 10 protein promoter [28]. Lung selective expression of IL-9 caused massive airway inflammation with eosinophils and lymphocytes as predominant infiltrating cell types [28]. Moreover, a striking finding was the manifestation of increased numbers of mast cells within the airway epithelium of IL-9 transgenic mice [28]. Physiologic evaluation of IL-9 transgenic mice demonstrated markedly increased airway hyperresponsiveness leading to increased airway resistance during challenge studies [28]. Another study investigated the contribution of IL-9 to the pathogenesis of asthma by examining the expression of IL-9 and its IL-9R in bronchial tissue from the human subjects [29]. There was a substantial increase in expression of *Il9* mRNA and IL-9 immunoreactive cells in asthmatic airways compared with healthy control subjects [29].

Scientists also have examined the role of IL-9 in eosinophil differentiation using the human cord blood CD34+ cells and human promyelocytic leukemia cells (HL-60) [30]. The addition of IL-9 to CD34+ cells cultured in the presence of IL-3 and IL-5 enhanced eosinophil development, and IL-9 alone induced the expression of IL-5Rα [30]. IL-9 also up-regulated the IL-5Rα chain cell surface expression during terminal eosinophil differentiation of HL-60 cell line [30]. These findings indicate that IL-9 may contribute in vivo eosinophil function by increasing their survival and IL-5-mediated differentiation and maturation. To find out the role of IL-9 pathway in eosinophilic tissue inflammation and its mechanism in affecting the eosinophils, a model of thioglycolate-induced peritoneal inflammation was developed in IL-9 transgenic (TG5) and background strain as control (FVB) mice [31]. In this model, a transient eosinophilic infiltration in the peritoneal cavity was observed in FVB mice, followed by IL-5 and IL-9 release [31]. In contrast, TG5 mice developed a massive eosinophilia (~80% of total cells) after thioglycolate injection. Thus, IL-9 may trigger and sustain eosinophilic inflammation via IL-5-driven eosinophilic maturation of precursors.

Expression of *Il9* mRNA and protein by human peripheral blood eosinophils was investigated to find out the role of IL-1β and TNF-α in releasing of IL-9 from cells [32]. Cultured human peripheral blood eosinophils isolated from asthmatic subjects produce IL-9 protein when stimulated with TNF-α and IL-1β [32]. These observations provide fundamental evidence for the role of eosinophils in regulating airway immune responses. To define the contribution of IL-9 during lung inflammation, transgenic mice was developed by Flavell and colleagues in which lung-specific expression of the IL-9 transgene is inducible by doxycycline [33]. IL-9 transgene induction caused in lymphocytic and eosinophilic infiltration in lung, formation of airway epithelial cell hypertrophy with mucus production, and mast cell hyperplasia, similar to what is observed in mice having constitutive expression of IL-9 in their lungs. Various cytokines, including IL-4, IL-5, and IL-13, were expressed in the lung in response to IL-9 [33]. Inhibition of IL-4 or IL-5 following IL-9 induction decreased airway eosinophilia without affecting mucus production [33]. In contrast, neutralization of IL-13 completely abolished both lung inflammation and mucus production. These results clearly informed that pathologic changes in the lung develop due to additional signals beyond IL-9, which could be provided by IL-4, IL-5, and IL-13, suggesting an indirect role of IL-9 [33]. One interesting report suggested that IL-9 contributes to the regulation of granulocytopoiesis in allergic inflammation [34]. Similarly, another study observed that the number of *Il9* mRNA expressing cells present in the airway of asthmatics significantly correlate with the numbers of eosinophils, but not mast cells or T cells [35]. This finding suggests that the elevated expression of IL-9 in allergic inflammation may contribute to local eosinophil infiltration and survival in asthma and other allergic atopic diseases. How IL-9 is critically involved in heart transplant eosinophilia in association with IL-4 and IL-5 has been defined in another study [36]. In this study, accumulation of *Il9* mRNA is observed in rejected graft of CD8-deficient mice. Expression of IL-9R and its effects on cultured human airway smooth muscle (ASM) cells were assessed by measuring the release of Th2-associated chemokines including eotaxin1/CCL11 and thymus and activation-regulated chemokine (TARC)/CCL17 [37].

Application of IL-9 on human ASM cells released eotaxin1/CCL11, but had no effect on the release of TARC/CCL17, in time- and dose-dependent manner [37]. One interesting study remarked that a single dose of a neutralizing IL-9 antibody is not adequate to diminish allergen-induced invasion of newly produced cells from bone marrow to airways [34]. Moreover, bone marrow cells express higher level of IL-9 in the presence of an allergen, which suggests that IL-9 might have a role in regulating of granulocytopoiesis in allergic inflammation. The eosinophils in the airways in asthma being accumulated directly react with house dust mite and produce IL-9 to further stimulate Th2-type immune responses [38]. The effects of IL-9 blockade on chronic airway inflammation have been evaluated [39]. The levels of IL-9, IL-17, and IFN-γ were significantly enhanced in the chronic group compared to the acute asthma group but the number of IL-9+ cells was not altered, with a diminution in Th17 cells in OVA-challenged caspase-1 knockout mice [39]. The numbers of eosinophils, neutrophils, B cells, mast cells, and Th17 cells were diminished, concomitant with lower levels of IgE, IL-5, IL-9, and IL-17 after administration of neutralizing IL-9 antibody [39]. This finding indicates that anti-IL-9 antibody treatment stops pulmonary infiltration of inflammatory cells and cytokine production, especially IL-17 [39]. Further studies have demonstrated that Th9 cells are responsible for inducing bronchial hyperresponsiveness independently of eosinophils and the signature cytokines, IL-9 and IL-10 [40].

5 IL-9 Signaling in Epithelial Cells

The epithelial tissue, one of the connective tissues that lines the respiratory and the gastrointestinal tracts, covers an area of several hundred square meters in adults. This surface of epithelial cells exposes the underlying mucosal immune system to potentially harmful substances including allergens and infectious agents. In particular, IL-9 is an important regulatory cytokine for mucosal immunity possibly due to expression of IL-9R on epithelial cells and surrounding immune cells at the large mucosal surfaces. As recent studies have identified an important regulatory role of

IL-9 in monitoring the critical balance between homeostasis and inflammation, the signaling of this cytokine in the mucosal immune system with special reference to allergic asthma, inflammatory bowel diseases, and others will be highlighted here.

Mucin production is inhibited when IL-9 signaling is blocked in the lavage of allergen challenged animals [41]. In contrast, administration of recombinant IL-9 augmented mucin gene transcription in respiratory epithelial cells [41]. The expression of another mucin gene, MUC4, is augmented by IL-9 in airway epithelial cells [42]. During lung eosinophilia in an allergic airway inflammation model, IL-9 plays a critical role by altering the expression of CC-type chemokines [43]. Recombinant IL-9 appears to enhance the expression of the chemokines in primary epithelial cell cultures [43]. IL-9 induces the release of T cell chemoattractant factors by airway epithelial cells [44]. IL-9 treatment aids in the release of IL-16 by HBE4-E6/E7 cells, but not BEAS-2B cells [44]. In primary epithelial cells, IL-9 induces the release of IL-16 and RANTES in a dose-dependent manner [44]. In epithelial cells, function of IL-9 in vivo could depend on other cytokines including IL-13 [45].

IL-9 affects changes in epithelial cell gene manifestation leading to goblet cell metaplasia, while IL-17C is functionally distinct in that its expression can be induced by bacterial challenge and inflammatory stimuli [46]. This study has compared the levels of IL-9 and IL-17C in chronic rhinosinusitis (CRS) with nasal polyps (CRSwNP) and without nasal polyps (CRSsNP) as well as atopy [46]. The level of expression of IL-9 was greater in CRSwNP than CRSsNP and control [46]. Similar findings were observed for IL-17C with higher expression in CRSwNP than CRSsNP and control. The average number of cells expressing IL-9 and IL-17C was higher in CRS with asthma and allergy compared to control [46]. This suggested that IL-9 and IL-17C were both involved in the pathogenesis of CRS, allergy, and asthma. Inflammatory cell expression of IL-9 and IL-17C was increased in CRS, particularly with allergy and asthma. These interleukins may play a vital role in pathogenesis of CRSwNP as well as atopy [46]. The expression of Th9-related cytokines and its distribution over the epithelial tissue in psoriatic arthritis (PsA) patients has also been studied. A significant up-regulation of IL-9 was observed in PsA gut, especially among infiltrating mononuclear cells, high endothelial venules, and Paneth cells [47]. IL-9-expressing mononuclear cells were observed in large part of Th9 cells [47]. IL-9 overexpression was accompanied by significant Paneth cell hyperplasia [47]. Paneth cells strongly overexpressed IL-9R, and stimulation of epithelial cells, isolated from PsA patients, with IL-9 resulted in overexpression of α-defensin 5 and IL-23p19. Peripheral and synovial expansion of α4β7+ Th9 cells was also observed in patients with PsA [47]. Increased expression of IL-9 and IL-9R was also found in synovial tissue. Resilient IL-9/Th9 polarization appears to be the predominant immunologic signature in patients in PsA [47].

A seminal study reported the crucial role for IL-9 in antitumor immunity generated via glucocorticoid-induced tumor necrosis factor receptor (TNFR)-related protein (GITR) with an agonistic GITR-specific antibody DTA-1 [48]. The IL-4 receptor knockout mice, which reduced the expression of IL-9, were resistant to tumor growth inhibition by DTA-1 [48]. Notably, neutralization of IL-9 considerably impaired tumor rejection induced by DTA-1 [48]. In particular, DTA-1-induced IL-9 promoted tumor-specific cytotoxic T lymphocyte (CTL) responses by enhancing the function of dendritic cells in vivo. Furthermore, GITR signaling enhanced the differentiation of IL-9-producing Th9 cells in a TNFR-associated factor-6 and NF-κB-dependent manner and inhibited the generation of induced regulatory T cells in vitro [48]. These findings demonstrate that GITR co-stimulation mediates anti-tumor immunity by promoting Th9 cell differentiation and enhancing CTL responses and thus provide a mechanism of action for GITR agonist-mediated cancer immunotherapies [48]. Th9 cells and IL-9 have also been detected in all Respiratory Epithelial Adenomatoid Hamartoma (REAH) patients [49]. The proportion of Th9 cells, the expression levels of *Il9* mRNA and IL-9 protein were significantly higher in the patients with REAH than in the controls. All these findings suggested that Th9 cells play a central role in the pathogenesis of this disease [49]. IL-9 induced a moderate production of IL-8 and formation of STIM1-puncta but not CXCL16, CCL22, TSLP, and IL-33 [50]. IL-9 induced ERK phosphorylation in both dose- and time-dependent manner, but not mTOR, S6K, p38, and STAT3 [50]. This is the first study to demonstrate the IL-9-STIM1-ERK-IL-8 axis in keratinocyte, and how this axis might play an important role in the pathophysiology of atopic dermatitis (AD) [50].

IL-9 could impact the junction protein in epithelial cells. It has been demonstrated that sealing proteins like occludin were up-regulated but the tight junction protein like Claudin1 showed less appearance in the inflamed colon of IL-9-deficient mice [50]. Additionally, the pore-forming molecule Claudin2 revealed expression in IL-9-deficient mice. These results illustrate the pleiotropic functions of IL-9 in changing intestinal permeability in experimental colitis. Thus, modulation of IL-9 function emerges as a new approach for regulating barrier function in intestinal inflammation [51].

On the same note, studies have been conducted to find out the role of IL-9 in inflammatory bowel disease. *Il9* mRNA expression was significantly augmented in patients with ulcerative colitis (UC). CD3+ T cells are the major IL-9-expressing cells in the inflamed gut. Surprisingly, IL-9 was co-localized with Th9 transcription factors, IRF4 and PU.1 [52]. Furthermore, IL-9R was overexpressed on gut epithelial cells and IL-9-induced STAT5 activation was also observed in these cells [52]. These results provide strong evidence that IL-9 could be involved in the pathogenesis of UC [52].

Higher quantities of T cells express the transcription factor PU.1 and IL-9 in ulcerative colitis patients [53]. Interestingly, mice with T cell specific deletion of PU.1 were safe from colitis, whereas treatment with antibody to IL-9 suppressed colitis [53]. Essentially, IL-9 reduced the intestinal barrier function and prevented mucosal wound healing in vivo. Thus, these findings suggest that the Th9 subset of helper T cells serves an important role in driving ulcerative colitis by regulating intestinal epithelial cells [53].

The proliferation of Th9 cell subset, up-regulation of PU.1, and increased secretion of IL-9 contributed to the pathogenesis of atopic dermatitis, which is also supported by the increased release of VEGF by keratinocytes after IL-9 stimulation [54].

IL-9 does not enhance goblet cell numbers in bronchial epithelial cell cultures from normal or asthmatic children. IL-9 and IL-13, either alone or in combination, attenuate ciliated cell numbers and transepithelial electrical resistance during differentiation of normal epithelium, which clinically stops mucociliary clearance [55]. Therefore, acting on epithelial cells, IL-9 can regulate various inflammatory responses.

6 IL-9 Signaling in Innate Lymphoid Cells

Innate lymphoid cells (ILCs), described very recently, belong to the lymphoid lineage, yet they do not have antigen-specific B- or T cell receptor [56]. These cells have been attributed to immune homeostasis, since their dysregulation could be responsible for the development of inflammatory disorders [56]. Based on the cytokines they produce and the transcription factors that are required for their development, ILCs can be divided into three broad groups among which type 2 innate lymphoid cells (ILC2s) produce IL-9 and play critical role against helminth infection [57].

In asthmatic patients the expression of IL-25, IL-22, IL-17, and IL-9 cytokines is increased, which also correlated with increased number of innate lymphoid cells in these patients [58]. Another study focused on the role of IL-9 and ILC2s in lung infection with *Nippostrongylus brasiliensis*, which resulted in considerable tissue injury [59]. Interestingly, in this study IL-9 receptor deficient mice showed lesser numbers of ILC2s present in the lung after infection [59]. Subsequently, a compromised IL-5, IL-13, and amphiregulin level was observed despite enhanced numbers of Th2 cells [59]. As a result, the reestablishment of tissue integrity was strongly inhibited due to the absence of IL-9 signaling. This outcome directs the unique role for IL-9 as an autocrine enhancer of ILC2 function, promoting tissue recovery from damage after helminth-induced lung inflammation [59]. In a different study during papain-induced lung inflammation, IL-9 production by ILCs depended on the presence of IL-2 from adaptive immune cells and was inhibited in the presence of IL-13 and IL-5. This result indicated that ILCs provide the missing link between the well-established functions of IL-9 in the regulation of type 2 helper T cell cytokines and its functions [60]. In response to helminths, alveolar cells produce IL-33 and TSLP, both of which are required to create IRF4-IL-9 axis in ILC2s [61]. IL-9 thus generated is critical to maintain epithelial cells during lung homeostasis [61]. IL-9 receptor highly expressed in eosinophilic esophagitis patients may require expanded ILC2s to mediate the pathophysiology of the disease. Thus, IL-9/IL-9 receptor signaling could impact multiple inflammatory processes with the help of ILCs.

7 Conclusion

In this chapter, we have summarized the effect of IL-9-mediated signaling on various cell types. The innate immune and nonimmune cells have been the focus of the IL-9 signaling in this chapter. Even though ILCs are very recently discovered lymphoid cell types, given the function of this cell type we anticipate further studies will unravel the impact of IL-9 on ILCs. Type 2 innate lymphoid cells strongly yield IL-9 in mice, not only those are challenged with papain and helminthic parasites, but also stimulated by IL-2 in vitro [60, 62]. Moreover, mast cells also produce IL-9 in immunological response to cytokine signals and Toll-like receptor [63–65]. CD8+ T cells have been demonstrated to produce IL-9 in animal model [66, 67]. IL-9 is also produced by natural killer T cells in allergic and autoimmune inflammation model [68, 69]. CD4+ T cells such as Th2, Th17, and regulatory T cells have been established to create IL-9, although the quantities of IL-9 expressed by each of these cell types are generally very small amount compared to production by a Th9 cell [70–72].

A number of studies published within the last few years demonstrated that IL-9 triggers the development of allergic and autoimmune diseases [73]. Many of these studies used different genetically manipulated murine models including IL-9-transgenic, IL-9-deficient, and IL-9-receptor-deficient mice but those studies occasionally led to contradictory results. The main reason of the discrepancies is due to the generation of a cocktail of different cytokines in any immunological response in the animal body resulting in the development of a heterogeneous population of immune cells. Further work is required to understand the implication and contribution of different levels of IL-9 in different cell types; those are not yet studied during various immune responses.

References

1. Wang X, Lupardus P, Laporte SL, Garcia KC (2009) Structural biology of shared cytokine receptors. Annu Rev Immunol 27:29–60. doi:10.1146/annurev.immunol.24.021605.090616
2. Goswami R, Kaplan MH (2011) A brief history of IL-9. J Immunol 186(6):3283–3288. doi:10.4049/jimmunol.1003049
3. Demoulin JB, Uyttenhove C, Van Roost E, DeLestre B, Donckers D, Van Snick J, Renauld JC (1996) A single tyrosine of the interleukin-9 (IL-9) receptor is required for STAT activation, antiapoptotic activity, and growth regulation by IL-9. Mol Cell Biol 16(9):4710–4716
4. Yin T, Keller SR, Quelle FW, Witthuhn BA, Tsang ML, Lienhard GE, Ihle JN, Yang YC (1995) Interleukin-9 induces tyrosine phosphorylation of insulin receptor substrate-1 via JAK tyrosine kinases. J Biol Chem 270(35):20497–20502
5. Demoulin JB, Louahed J, Dumoutier L, Stevens M, Renauld JC (2003) MAP kinase activation by interleukin-9 in lymphoid and mast cell lines. Oncogene 22(12):1763–1770. doi:10.1038/sj.onc.1206253
6. Richard M, Louahed J, Demoulin JB, Renauld JC (1999) Interleukin-9 regulates NF-kappaB activity through BCL3 gene induction. Blood 93(12):4318–4327
7. Xiao H, Yin T, Wang XY, Uchida T, Chung J, White MF, Yang YC (2002) Specificity of interleukin-2 receptor gamma chain superfamily cytokines is mediated by insulin receptor substrate-dependent pathway. J Biol Chem 277(10):8091–8098. doi:10.1074/jbc.M106650200
8. Lejeune D, Demoulin JB, Renauld JC (2001) Interleukin 9 induces expression of three cytokine signal inhibitors: cytokine-inducible SH2-containing protein, suppressor of cytokine signalling (SOCS)-2 and SOCS-3, but only SOCS-3 overexpression suppresses interleukin 9 signalling. Biochem J 353(Pt 1):109–116
9. Dugas B, Renauld JC, Pene J, Bonnefoy JY, Peti-Frere C, Braquet P, Bousquet J, Van Snick J, Mencia-Huerta JM (1993) Interleukin-9 potentiates the interleukin-4-induced immunoglobulin (IgG, IgM and IgE) production by normal human B lymphocytes. Eur J Immunol 23(7):1687–1692. doi:10.1002/eji.1830230743
10. Wiener Z, Falus A, Toth S (2004) IL-9 increases the expression of several cytokines in activated mast cells, while the IL-9-induced IL-9 production is inhibited in mast cells of histamine-free transgenic mice. Cytokine 26(3):122–130. doi:10.1016/j.cyto.2004.01.006
11. Lu L, Leemhuis T, Srour EF, Yang YC (1992) Human interleukin (IL)-9 specifically stimulates proliferation of CD34+++DR+CD33− erythroid progenitors in normal human bone marrow in the absence of serum. Exp Hematol 20(4):418–424
12. Lemoli RM, Fortuna A, Fogli M, Motta MR, Rizzi S, Benini C, Tura S (1994) Stem cell factor (c-kit ligand) enhances the interleukin-9-dependent proliferation of human CD34+ and CD34+CD33−DR− cells. Exp Hematol 22(9):919–923
13. Fawaz LM, Sharif-Askari E, Hajoui O, Soussi-Gounni A, Hamid Q, Mazer BD (2007) Expression of IL-9 receptor alpha chain on human germinal center B cells modulates IgE secretion. J Allergy Clin Immunol 120(5):1208–1215. doi:10.1016/j.jaci.2007.08.022
14. Vink A, Warnier G, Brombacher F, Renauld JC (1999) Interleukin 9-induced in vivo expansion of the B-1 lymphocyte population. J Exp Med 189(9):1413–1423
15. Knoops L, Louahed J, Renauld JC (2004) IL-9-induced expansion of B-1b cells restores numbers but not function of B-1 lymphocytes in xid mice. J Immunol 172(10):6101–6106
16. Petit-Frere C, Dugas B, Braquet P, Mencia-Huerta JM (1993) Interleukin-9 potentiates the interleukin-4-induced IgE and IgG1 release from murine B lymphocytes. Immunology 79(1):146–151
17. Lv X, Feng L, Ge X, Lu K, Wang X (2016) Interleukin-9 promotes cell survival and drug resistance in diffuse large B-cell lymphoma. J Exp Clin Cancer Res 35(1):106. doi:10.1186/s13046-016-0374-3
18. Lv X, Feng L, Fang X, Jiang Y, Wang X (2013) Overexpression of IL-9 receptor in diffuse large B-cell lymphoma. Int J Clin Exp Pathol 6(5):911–916
19. Feng LL, Gao JM, Li PP, Wang X (2011) IL-9 contributes to immunosuppression mediated by regulatory T cells and mast cells in B-cell non-hodgkin's lymphoma. J Clin Immunol 31(6):1084–1094. doi:10.1007/s10875-011-9584-9
20. Arras M, Huaux F, Vink A, Delos M, Coutelier JP, Many MC, Barbarin V, Renauld JC, Lison D (2001) Interleukin-9 reduces lung fibrosis and type 2 immune polarization induced by silica particles in a murine model. Am J Respir Cell Mol Biol 24(4):368–375. doi:10.1165/ajrcmb.24.4.4249
21. Arras M, Louahed J, Simoen V, Barbarin V, Misson P, van den Brule S, Delos M, Knoops

L, Renauld JC, Lison D, Huaux F (2006) B Lymphocytes are critical for lung fibrosis control and prostaglandin E2 regulation in IL-9 transgenic mice. Am J Respir Cell Mol Biol 34(5):573–580. doi:10.1165/rcmb.2004-0383OC
22. Matsuzawa S, Sakashita K, Kinoshita T, Ito S, Yamashita T, Koike K (2003) IL-9 enhances the growth of human mast cell progenitors under stimulation with stem cell factor. J Immunol 170(7):3461–3467
23. Lora JM, Al-Garawi A, Pickard MD, Price KS, Bagga S, Sicoli J, Hodge MR, Gutierrez-Ramos JC, Briskin MJ, Boyce JA (2003) FcepsilonRI-dependent gene expression in human mast cells is differentially controlled by T helper type 2 cytokines. J Allergy Clin Immunol 112(6):1119–1126. doi:10.1016/j.jaci.2003.08.042
24. Khalil RM, Luz A, Mailhammer R, Moeller J, Mohamed AA, Omran S, Dormer P, Hultner L (1996) *Schistosoma mansoni* infection in mice augments the capacity for interleukin 3 (IL-3) and IL-9 production and concurrently enlarges progenitor pools for mast cells and granulocytes-macrophages. Infect Immun 64(12):4960–4966
25. Pilette C, Ouadrhiri Y, Van Snick J, Renauld JC, Staquet P, Vaerman JP, Sibille Y (2002) Oxidative burst in lipopolysaccharide-activated human alveolar macrophages is inhibited by interleukin-9. Eur Respir J 20(5):1198–1205
26. Panzner P, Lafitte JJ, Tsicopoulos A, Hamid Q, Tulic MK (2003) Marked up-regulation of T lymphocytes and expression of interleukin-9 in bronchial biopsies from patients With chronic bronchitis with obstruction. Chest 124(5):1909–1915
27. Tulic MK, Hamid Q (2003) Contribution of the distal lung to the pathologic and physiologic changes in asthma: potential therapeutic target Roger S. Mitchell lecture. Chest 123(3 Suppl):348S–355S
28. Temann UA, Geba GP, Rankin JA, Flavell RA (1998) Expression of interleukin 9 in the lungs of transgenic mice causes airway inflammation, mast cell hyperplasia, and bronchial hyperresponsiveness. J Exp Med 188(7):1307–1320
29. Shimbara A, Christodoulopoulos P, Soussi-Gounni A, Olivenstein R, Nakamura Y, Levitt RC, Nicolaides NC, Holroyd KJ, Tsicopoulos A, Lafitte JJ, Wallaert B, Hamid QA (2000) IL-9 and its receptor in allergic and nonallergic lung disease: increased expression in asthma. J Allergy Clin Immunol 105(1 Pt 1):108–115
30. Gounni AS, Gregory B, Nutku E, Aris F, Latifa K, Minshall E, North J, Tavernier J, Levit R, Nicolaides N, Robinson D, Hamid Q (2000) Interleukin-9 enhances interleukin-5 receptor expression, differentiation, and survival of human eosinophils. Blood 96(6):2163–2171
31. Louahed J, Zhou Y, Maloy WL, Rani PU, Weiss C, Tomer Y, Vink A, Renauld J, Van Snick J, Nicolaides NC, Levitt RC, Haczku A (2001) Interleukin 9 promotes influx and local maturation of eosinophils. Blood 97(4):1035–1042
32. Gounni AS, Nutku E, Koussih L, Aris F, Louahed J, Levitt RC, Nicolaides NC, Hamid Q (2000) IL-9 expression by human eosinophils: regulation by IL-1beta and TNF-alpha. J Allergy Clin Immunol 106(3):460–466. doi:10.1067/mai.2000.109172
33. Temann UA, Ray P, Flavell RA (2002) Pulmonary overexpression of IL-9 induces Th2 cytokine expression, leading to immune pathology. J Clin Invest 109(1):29–39. doi:10.1172/JCI13696
34. Sitkauskiene B, Radinger M, Bossios A, Johansson AK, Sakalauskas R, Lotvall J (2005) Airway allergen exposure stimulates bone marrow eosinophilia partly via IL-9. Respir Res 6:33. doi:10.1186/1465-9921-6-33
35. Ying S, Meng Q, Kay AB, Robinson DS (2002) Elevated expression of interleukin-9 mRNA in the bronchial mucosa of atopic asthmatics and allergen-induced cutaneous late-phase reaction: relationships to eosinophils, mast cells and T lymphocytes. Clin Exp Allergy 32(6):866–871
36. Poulin LF, Richard M, Le Moine A, Kiss R, McKenzie AN, Goldman M, Renauld JC, Van Snick J, Braun MY (2003) Interleukin-9 promotes eosinophilic rejection of mouse heart allografts. Transplantation 76(3):572–577. doi:10.1097/01.TP.0000071201.32424.D2
37. Gounni AS, Hamid Q, Rahman SM, Hoeck J, Yang J, Shan L (2004) IL-9-mediated induction of eotaxin1/CCL11 in human airway smooth muscle cells. J Immunol 173(4):2771–2779
38. Fujisawa T, Katsumata H, Kato Y (2008) House dust mite extract induces interleukin-9 expression in human eosinophils. Allergol Int 57(2):141–146. doi:10.2332/allergolint.O-07-498
39. Kim MS, Cho KA, Cho YJ, Woo SY (2013) Effects of interleukin-9 blockade on chronic airway inflammation in murine asthma models. Allerg Asthma Immunol Res 5(4):197–206. doi:10.4168/aair.2013.5.4.197
40. Saeki M, Kaminuma O, Nishimura T, Kitamura N, Mori A, Hiroi T (2016) Th9 cells elicit eosinophil-independent bronchial hyperresponsiveness in mice. Allergol Int 65S:S24–S29. doi:10.1016/j.alit.2016.05.003
41. Longphre M, Li D, Gallup M, Drori E, Ordonez CL, Redman T, Wenzel S, Bice DE,

Fahy JV, Basbaum C (1999) Allergen-induced IL-9 directly stimulates mucin transcription in respiratory epithelial cells. J Clin Invest 104(10):1375–1382. doi:10.1172/JCI6097

42. Damera G, Xia B, Ancha HR, Sachdev GP (2006) IL-9 modulated MUC4 gene and glycoprotein expression in airway epithelial cells. Biosci Rep 26(1):55–67. doi:10.1007/s10540-006-9000-5
43. Dong Q, Louahed J, Vink A, Sullivan CD, Messler CJ, Zhou Y, Haczku A, Huaux F, Arras M, Holroyd KJ, Renauld JC, Levitt RC, Nicolaides NC (1999) IL-9 induces chemokine expression in lung epithelial cells and baseline airway eosinophilia in transgenic mice. Eur J Immunol 29(7):2130–2139
44. Little FF, Cruikshank WW, Center DM (2001) Il-9 stimulates release of chemotactic factors from human bronchial epithelial cells. Am J Respir Cell Mol Biol 25(3):347–352. doi:10.1165/ajrcmb.25.3.4349
45. Steenwinckel V, Louahed J, Orabona C, Huaux F, Warnier G, McKenzie A, Lison D, Levitt R, Renauld JC (2007) IL-13 mediates in vivo IL-9 activities on lung epithelial cells but not on hematopoietic cells. J Immunol 178(5):3244–3251
46. Olcott CM, Han JK, Cunningham TD, Franzese CB (2016) Interleukin-9 and interleukin-17C in chronic rhinosinusitis. Int Forum Allerg Rhinol 6(8):841–847. doi:10.1002/alr.21745
47. Ciccia F, Guggino G, Ferrante A, Raimondo S, Bignone R, Rodolico V, Peralta S, Van Tok M, Cannizzaro A, Schinocca C, Ruscitti P, Cipriani P, Giacomelli R, Alessandro R, Dieli F, Rizzo A, Baeten D, Triolo G (2016) Interleukin-9 overexpression and Th9 polarization characterize the inflamed gut, the synovial tissue, and the peripheral blood of patients with psoriatic arthritis. Arthritis Rheumatol 68(8):1922–1931. doi:10.1002/art.39649
48. Kim IK, Kim BS, Koh CH, Seok JW, Park JS, Shin KS, Bae EA, Lee GE, Jeon H, Cho J, Jung Y, Han D, Kwon BS, Lee HY, Chung Y, Kang CY (2015) Glucocorticoid-induced tumor necrosis factor receptor-related protein co-stimulation facilitates tumor regression by inducing IL-9-producing helper T cells. Nat Med 21(9):1010–1017. doi:10.1038/nm.3922
49. Gu ZW, Wang YX, Cao ZW (2015) T-Helper type 9 cells play a central role in the pathogenesis of respiratory epithelial adenomatoid hamartoma. Medicine (Baltimore) 94(26):e1050. doi:10.1097/MD.0000000000001050
50. Hong CH, Chang KL, Wang HJ, Yu HS, Lee CH (2015) IL-9 induces IL-8 production via STIM1 activation and ERK phosphorylation in epidermal keratinocytes: a plausible mechanism of IL-9R in atopic dermatitis. J Dermatol Sci 78(3):206–214. doi:10.1016/j.jdermsci.2015.03.004
51. Gerlach K, McKenzie AN, Neurath MF, Weigmann B (2015) IL-9 regulates intestinal barrier function in experimental T cell-mediated colitis. Tissue barriers 3(1-2):e983777. doi:10.4161/21688370.2014.983777
52. Nalleweg N, Chiriac MT, Podstawa E, Lehmann C, Rau TT, Atreya R, Krauss E, Hundorfean G, Fichtner-Feigl S, Hartmann A, Becker C, Mudter J (2015) IL-9 and its receptor are predominantly involved in the pathogenesis of UC. Gut 64(5):743–755. doi:10.1136/gutjnl-2013-305947
53. Gerlach K, Hwang Y, Nikolaev A, Atreya R, Dornhoff H, Steiner S, Lehr HA, Wirtz S, Vieth M, Waisman A, Rosenbauer F, McKenzie AN, Weigmann B, Neurath MF (2014) TH9 cells that express the transcription factor PU.1 drive T cell-mediated colitis via IL-9 receptor signaling in intestinal epithelial cells. Nat Immunol 15(7):676–686. doi:10.1038/ni.2920
54. Ma L, Xue HB, Guan XH, Shu CM, Zhang JH, Yu J (2014) Possible pathogenic role of T helper type 9 cells and interleukin (IL)-9 in atopic dermatitis. Clin Exp Immunol 175(1):25–31. doi:10.1111/cei.12198
55. Parker JC, Thavagnanam S, Skibinski G, Lyons J, Bell J, Heaney LG, Shields MD (2013) Chronic IL9 and IL-13 exposure leads to an altered differentiation of ciliated cells in a well-differentiated paediatric bronchial epithelial cell model. PLoS One 8(5):e61023. doi:10.1371/journal.pone.0061023
56. Walker JA, Barlow JL, McKenzie AN (2013) Innate lymphoid cells—how did we miss them? Nat Rev Immunol 13(2):75–87. doi:10.1038/nri3349
57. Neill DR, Wong SH, Bellosi A, Flynn RJ, Daly M, Langford TK, Bucks C, Kane CM, Fallon PG, Pannell R, Jolin HE, McKenzie AN (2010) Nuocytes represent a new innate effector leukocyte that mediates type-2 immunity. Nature 464(7293):1367–1370. doi:10.1038/nature08900
58. Sherkat R, Yazdani R, Ganjalikhani Hakemi M, Homayouni V, Farahani R, Hosseini M, Rezaei A (2014) Innate lymphoid cells and cytokines of the novel subtypes of helper T cells in asthma. Asia Pac Allergy 4(4):212–221. doi:10.5415/apallergy.2014.4.4.212
59. Turner JE, Morrison PJ, Wilhelm C, Wilson M, Ahlfors H, Renauld JC, Panzer U, Helmby H, Stockinger B (2013) IL-9-mediated survival of type 2 innate lymphoid cells promotes damage control in helminth-induced lung inflammation. J Exp Med 210(13):2951–2965. doi:10.1084/jem.20130071

60. Wilhelm C, Hirota K, Stieglitz B, Van Snick J, Tolaini M, Lahl K, Sparwasser T, Helmby H, Stockinger B (2011) An IL-9 fate reporter demonstrates the induction of an innate IL-9 response in lung inflammation. Nat Immunol 12(11):1071–1077. doi:10.1038/ni.2133
61. Mohapatra A, Van Dyken SJ, Schneider C, Nussbaum JC, Liang HE, Locksley RM (2016) Group 2 innate lymphoid cells utilize the IRF4-IL-9 module to coordinate epithelial cell maintenance of lung homeostasis. Mucosal Immunol 9(1):275–286. doi:10.1038/mi.2015.59
62. Licona-Limon P, Henao-Mejia J, Temann AU, Gagliani N, Licona-Limon I, Ishigame H, Hao L, Herbert DR, Flavell RA (2013) Th9 cells drive host immunity against gastrointestinal worm infection. Immunity 39(4):744–757. doi:10.1016/j.immuni.2013.07.020
63. Stassen M, Arnold M, Hultner L, Muller C, Neudorfl C, Reineke T, Schmitt E (2000) Murine bone marrow-derived mast cells as potent producers of IL-9: costimulatory function of IL-10 and kit ligand in the presence of IL-1. J Immunol 164(11):5549–5555
64. Stassen M, Muller C, Arnold M, Hultner L, Klein-Hessling S, Neudorfl C, Reineke T, Serfling E, Schmitt E (2001) IL-9 and IL-13 production by activated mast cells is strongly enhanced in the presence of lipopolysaccharide: NF-kappa B is decisively involved in the expression of IL-9. J Immunol 166(7):4391–4398
65. Stassen M, Klein M, Becker M, Bopp T, Neudorfl C, Richter C, Heib V, Klein-Hessling S, Serfling E, Schild H, Schmitt E (2007) p38 MAP kinase drives the expression of mast cell-derived IL-9 via activation of the transcription factor GATA-1. Mol Immunol 44(5):926–933. doi:10.1016/j.molimm.2006.03.019
66. Lu Y, Hong B, Li H, Zheng Y, Zhang M, Wang S, Qian J, Yi Q (2014) Tumor-specific IL-9-producing CD8+ Tc9 cells are superior effector than type-I cytotoxic Tc1 cells for adoptive immunotherapy of cancers. Proc Natl Acad Sci U S A 111(6):2265–2270. doi:10.1073/pnas.1317431111
67. Visekruna A, Ritter J, Scholz T, Campos L, Guralnik A, Poncette L, Raifer H, Hagner S, Garn H, Staudt V, Bopp T, Reuter S, Taube C, Loser K, Huber M (2013) Tc9 cells, a new subset of CD8(+) T cells, support Th2-mediated airway inflammation. Eur J Immunol 43(3):606–618. doi:10.1002/eji.201242825
68. Kim HS, Chung DH (2013) IL-9-producing invariant NKT cells protect against DSS-induced colitis in an IL-4-dependent manner. Mucosal Immunol 6(2):347–357. doi:10.1038/mi.2012.77
69. Jones TG, Hallgren J, Humbles A, Burwell T, Finkelman FD, Alcaide P, Austen KF, Gurish MF (2009) Antigen-induced increases in pulmonary mast cell progenitor numbers depend on IL-9 and CD1d-restricted NKT cells. J Immunol 183(8):5251–5260. doi:10.4049/jimmunol.0901471
70. Chang HC, Sehra S, Goswami R, Yao W, Yu Q, Stritesky GL, Jabeen R, McKinley C, Ahyi AN, Han L, Nguyen ET, Robertson MJ, Perumal NB, Tepper RS, Nutt SL, Kaplan MH (2010) The transcription factor PU.1 is required for the development of IL-9-producing T cells and allergic inflammation. Nat Immunol 11(6):527–534. doi:10.1038/ni.1867
71. Nowak EC, Weaver CT, Turner H, Begum-Haque S, Becher B, Schreiner B, Coyle AJ, Kasper LH, Noelle RJ (2009) IL-9 as a mediator of Th17-driven inflammatory disease. J Exp Med 206(8):1653–1660. doi:10.1084/jem.20090246
72. Lu LF, Lind EF, Gondek DC, Bennett KA, Gleeson MW, Pino-Lagos K, Scott ZA, Coyle AJ, Reed JL, Van Snick J, Strom TB, Zheng XX, Noelle RJ (2006) Mast cells are essential intermediaries in regulatory T-cell tolerance. Nature 442(7106):997–1002. doi:10.1038/nature05010
73. Stassen M, Schmitt E, Bopp T (2012) From interleukin-9 to T helper 9 cells. Ann N Y Acad Sci 1247:56–68. doi:10.1111/j.1749-6632.2011.06351.x

Chapter 4

A Method to In Vitro Differentiate Th9 Cells from Mouse Naïve CD4+ T Cells

Duy Pham

Abstract

CD4⁺ T helper cells with the ability to produce effector cytokines provide host protection by regulating immune responses against pathogens. In contrast, these cells are responsible for the development of various inflammatory disease if not regulated properly. Previous studies using naïve CD4⁺ T cell activation in vitro have shown the requirement of various cytokine combinations in addition to TCR activation to differentiate naïve CD4⁺ T cells into various effector T helper lineages. The recently discovered CD4⁺ T helper subset is IL-9-producing Th9 cells. Since Th9 cell differentiation in vitro is essential in understanding the molecular mechanism in regulating Th9 cell development, it is critical to develop a basic protocol in polarizing naïve CD4⁺ T cells to Th9 cells in vitro. Here we describe a simple method for Th9 cell culture conditions in vitro that can be used for other molecular analyses.

Key words Naïve CD4+ T cells, Th9, IL-9, IL-4, TGF-β

1 Introduction

T helper type 9 (Th9) cells are a new T helper cell subset that has been characterized in recent years. Th9 cells secrete IL-9 that is instrumental in modulating allergic inflammation and immunity to intestinal parasites. IL-9 has broad effects on various cell types that play pivotal role for the development of immunity and inflammation, including smooth muscle cells, stem cells, lymphocytes, mast cells, and epithelial cells [1, 2]. Th9 cell differentiation requires both IL-4 and TGF-β signaling [3, 4]. IL-4 activates STAT6 to induce the expression of transcription factors GATA3, IRF4, and BATF, while TFG-β induces PU.1 expression. These transcription factors are required for optimal Th9 cell differentiation [5–9]. Other cytokines such as TSLP, IL-2, and IL-1 have been shown to modulate IL-9 production by CD4⁺ T cells, yet less is known about their downstream modulators [6, 10]. Thus, proper Th9 cell differentiation protocol in vitro is essential in shedding light on the signaling pathways that control Th9 cell development.

Ritobrata Goswami (ed.), *Th9 Cells: Methods and Protocols*, Methods in Molecular Biology, vol. 1585, DOI 10.1007/978-1-4939-6877-0_4, © Springer Science+Business Media LLC 2017

2 Materials

Prepare and store all reagents at 4 °C (unless indicated otherwise).

2.1 Equipment

1. Centrifuge.
2. Fluorescence-activated cell sorter (FACS).
3. MultiStand for magnetic separators.
4. MiniMACS and MidiMACS magnetic separators.
5. MS and LS columns.
6. 15 and 50 mL polypropylene conical tubes.
7. Polystyrene tissue culture plates.
8. 60 × 15 mm culture dish.
9. Frosted slides.
10. 0.22 and 0.45 μm filters.

2.2 Additional Reagents

1. T cells restimulation cocktail: GolgiStop (Monensin solution), Phorbol 12-myristate 13-acetate (PMA), and ionomycin.
2. Formaldehyde solution for molecular biology, 36.5–38% in H_2O.
3. Trypan blue.

2.3 Mice

We use spleen and lymph nodes from 6–12-week-old C57BL/6 mice. All experiments should be performed with the approval of the University Institutional Animal Care and Use Committee.

2.4 Buffers

1. Prepare RPMI complete medium for mouse under sterile conditions: 500 mL RPMI 1640 with 10% (v/v) fetal bovine serum (FBS), 200 U/mL penicillin–streptomycin, 1 mM sodium pyruvate, 1 mM L-glutamine, 50 μM MEM nonessential amino acids solution, 5 mM HEPES, 50 μM 2-mercaptoethanol. Filter using 0.22 μm filter. Store at 4 °C.
2. Prepare magnetic cell sorting (MACS) buffer under sterile conditions: phosphate-buffered saline (PBS), 0.5% bovine serum albumin (BSA) (w/v), 2 mM EDTA, pH 8. Filter using 0.22 μm filter. Store at 4 °C.
3. Red blood cell lysis buffer: Add 8.3 g NH_4Cl and 10 mM Tris–HCl pH 7.4 to 900 mL water. Mix and adjust pH to 7.5 ± 0.2. Add water to make 1 l, filter using 0.22 μm filter. Autoclave and store at 4 °C.
4. Fluorescence-activated cell sorting (FACS) buffer: PBS, 0.02% BSA (w/v), 0.01% sodium azide (w/v). Mix well and filter using 0.22 μm filter. Store at 4 °C.
5. Permeabilization buffer: PBS, 2% BSA (w/v), 0.1% saponin (w/v), 0.01% sodium azide (w/v). Mix well and filter using 0.22 μm filter. Store at 4 °C.

2.5 Antibodies and Cytokines

See manufacture's datasheet for antibody and cytokine storage conditions. Small aliquots should be made to prevent multiple freeze–thaw cycles.

1. $CD4^+$ T cell biotin-antibody cocktail for depletion of non-$CD4^+$ T cells: biotin-conjugated monoclonal anti-mouse CD8a, CD45R, CD11b, CD25, CD49b, TCRγ/δ, and Ter-119 antibodies.
2. Monoclonal (mouse IgG1) anti-biotin microbeads antibody.
3. Monoclonal anti-CD62L (L-selectin, rat IgG2a) microbeads antibody.
4. Th9-polarizing antibody and cytokine cocktail: purified anti-mouse CD3 (clone 145-2C11), purified anti-mouse CD28 (clone 37.51), and anti-mouse IFN-γ (clone XMG) antibodies; recombinant human TGF-β and IL-2, and recombinant mouse IL-4.
5. Fluorochrome-conjugated anti-mouse IL-9 and IL-4 antibodies.

3 Methods

3.1 Naïve CD4+ T Cell Isolation (See Notes 1–3).

3.1.1 Sample Preparation

1. Process the spleen and/or lymph nodes by the frosted slide method. Place the spleen in the bottom half of a 60 × 15 mm culture dish with 5 mL of PBS.
2. Press the spleen and/or lymph nodes between the frosted ends of two microscope slides using a gentle circular motion until only the empty capsule remains.
3. Wash the frosted slide with 10 mL of MACS buffer and transfer to a new 15 mL tube through a 0.45 μm filter.
4. Rinse the culture dish and filter with 5 mL of MACS buffer.
5. Centrifuge samples at 300 × *g* for 5 min at 4 °C.
6. Resuspend cell pellets in 3 mL of red blood cell lysis buffer and incubate at room temperature for 3 min.
7. Add 20 mL of RPMI complete medium to stop the reaction.
8. Centrifuge samples at 300 × *g* for 5 min at 4 °C.
9. Wash once with 10 mL of MACS buffer by centrifuging at 300 × *g* for 5 min at 4 °C.
10. Perform live cell count using Trypan Blue.
11. Aspirate supernatant completely and resuspend cell pellet in 400 μL of MACS buffer per 10^8 total cells.
12. Add 100 μL of $CD4^+$ T cells biotin-antibody cocktail (including biotin-conjugated monoclonal anti-mouse CD8a, CD45R, CD11b, CD25, CD49b, TCRγ/δ, and Ter-119 antibodies per 10^8 cells (*see* **Note 4**).

13. Mix well and incubate at 4 °C for 10 min.
14. Add 300 μL of MACS buffer and 200 μL of anti-biotin microbeads per 10^8 cells.
15. Mix well and incubate at 4 °C for 15 min.
16. Centrifuge samples at 300 × *g* for 5 min at 4 °C.
17. Aspirate supernatant and resuspend pellet in 500 μL of MACS buffer.

3.1.2 Magnetic Separation—Depletion of Non-CD4$^+$ T Cells

1. Place LS column in the MidiMACS Separator.
2. Rinse column with 3 mL of MACS buffer.
3. Apply cell suspension onto the column.
4. Wash the column three times with 3 mL of MACS buffer.
5. The flow-through contains the unlabeled pre-enriched CD4$^+$ T cells.
6. Centrifuge samples at 300 × *g* for 5 min at 4 °C.
7. Aspirate supernatant completely and resuspend cell pellet in 800 μL of MACS buffer.
8. Add 200 μL of CD62L (L-selectin) microbeads to samples.
9. Mix well and incubate at 4 °C for 15 min.
10. Centrifuge samples at 300 × *g* for 5 min at 4 °C.
11. Aspirate supernatant and resuspend pellet in 500 μL of MACS buffer.

3.1.3 Magnetic Separation—Positive Selection of CD4$^+$CD62L$^+$ T Cells

1. Place MS column in the MiniMACS Separator.
2. Rinse column with 500 μL of MACS buffer.
3. Apply cell suspension onto the column.
4. Wash the column three times with 500 μL of MACS buffer.
5. Remove column and place it on a fresh collection tube.
6. Add 1 mL of MACS buffer onto the column. Push the plunger into the column to flush out the magnetically labeled CD4$^+$CD62L$^+$ T Cells.
7. Wash cells once with 10 mL of MACS buffer by centrifuging samples at 300 × *g* for 5 min at 4 °C.
8. Perform live cell count using Trypan Blue exclusion method.
9. Resuspend pellet in RPMI compete medium to 1 × 10^6 cells per mL.

3.2 Setup Th9 Cell Culture

1. Coat tissue culture plate with 2 μg/mL of anti-CD3 diluted in PBS for at least 2 h at 37 °C or overnight at 4 °C.
2. Aspirate PBS from tissue culture plate.
3. Seed cell suspension onto antibody-coated tissue culture plate at the concentration of 1 million cells per mL.

4. Prepare and add Th9 cytokine and antibody cocktail containing 1 μg/mL of anti-CD28, 20 ng/mL of IL-4, 2 ng/mL of human TGF-β, and 10 μg/mL of anti-IFN-γ (*see* **Note 5**).
5. Gently rock tissue culture plate back and forth to mix properly.
6. Incubate plate in the incubator at 37 °C and 5% CO_2 for 72 h.
7. After 72 h, expand cell culture by gently pipetting up and down to remove cells and transfer to a new plate with fresh RPMI compete medium (three times the original volume).
8. Add 20 ng/mL of IL-4, 2 ng/mL of human TGF-β, and 50 U/mL of human IL-2 to the culture medium.
9. Incubate plate in the incubator at 37 °C and 5% CO_2 for an additional 48 h.

3.3 Assess Th9 Differentiation by Flow Cytometry (See Notes 6 and 7)

3.3.1 Restimulation of Th9 Cells

1. Harvest and count cells.
2. Wash cells once with 10 mL of RPMI compete medium.
3. Centrifuge samples at 300 × *g* for 5 min at 4 °C.
4. Aspirate supernatant and resuspend pellet in RPMI compete medium to 1 × 10^6 cells per mL.
5. Seed cell suspension onto tissue plate at 1 × 10^6 cells per mL.
6. Add 50 ng/mL of PMA and 750 ng/mL of ionomycin to the medium.
7. Incubate cells in the incubator at 37 °C and 5% CO_2 for 2 h.
8. Add GolgiStop solution at 1:1000 dilution and culture for an additional 4 h.

3.3.2 Cell Staining for Flow Cytometry

1. Wash cells twice with 3 mL of FACS Buffer. Centrifuge samples at 300 × *g* for 5 min at 4 °C.
2. Add surface antibodies if needed and incubate for 30 min at 4 °C (*see* **Note 8**).
3. Wash cells once with 3 mL of FACS buffer. Centrifuge samples at 300 × *g* for 5 min at 4 °C.
4. Fix cells in 100 μL per 1 × 10^6 cells using 2% formaldehyde for 10 min at room temperature. Keep in the dark.
5. Wash cells once with 3 mL of FACS buffer and once with 1 mL of permeabilization buffer by centrifuging samples at 300x*g* for 5 min at 4 °C.
6. Add fluorochrome conjugated antibodies diluted in FACS buffer to the samples and incubate 30 min at 4 °C in the dark (*see* **Note 8**).
7. Wash cells once with 3 mL of FACS buffer. Centrifuge samples at 300 × *g* for 5 min at 4 °C.
8. Resuspend cells in 500 μL of FACS buffer.
9. Cells can be kept at 4 °C in the dark for flow cytometry analysis (Fig. 1).

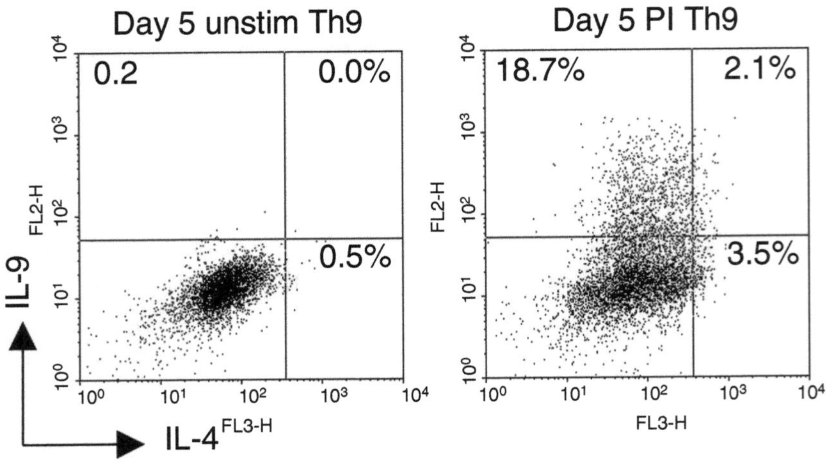

Fig. 1 Flow cytometry obtained from day 5 unstimulated or PMA and ionomycin (PI) restimulated Th9 cells that were stained using fluorochrome-conjugated anti-mouse IL-9 and IL-4 antibodies

4 Notes

1. Perform fast and keep reagents cold during naïve $CD4^{+}$ T cell isolation.
2. AutoMACS Separator can be used for negative and positive selection during the isolation of $CD4^{+}CD62L^{+}$ T cells. Using either magnetic separator or autoMACS Separator, one can achieve greater than 90% cell purity.
3. Alternatively, total splenocytes can be labeled with fluorochrome antibodies and sorted by flow cytometry for naïve $CD4^{+}CD25^{-}CD44^{low}CD62L^{high}$ T cells.
4. MS and LS columns can retain up to 1×10^{7} and 1×10^{8} labeled cells, respectively. The amount of antibodies should be optimized for up to 10^{8} cells. Scale up buffer volume and antibodies for higher cell numbers.
5. The concentration of IL-4 and TGF-β can be adjusted to optimize Th9 cell culture conditions. We have observed that increasing the concentration of anti-CD3 and anti-CD28 antibodies up to 10 μg/mL can enhance IL-9 production. This can vary depending on the purpose of the experiment.
6. Resting differentiated Th9 cells or other T helper cells (Th1 or Th2) can be used as negative controls for evaluating Th9 differentiation using flow cytometry.
7. Th9 cell differentiation can be assessed using qPCR or ELISA. Day 5 differentiated cells are washed and restimulated for 6 or 24 h with 2 μg/mL of anti-CD3 for qPCR or ELISA, respectively.

8. Dead cells can interfere with the interpretation of flow cytometry data. The Live/Dead cell viability kit that stains dead cells can be performed to differentiate between live and dead cells in flow cytometry. Live/Dead cell viability kit is comparable with common fixation and permeabilization methods for intracellular staining. Fluorochrome conjugated antibodies should be used at proper concentration according to manufacturer's protocol. Alternatively, antibodies can be titrated to determine optimal concentration.

Acknowledgment

This work was supported by NIH grants T32 HL007910 and T32 AI007051 to D.P.

References

1. Goswami R, Kaplan MH (2011) A brief history of IL-9. J Immunol 186(6):3283–3288. doi:10.4049/jimmunol.1003049
2. Kaplan MH (2013) Th9 cells: differentiation and disease. Immunol Rev 252(1):104–115. doi:10.1111/imr.12028
3. Dardalhon V, Awasthi A, Kwon H, Galileos G, Gao W, Sobel RA, Mitsdoerffer M, Strom TB, Elyaman W, Ho IC, Khy S, Oukka M, Kuchroo VK (2008) IL-4 inhibits TGF-beta-induced Foxp3+ T cells and, together with TGF-beta, generates IL-9+ IL-10+ Foxp3(−) effector T cells. Nat Immunol (12):1347–1355. doi:10.1038/ni.1677
4. Veldhoen M, Uyttenhove C, van Snick J, Helmby H, Westendorf A, Buer J, Martin B, Wilhelm C, Stockinger B (2008) Transforming growth factor-beta 'reprograms' the differentiation of T helper 2 cells and promotes an interleukin 9-producing subset. Nat Immunol 9(12):1341–1346. doi:10.1038/ni.1659
5. Goswami R, Jabeen R, Yagi R, Pham D, Zhu J, Goenka S, Kaplan MH (2012) STAT6-dependent regulation of Th9 development. J Immunol 188(3):968–975. doi:10.4049/jimmunol.1102840
6. Jabeen R, Goswami R, Awe O, Kulkarni A, Nguyen ET, Attenasio A, Walsh D, Olson MR, Kim MH, Tepper RS, Sun J, Kim CH, Taparowsky EJ, Zhou B, Kaplan MH (2013) Th9 cell development requires a BATF-regulated transcriptional network. J Clin Invest 123(11):4641–4653. doi:10.1172/JCI69489
7. Staudt V, Bothur E, Klein M, Lingnau K, Reuter S, Grebe N, Gerlitzki B, Hoffmann M, Ulges A, Taube C, Dehzad N, Becker M, Stassen M, Steinborn A, Lohoff M, Schild H, Schmitt E, Bopp T (2010) Interferon-regulatory factor 4 is essential for the developmental program of T helper 9 cells. Immunity 33(2):192–202. doi:10.1016/j.immuni.2010.07.014
8. Mantel PY, Kuipers H, Boyman O, Rhyner C, Ouaked N, Ruckert B, Karagiannidis C, Lambrecht BN, Hendriks RW, Crameri R, Akdis CA, Blaser K, Schmidt-Weber CB (2007) GATA3-driven Th2 responses inhibit TGF-beta1-induced FOXP3 expression and the formation of regulatory T cells. PLoS Biol 5(12):e329. doi:10.1371/journal.pbio.0050329
9. Chang HC, Sehra S, Goswami R, Yao W, Yu Q, Stritesky GL, Jabeen R, McKinley C, Ahyi AN, Han L, Nguyen ET, Robertson MJ, Perumal NB, Tepper RS, Nutt SL, Kaplan MH (2010) The transcription factor PU.1 is required for the development of IL-9-producing T cells and allergic inflammation. Nat Immunol 11(6):527–534. doi:10.1038/ni.1867
10. Yao W, Zhang Y, Jabeen R, Nguyen ET, Wilkes DS, Tepper RS, Kaplan MH, Zhou B (2013) Interleukin-9 is required for allergic airway inflammation mediated by the cytokine TSLP. Immunity 38(2):360–372. doi:10.1016/j.immuni.2013.01.007

Chapter 5

T Cell Receptor and Co-Stimulatory Signals for Th9 Generation

Françoise Meylan and Julio Gomez-Rodriguez

Abstract

In vitro polarization of naïve CD4⁺ T cells toward distinct T helper lineages is crucial for establishing the factors and features that determine the differentiation, stability, and effector function for each T helper subsets. In this regard, the recently defined Th9 subset has been reported with two essential cytokines requirement for their generation. Generating Th9 cells in vitro from naïve CD4⁺ T cells requires the combination of TGF-β and IL-4. However, the amount of IL-9 producing under these minimal conditions is often small. The intent of this chapter is to provide examples to increase the generation of IL-9 producing T cells in vitro by modulating TCR strength and co-stimulation through the TNF family member TL1A. We hope that these methods to efficiently differentiate naïve CD4⁺ T cells toward IL-9 producing cells will facilitate understanding the differentiation and function of Th9 cells and their pathogenesis in various inflammatory and autoimmune diseases.

Key words Th9 lymphocytes, TCR, TGF-β, IL-4, TL1A, Ex vivo generation

1 Introduction

CD4⁺ T cells are an important part of the adaptive immune system and have a critical role in responses against invading pathogens. Naïve CD4⁺ T cells can differentiate into distinctive effector T helper (Th) cells upon T cell receptor (TCR) engagement and cytokines secreted by innate cells. These subsets include Th1, Th2, Th17, T follicular, and T regulatory cells that upon differentiation secrete their given signature cytokines [1]. Recently, Th9 cells have been added to the T helper subset family [2]. These cells predominantly secrete IL-9, originally named p40 [3], as their signature cytokine. Number of studies has shown that IL-9 has both detrimental and beneficial roles in response to the immune system [4]. In fact, IL-9 has been shown to have harmful effects in a numbers of human and mouse studies. IL-9 was found in the bronchoalveolar lavage of asthmatic patients [5]; and in mouse IL-9 exacerbated asthma pathology in both papain and OVA-induced lung inflammation [6].

Ritobrata Goswami (ed.), *Th9 Cells: Methods and Protocols*, Methods in Molecular Biology, vol. 1585,
DOI 10.1007/978-1-4939-6877-0_5, © Springer Science+Business Media LLC 2017

These observations were also supported by mouse models with transgenic expression of IL-9 resulting in the development of allergic inflammation [7–9]. However, the pulmonary inflammation caused by IL-9 is more complex because the IL-9-induced asthma pathology is in part mediated by IL-13, which can be produced by Th2 cells. It became more complicated as studies have demonstrated that type 2 innate lymphocytes (ILC2) are the main source of IL-9 in papain induced pulmonary inflammation [10]. Several studies also involved a pathogenic role for Th9 cells in mouse model of EAE [11–13]. Interestingly, IL-9 has been shown to have also an important role in the eradication of pathogens such as *Leishmania major* [14], *Trichuris muris* [15], *Schistosoma mansoni* [16], and *Nippostrongylus brasiliensis* [17]. Furthermore, Th9 cells have been implicated in cancer biology with mouse studies showing that IL-9 produced by Th9 cells have antitumor activity especially in melanoma by increasing lymphocyte infiltration in the tumor and enhancing the antitumor activity of mast cells [18, 19].

The production of IL-9 by $CD4^+$ T cells was reported over 2 decades ago [20], but it was only recently shown that TGF-β could reprogram T helper 2 cells to lose their signature cytokine profiles and promote them into producing IL-9 [2, 21]. These studies paved the basis to generate Th9 directly from naïve $CD4^+$ T lymphocytes. In mice, TGF-β and IL-4, in the presence of TCR and CD28 signals (antigen-presenting cells, plate-bound anti-CD3 plus anti-CD28 or anti-CD3/CD28 coated microbeads) are required and sufficient to induce the differentiation of naïve $CD4^+CD25^-$ T cells into Th9 cells. However, one of the major issues toward recognizing these IL-9 producing cells as a new T helper population came from the modest fraction of differentiated cells secreting IL-9 that initial studies were able to generate from primary naïve $CD4^+$ T cells. In this context, the ability to generate large amount of T helper cells capable of producing IL-9 will improve the ability to study their function in more detail.

Recent studies have identified a plethora of molecules that can modulate the potential of in vitro differentiation toward Th9 both in mouse and in human $CD4^+$ T cells. In this regard, IL-2 [20], the TNF family members OX40 ligand (OX40L) [22], and TNF-like factor 1A (TL1A) [6], IL-25 [23], IL-1 [24, 25] , TSLP [26], Jagged-2 [27], calcitonin gene-related peptide [28], nitric oxide [29], have been shown to enhance Th9 differentiation. In contrast, other molecules such as cyclooxygenase-2 [30], programmed cell death ligand 2 [31], 1,25-dihidroxyvitamin D3 [32], and IFN-γ [20] have inhibitory effects toward Th9 generation. Interestingly, IL-21 can have either a positive or a negative effect on Th9 generation depending on the culture milieu [33]. Numbers of transcription factors has been identified to be required for Th9 differentiation, such as STAT6, GATA3, IRF4 [21, 34, 35], STAT5 [6, 36], BATF [37], PU.1 [38]. TL1A is highly potent in enhancing Th9 differentiation through an IL-2 and STAT5-dependent mechanism and Th9 differentiated in the presence of TL1A are more pathogenic. These data identified TL1A–DR3 interactions as a novel pathway that promotes Th9 differentiation and pathogenicity [6]. TL1A is a cytokine belonging to the tumor necrosis factor (TNF) family, which was identified in 2002 [39] as the ligand for the TNFR receptor DR3. Like other TNF ligands, TL1A can be produced as a membrane-bound protein or cleaved into a soluble form and its expression is controlled tightly [40]. TL1A can also co-stimulate T cell proliferation and cytokine production similar to other TNF family members [39, 40]. While endogenous TL1A-DR3 interactions are not required for murine in vitro T cell differentiation into Th1, Th2, Th9, or Th17, addition of exogenous TL1A can regulate their specific cytokine secretion with the most effect on IL-9 secretion [6, 40–42].

In addition, the strength of the TCR plays a fundamental role in the generation of Th9 cells, as the amount of IL-9 produced correlates with the strength of the TCR signal [36]. Indeed, these results were further confirmed by studies using Itk deficient cells. Itk, which belongs to the Tec family of cytosolic tyrosine kinases is a crucial component of the TCR signaling [43], and has been recently shown to be a positive modulator of Th9 cell differentiation [36]. These studies further emphasize the cross talk between TCR and IL-2 signaling on IL-9 expression.

As Th9 cells have gained increasing attention in the field of asthma, IBD, EAE and tumor immunology, the development of protocols allowing for large numbers of IL-9 producing cells ex vivo is essential to further study and understand their function, regulation, lineage stability, and impact in various pathologies. For instance the pathological fate of antigen-specific Th9 cells adoptively transferred to a host and their pro-inflammatory properties need to be further elucidated. In this regard, the ability to generate highly polarized Th9 cells is crucial. In this chapter, we describe protocols suitable to increase the in vitro generation of mouse and human Th9 lymphocytes from naïve $CD4^{+}CD25^{-}$ T cells.

2 Materials

2.1 Generation of Murine Th9 Cells in Presence of the TNF Family Member TL1A

1. Eight- to ten-week-old C57BL/6 mice.
2. Complete medium IMDM+ glutaMAX supplemented with 10% FBS (v/v), 100 U/mL penicillin–streptomycin, 55 μM 2-mercaptoethanol (*see* **Note 1**).
3. $CD4^{+}$ T Cell Isolation Kit II, mouse (Miltenyi Biotech) (*see* **Notes 2** and **3**).
4. Anti-mouse CD4 and CD8 labeled antibodies.

5. Magnetic microbeads against the label of anti-mouse CD4 and CD8 antibodies.
6. Cytokines: recombinant mouse TL1A, recombinant murine IL-4 and recombinant human TGF-β1.
7. Blocking antibodies: anti-mouse IFN-γ antibodies (clone XMG1.2), anti-mouse CD3 (clone 2C11), anti-mouse CD28 (clone 37.51).
8. LS Columns and magnets (MidiMACS, Miltenyi) (*see* **Notes 2** and **3**).
9. Mitomycin C (*see* **Note 4**).

2.2 Generation of Human Th9 Cells in Presence of the TNF Family Member TL1A

1. Human whole blood from healthy donors or patients.
2. Ficoll-Paque™ PLUS (GE Healthcare Bio-Sciences AB).
3. Complete medium IMDM+ glutaMAX supplemented with 10% FBS (v/v), 100 U/mL penicllin–streptomycin, 55 μM 2-mercaptoethanol (*see* **Note 1**).
4. Miltenyi Human Naïve CD4$^+$ T Cell Isolation Kit II.
5. Cytokines: recombinant human TL1A (R&D), recombinant human IL-4, recombinant human TGF-β1, recombinant human IL-2.
6. LS Columns and magnets (MidiMACS, Miltenyi).
7. Dynabeads Human T-activator CD3/CD28 (ThermoFisher).
8. Magna GrlP Rack Magnet to recover washed microbeads.

2.3 Flow Cytometry

1. FACS buffer (1× PBS, 0.5% BSA).
2. 4% paraformaldehyde in PBS.
3. Permeabilization buffer: Triton buffer with 0.05% BSA.
4. Anti-mouse IL-9 (clone RM9A4), anti-mouse CD4 (clone RM4-5), anti-human/mouse CD44 (clone IM7), anti-mouse CD25 (clone PC61.5), anti-mouse CD4 (clone RM9A4), anti-mouse CD62L (clone MEL-14), anti-human IL-9 (clone MH9A4), and anti-human CD4 (clone RPA-T4).
5. Cell viability dye and LiveDead fluorescent reactive dye.
6. Flow cytometer (i.e., LSRII, Fortessa, BD Biosciences).
7. Cell Sorter (i.e., FACSAria, BD Bioscences).
8. Flowjo software (Version 9.9 Tree Star, Inc).

2.4 Luminex

1. Bio-Plex Pro Mouse Cytokine IL-9 Set (Bio Rad) (*see* **Note 5**).
2. Bio-Plex Pro Human Cytokine IL-9 Set (Bio Rad catalogue) (*see* **Note 5**).
3. Luminex® 100/200™ or MAGPIX® system.

2.5 Real-Time PCR

1. RNeasy mini kit (QIAGEN).
2. TaqMan probes for *Il9*, Human 18S rRNA as internal control (*see* **Note 6**).
3. NanoDrop ND1000 spectrophotometer (NanoDrop).
4. TaqMan Reverse Transcription Reagents (*see* **Note 7**).
5. TaqMan Universal PCR Master Mix (*see* **Note 7**).
6. 96-well plates.
7. Real-time PCR machine.

2.6 General Reagents.

1. Sterile PBS.
2. Red blood cell lysis buffer.
3. Cell strainers 70 μm.
4. Centrifuge tubes 15, 50 mL.
5. Hemocytometer.
6. 0.4% trypan blue.
7. MACS sorting buffer (0.5% BSA, 2 mM EDTA).
8. 48-well plates and petri dishes.
9. Phorbol 12-myristate 13-acetate (PMA).
10. Ionomycin.
11. Golgi Stop (BD Biosciences) (*see* **Note 8**).
12. TRIzol® RNA isolation reagent.
13. Chloroform.

3 Methods

3.1 Generation of Murine Th9 Cells in Presence of the TNF Family Member TL1A

1. Euthanize mice, spray fur with 70% ethanol and harvest spleens and lymph nodes. The remaining steps should be performed in a tissue culture hood to maintain aseptic conditions. Separate splenocytes into a single cell suspension by smashing the spleens with a 3 mL syringe plunger in a petri dish and lyse the red blood cells with the lysis buffer for 2 min at room temperature. Pass the cell suspension through a 70 μm cell strainer with complete medium into a 50 mL tube to wash the cells through. Fill the tube complete medium and centrifuge for 5 min at 544 × *g* at 4 °C. Remove the supernatant and proceed to **step 3**.
2. Transfer the lymph nodes into a cell strainer placed into a 50 mL tube and smash the lymph nodes with a syringe plunger and let the cells pass through with enough complete medium to wash the cells through. Centrifuge for 5 min at 544 × *g* at 4 °C. Remove the supernatant.

3. Resuspend the pellets from spleens obtained from **step 1** and the lymph nodes with complete medium, count the cells using trypan blue to exclude dead cells counts in a hemocytometer and centrifuge as before.
4. Purify the cells by negative selection using the CD4$^+$ T Cell Isolation Kit II (as directed by the manufacturer's instructions). Resuspend cells in 40 μL of MACS sorting buffer per 10^8 cells and incubate with 10 μL of biotin–antibody cocktail per 10^8 cells for 10 min at 4 °C.
5. Add 30 μL of MACS buffer per 10^8 cells and incubate for 15 min at 4 °C with 20 μL of anti-biotin microbeads per 10^8 cells.
6. Wash cells with MACS buffer and resuspend at 10^8 cells/mL and pass through magnetic cell separation using LS columns.
7. Centrifuge naïve CD4$^+$ T cells (negative selection) with IMDM complete medium, resuspend in 2 mL, and count the cells (*see* **Note 2** and **3**).
8. Resuspend cells in FACS buffer at 2×10^7 cells/mL and stain for 15 min at 4 °C with a cocktail of antibodies including anti-mouse CD4, anti-mouse CD25, anti-mouse CD44, and anti-mouse CD62L (each antibody should be titrated to obtain the highest signal for positive population and the lowest signal for the negative population). Wash with FACS buffer and centrifuge.
9. The labeled cells are prepared for sorting for naïve CD4$^+$ T cells by resuspending the cells at 2×10^7/mL in FACS buffer. In order to exclude dead cells, add a cell viability dye prior to sort for CD4^{pos} CD25^{neg} CD44^{low} CD62L^{high} gating on the live singlets population into IMDM complete medium.
10. Wash the sorted cells once with IMDM complete medium and resuspend in 1 mL of complete medium (approximately 3–4 × 10^6 naïve CD4$^+$ T cells/mouse).
11. Prepare T-depleted splenocytes as antigen presenting cells (APCs). Isolate three spleens as mentioned in one and incubate the splenocytes resuspended at 2×10^7 cells/mL in FACS buffer with labeled anti-CD4 and anti-CD8 mouse antibodies (magnetic microbeads against the label of CD4 and CD8 antibodies) at 4 °C for 20 min.
12. Wash the cells and resuspend in MACS buffer with 40 μL of MACS sorting buffer per 10^8 cells and incubate with 10 μL of microbeads against the fluorochrome selected to stain the splenocytes above per 10^8 cells for 10 min at 4 °C.
13. Wash cells with MACS buffer and resuspend at 10^8 cells/mL and pass through magnetic cell separation using LS columns. Centrifuge T-depleted fraction (negative selection) with PBS and resuspend at 4×10^7 cells/mL in PBS and incubate for 30 min at 37 °C with mitomycin C (50 μg/mL). Wash the

cells four times with PBS, resuspend in 4 mL of IMDM complete medium and count (*see* **Note 4**).

14. Sorted naïve CD4+ T cells (2×10^5) are cocultured with antigen presenting cells at a ratio of 1:5 in 48-well plates in a total volume of 1 mL (*see* **Note 9**).
15. Add the Th9 differentiation cocktail: 1 μg/mL anti-CD3, 3 μg/mL anti-CD28, 20 ng/mL IL-4, 5 ng/mL TGF-β1, 10 ng/mL mouse TL1A, and 10 μg/mL anti-IFN-γ.

On day 3, cells producing IL-9 can be obtained. These cells can be characterized by real-time PCR, flow cytometry, or Luminex for expression of specific markers and cytokine production (*see* Fig. 1). These polarized cells can also be used for further analysis in vitro or to evaluate their potential in vivo.

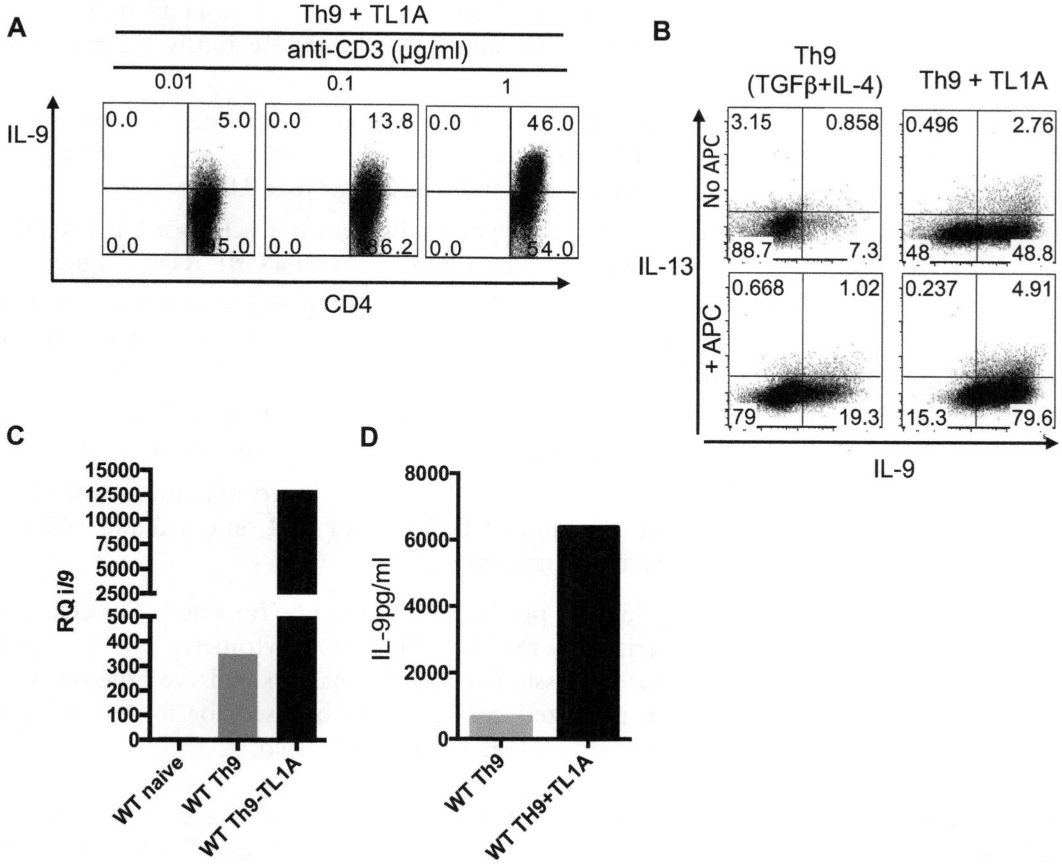

Fig. 1 Example of mouse Th9 generation in vitro. (**a**) Mouse naïve CD4+ T cells were cultured in Th9-polarizing conditions with mouse TL1A and in presence of APCs and varying concentration of anti-CD3 antibody and were analyzed for their IL-9 expression by flow cytometry. (**b**) Mouse naïve CD4+ T cells were cultured on coated plate with anti-CD3 and anti-CD28 in absence (No APC) or in presence of APC in Th9-polarizing condition with or without mouse TL1A and analyzed for IL-9 expression by flow cytometry. (**c**) Cells were collected after 3 days of differentiation conditions and analyzed for *Il9* mRNA by RT-PCR, naïve CD4+ T cells were used as comparison. (**d**) Supernatants from cells differentiated for 3 days were collected and analyzed for IL-9 protein levels

3.2 Generation of Human Th9 Cells in Presence of the TNF Family Member TL1A

1. Obtain blood from healthy donors or patients.
2. Dilute whole blood with PBS (1:1) in a 50 mL tube.
3. Carefully pipet the diluted blood to a 50 mL tube containing 15 mL of Ficoll medium.
4. Spin at 863 × *g* for 20 min at room temperature without brake.
5. Carefully remove the PBMC layer to a 50 mL tube and wash the cells with PBS twice at 486 × *g* at 4 °C for 6 min.
6. Isolate naïve $CD4^+$ T cells from PBMC by negative selection using Human naïve $CD4^+$ T Cell Isolation Kit II. Resuspend total PBMC in 40 μL of MACS buffer per 10^8 cells and incubate with 10 μL of biotin–antibody cocktail per 10^8 cells for 10 min at 4 °C.
7. Add 30 μL of MACS buffer per 10^8 cells and incubate for 15 min at 4 °C with 20 μL of anti-biotin microbeads per 10^8 cells. Wash cells with MACS buffer and resuspend at 10^8 cells/mL and pass through LS column previously washed with MACS buffer.
8. Centrifuge the sorted naïve $CD4^+$ T cells (negative selection) with IMDM complete medium at 486 × *g* at 4 °C for 6 min, resuspend in 2 mL and count (*see* **Note 10**).
9. Prepare sufficient Dynabeads Human T-activator CD3/CD28 by washing with PBS using the Magna GrIP Rack Magnet.
10. Sorted naïve $CD4^+$ T cells (2×10^5) are plated in 48-well plate in a total volume of 1 mL and the recommended amount of CD3/CD28 Dynabeads.
11. Add the Th9 differentiation cocktail: 20 ng/mL human IL-4, 5 ng/mL human TGF-β1, and 10 ng/mL human TL1A.
12. Differentiated cells were expanded into new media with half the initial amount of TGF-β1 and IL-4 on day 3 and 100 U/mL of recombinant hIL-2.

On day 5, cells producing IL-9 can be obtained. The cells can be characterized by real-time PCR, flow cytometry (*see* Fig. 2) or Luminex for expression of specific markers and cytokine production. These polarized cells can also be used for further analysis in vitro or to evaluate their potential in vivo.

3.3 Characterization of Ex Vivo Generated Th9 Cells

3.3.1 Real-Time PCR

1. Differentiated Th9 cells and naïve $CD4^+$ T cells are collected in 1 mL of TRIzol and proceed to purify total RNA or freeze at −80 °C.
2. Add 200 μL of chloroform, mix for 2 min leave at room temperature and centrifuge for 15 min at 9300 × *g* at 4 °C.
3. Carefully collect the aqueous phase and mix 1:1 with alcohol 70% at 1:1 ratio.

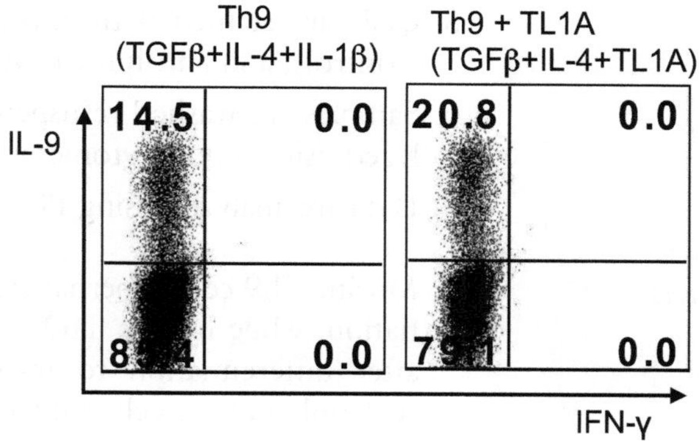

Fig. 2 Example of humanTh9 generation in vitro. Human naïve CD4⁺ T cells were cultured in Th9-polarizing conditions with CD3/CD28 Dynabeads in presence of 10 ng/ml of human IL-1β or 10 ng/ml of human TL1A and were analyzed for their IL-9 expression by flow cytometry

4. Extract total RNA using Qiagen RNeasy Mini kit following the manufacturer's protocol.
5. Measure RNA concentration using NanoDrop ND1000 spectrophotometer.
6. Synthesize cDNA using 200 ng of total RNA in a 20 μL reaction using TaqMan Reverse Transcription Reagent kit, following the manufacturer's instructions. Dilute the cDNA with water at 1:6 (*see* **Note** 7).
7. Real-time PCR reaction: PCR reactions are set up in 20 μL volume using 96-well plates containing 10 μL TaqMan Universal PCR Master Mix, 1 μL IL-9 TaqMan® primer/probe set, 4 μL of the cDNA synthesis reaction, and 5 μL of nuclease-free water. Use 0.2 μL of 18S RNA. Reactions are run on a real-time PCR Instrument.
8. Data are analyzed by using the comparative Ct method as means of relative quantification, normalized to 18S rRNA as an endogenous reference and relative to naïve CD4⁺ T cells. Relative quantification values are expressed as $2^{-\Delta\Delta Ct}$.

3.3.2 Flow Cytometry

1. After 72 h (mouse) or 5 days (human), differentiated Th9 cells are restimulated with PMA + Ionomycin in presence of Golgi Stop for 4 h.
2. Wash cells with PBS and stain with LiveDead fluorescent reactive dye (1/1000) for 30 min at 4 °C.
3. Incubate cells with 4% paraformaldehyde at room temperature for 15 min.
4. Permeabilize cells with Triton buffer for 10 min at room temperature.

5. Cells are stained with anti-human/mouse IL-9 (or Isotype Control), and anti-human/mouse CD4 at 4 °C for 30 min.
6. Samples are washed resuspended with Triton buffer and analyzed using a flow cytometer.
7. Data are analyzed using FlowJo software.

3.3.3 Luminex

1. Murine Th9 cells supernatant is collected 3 days after differentiation, while human Th9 cell supernatant is collected 5 days after differentiation to measure IL-9 protein level before restimulating the cells with PMA/ionomycin (for the analysis by flow cytometry).
2. Mouse IL-9 is detected in the supernatant using Bio-Plex Pro Mouse Cytokine IL-9 Set (Bio-Rad) (*see* **Note 5**) following the manufacturer's instructions.
3. Human IL-9 is detected in the supernatant using Bio-Plex Pro Human Cytokine IL-9 Set (Bio-Rad) (*see* **Note 5**) following the manufacturer's instructions.

4 Notes

1. While RPMI medium is widely used for primary T cell culture, IMDM medium has been shown to give slightly better Th9 polarization.
2. An alternative kit to this procedure is the EasySep™ mouse $CD4^{+}T$ cell Isolation kit.
3. An alternative method for purifying naïve $CD4^{+}$ T cells without a cell sorter is the use of a naïve $CD4^{+}$ T cell isolation kit. However, the purity obtained with a kit is lower than the purity obtained using a cell sorter.
4. An alternative to mitomycin C treatment to inhibit APCs growth is the use of irradiation with 3000 rad (30 Gy).
5. An alternative to the use of beads-based multiplex assay (Luminex) to measure protein in the supernatants is ELISA.
6. Human 18S rRNA can be used for endogenous control for both human and mouse.
7. An alternative to the two-step protocol presented here is to use the iTaq™ Universal Probes One-Step Kit.
8. Alternative to BD GolgiStop™ are monensin and brefeldin A.
9. Purified naïve $CD4^{+}$ T cells can be plated on anti-CD3 and anti-CD28 coated plate (5 μg/mL in PBS and incubate at 4 °C overnight or 2 h at 37 °C, wash three times with PBS). In the presence of APCs, soluble anti-CD3 can be used at a concentration of 1 μg/mL and soluble anti-CD28 at a concentration of 3 μg/mL

to stimulate $CD4^+$ T cells. When using antigen specific stimulation in presence of APCs, 0.5 μg/mL of anti-CD28 can be used, and 2 ng/mL of TGF-β1.

10. An alternative to the Miltenyi or EasySep kits to purify human naïve $CD4^+$ T cells is to sort the naïve T cells by flow cytometry similar to the protocol described for mouse; however, for human naïve T cells the surface markers used are CD4, CD45RA, CD45RO, CCR7, and CD62L. The naïve population is defined as $CD4^+CD45RA^+CD45RO^-$ $CCR7^+CD62L^+$.

References

1. DuPage M, Bluestone JA (2016) Harnessing the plasticity of CD4(+) T cells to treat immune-mediated disease. Nat Rev Immunol 16(3):149–163. doi:10.1038/nri.2015.18
2. Veldhoen M, Uyttenhove C, van Snick J, Helmby H, Westendorf A, Buer J, Martin B, Wilhelm C, Stockinger B (2008) Transforming growth factor-beta 'reprograms' the differentiation of T helper 2 cells and promotes an interleukin 9-producing subset. Nat Immunol 9(12):1341–1346. doi:10.1038/ni.1659
3. Hultner L, Druez C, Moeller J, Uyttenhove C, Schmitt E, Rude E, Dormer P, Van Snick J (1990) Mast cell growth-enhancing activity (MEA) is structurally related and functionally identical to the novel mouse T cell growth factor P40/TCGFIII (interleukin 9). Eur J Immunol 20(6):1413–1416. doi:10.1002/eji.1830200632
4. Kaplan MH, Hufford MM, Olson MR (2015) The development and in vivo function of T helper 9 cells. Nat Rev Immunol 15(5):295–307. doi:10.1038/nri3824
5. Shimbara A, Christodoulopoulos P, Soussi-Gounni A, Olivenstein R, Nakamura Y, Levitt RC, Nicolaides NC, Holroyd KJ, Tsicopoulos A, Lafitte JJ, Wallaert B, Hamid QA (2000) IL-9 and its receptor in allergic and nonallergic lung disease: increased expression in asthma. J Allergy Clin Immunol 105 (1 Pt 1):108–115
6. Richard AC, Tan C, Hawley ET, Gomez-Rodriguez J, Goswami R, Yang XP, Cruz AC, Penumetcha P, Hayes ET, Pelletier M, Gabay O, Walsh M, Ferdinand JR, Keane-Myers A, Choi Y, O'Shea JJ, Al-Shamkhani A, Kaplan MH, Gery I, Siegel RM, Meylan F (2015) The TNF-family ligand TL1A and its receptor DR3 promote T cell-mediated allergic immunopathology by enhancing differentiation and pathogenicity of IL-9-producing T cells. J Immunol 194(8):3567–3582. doi:10.4049/jimmunol.1401220
7. Forbes EE, Groschwitz K, Abonia JP, Brandt EB, Cohen E, Blanchard C, Ahrens R, Seidu L, McKenzie A, Strait R, Finkelman FD, Foster PS, Matthaei KI, Rothenberg ME, Hogan SP (2008) IL-9- and mast cell-mediated intestinal permeability predisposes to oral antigen hypersensitivity. J Exp Med 205(4):897–913. doi:10.1084/jem.20071046
8. Temann UA, Geba GP, Rankin JA, Flavell RA (1998) Expression of interleukin 9 in the lungs of transgenic mice causes airway inflammation, mast cell hyperplasia, and bronchial hyperresponsiveness. J Exp Med 188(7):1307–1320
9. Temann UA, Ray P, Flavell RA (2002) Pulmonary overexpression of IL-9 induces Th2 cytokine expression, leading to immune pathology. J Clin Invest 109(1):29–39. doi:10.1172/JCI13696
10. Wilhelm C, Hirota K, Stieglitz B, Van Snick J, Tolaini M, Lahl K, Sparwasser T, Helmby H, Stockinger B (2011) An IL-9 fate reporter demonstrates the induction of an innate IL-9 response in lung inflammation. Nat Immunol 12(11):1071–1077. doi:10.1038/ni.2133
11. Jager A, Dardalhon V, Sobel RA, Bettelli E, Kuchroo VK (2009) Th1, Th17, and Th9 effector cells induce experimental autoimmune encephalomyelitis with different pathological phenotypes. J Immunol 183(11):7169–7177. doi:10.4049/jimmunol.0901906
12. Murugaiyan G, Beynon V, Pires Da Cunha A, Joller N, Weiner HL (2012) IFN-gamma limits Th9-mediated autoimmune inflammation through dendritic cell modulation of IL-27. J Immunol 189(11):5277–5283. doi:10.4049/jimmunol.1200808
13. Zhou Y, Sonobe Y, Akahori T, Jin S, Kawanokuchi J, Noda M, Iwakura Y, Mizuno T, Suzumura A (2011) IL-9 promotes Th17 cell migration into the central nervous system via CC chemokine ligand-20 produced by astrocytes. J Immunol 186(7):4415–4421. doi:10.4049/jimmunol.1003307

14. Gessner A, Blum H, Rollinghoff M (1993) Differential regulation of IL-9-expression after infection with Leishmania major in susceptible and resistant mice. Immunobiology 189(5):419–435. doi:10.1016/S0171-2985(11)80414-6

15. Faulkner H, Renauld JC, Van Snick J, Grencis RK (1998) Interleukin-9 enhances resistance to the intestinal nematode Trichuris muris. Infect Immun 66(8):3832–3840

16. Fallon PG, Smith P, Richardson EJ, Jones FJ, Faulkner HC, Van Snick J, Renauld JC, Grencis RK, Dunne DW (2000) Expression of interleukin-9 leads to Th2 cytokine-dominated responses and fatal enteropathy in mice with chronic *Schistosoma mansoni* infections. Infect Immun 68(10):6005–6011

17. Licona-Limon P, Henao-Mejia J, Temann AU, Gagliani N, Licona-Limon I, Ishigame H, Hao L, Herbert DR, Flavell RA (2013) Th9 cells drive host immunity against gastrointestinal worm infection. Immunity 39(4):744–757. doi:10.1016/j.immuni.2013.07.020

18. Lu Y, Hong S, Li H, Park J, Hong B, Wang L, Zheng Y, Liu Z, Xu J, He J, Yang J, Qian J, Yi Q (2012) Th9 cells promote antitumor immune responses in vivo. J Clin Invest 122(11):4160–4171. doi:10.1172/JCI65459

19. Purwar R, Schlapbach C, Xiao S, Kang HS, Elyaman W, Jiang X, Jetten AM, Khoury SJ, Fuhlbrigge RC, Kuchroo VK, Clark RA, Kupper TS (2012) Robust tumor immunity to melanoma mediated by interleukin-9-producing T cells. Nat Med 18(8):1248–1253. doi:10.1038/nm.2856

20. Schmitt E, Germann T, Goedert S, Hoehn P, Huels C, Koelsch S, Kuhn R, Muller W, Palm N, Rude E (1994) IL-9 production of naive CD4+ T cells depends on IL-2, is synergistically enhanced by a combination of TGF-beta and IL-4, and is inhibited by IFN-gamma. J Immunol 153(9):3989–3996

21. Dardalhon V, Awasthi A, Kwon H, Galileos G, Gao W, Sobel RA, Mitsdoerffer M, Strom TB, Elyaman W, Ho IC, Khoury S, Oukka M, Kuchroo VK (2008) IL-4 inhibits TGF-beta-induced Foxp3+ T cells and, together with TGF-beta, generates IL-9+ IL-10+ Foxp3(−) effector T cells. Nat Immunol (12):1347–1355. doi:10.1038/ni.1677

22. Xiao X, Balasubramanian S, Liu W, Chu X, Wang H, Taparowsky EJ, Fu YX, Choi Y, Walsh MC, Li XC (2012) OX40 signaling favors the induction of T(H)9 cells and airway inflammation. Nat Immunol. doi:10.1038/ni.2390

23. Angkasekwinai P, Chang SH, Thapa M, Watarai H, Dong C (2010) Regulation of IL-9 expression by IL-25 signaling. Nat Immunol 11(3):250–256. doi:10.1038/ni.1846

24. Uyttenhove C, Brombacher F, Van Snick J (2010) TGF-beta interactions with IL-1 family members trigger IL-4-independent IL-9 production by mouse CD4(+) T cells. Eur J Immunol 40(8):2230–2235. doi:10.1002/eji.200940281

25. Vegran F, Berger H, Boidot R, Mignot G, Bruchard M, Dosset M, Chalmin F, Rebe C, Derangere V, Ryffel B, Kato M, Prevost-Blondel A, Ghiringhelli F, Apetoh L (2014) The transcription factor IRF1 dictates the IL-21-dependent anticancer functions of TH9 cells. Nat Immunol 15(8):758–766. doi:10.1038/ni.2925

26. Yao W, Zhang Y, Jabeen R, Nguyen ET, Wilkes DS, Tepper RS, Kaplan MH, Zhou B (2013) Interleukin-9 is required for allergic airway inflammation mediated by the cytokine TSLP. Immunity 38(2):360–372. doi:10.1016/j.immuni.2013.01.007

27. Elyaman W, Bassil R, Bradshaw EM, Orent W, Lahoud Y, Zhu B, Radtke F, Yagita H, Khoury SJ (2012) Notch receptors and Smad3 signaling cooperate in the induction of interleukin-9-producing T cells. Immunity 36(4):623–634. doi:10.1016/j.immuni.2012.01.020

28. Mikami N, Miyagi Y, Sueda K, Takatsuji M, Fukada S, Yamamoto H, Tsujikawa K (2013) Calcitonin gene-related peptide and cyclic adenosine 5'-monophosphate/protein kinase A pathway promote IL-9 production in Th9 differentiation process. J Immunol 190(8):4046–4055. doi:10.4049/jimmunol.1203102

29. Niedbala W, Besnard AG, Nascimento DC, Donate PB, Sonego F, Yip E, Guabiraba R, Chang HD, Fukada SY, Salmond RJ, Schmitt E, Bopp T, Ryffel B, Liew FY (2014) Nitric oxide enhances Th9 cell differentiation and airway inflammation. Nat Commun 5:4575. doi:10.1038/ncomms5575

30. Li H, Edin ML, Bradbury JA, Graves JP, DeGraff LM, Gruzdev A, Cheng J, Dackor RT, Wang PM, Bortner CD, Garantziotis S, Jetten AM, Zeldin DC (2013) Cyclooxygenase-2 inhibits T helper cell type 9 differentiation during allergic lung inflammation via down-regulation of IL-17RB. Am J Respir Crit Care Med 187(8):812–822. doi:10.1164/rccm.201211-2073OC

31. Kerzerho J, Maazi H, Speak AO, Szely N, Lombardi V, Khoo B, Geryak S, Lam J, Soroosh P, Van Snick J, Akbari O (2013) Programmed cell death ligand 2 regulates TH9 differentiation and induction of chronic airway hyperreactivity. J Allergy Clin Immunol 131(4):1041–1057. doi:10.1016/j.jaci.2012.09.027

32. Palmer MT, Lee YK, Maynard CL, Oliver JR, Bikle DD, Jetten AM, Weaver CT (2011)

Lineage-specificeffectsof1,25-dihydroxyvitamin D(3) on the development of effector CD4 T cells. J Biol Chem 286(2):997–1004. doi:10.1074/jbc.M110.163790

33. Liao W, Spolski R, Li P, Du N, West EE, Ren M, Mitra S, Leonard WJ (2014) Opposing actions of IL-2 and IL-21 on Th9 differentiation correlate with their differential regulation of BCL6 expression. Proc Natl Acad Sci U S A 111(9):3508–3513. doi:10.1073/pnas.1301138111
34. Goswami R, Jabeen R, Yagi R, Pham D, Zhu J, Goenka S, Kaplan MH (2012) STAT6-dependent regulation of Th9 development. J Immunol 188(3):968–975. doi:10.4049/jimmunol.1102840
35. Staudt V, Bothur E, Klein M, Lingnau K, Reuter S, Grebe N, Gerlitzki B, Hoffmann M, Ulges A, Taube C, Dehzad N, Becker M, Stassen M, Steinborn A, Lohoff M, Schild H, Schmitt E, Bopp T (2010) Interferon-regulatory factor 4 is essential for the developmental program of T helper 9 cells. Immunity 33(2):192–202. doi:10.1016/j.immuni.2010.07.014
36. Gomez-Rodriguez J, Meylan F, Handon R, Hayes ET, Anderson SM, Kirby MR, Siegel RM, Schwartzberg PL (2016) Itk is required for Th9 differentiation via TCR-mediated induction of IL-2 and IRF4. Nat Commun 7:10857. doi:10.1038/ncomms10857
37. Jabeen R, Goswami R, Awe O, Kulkarni A, Nguyen ET, Attenasio A, Walsh D, Olson MR, Kim MH, Tepper RS, Sun J, Kim CH, Taparowsky EJ, Zhou B, Kaplan MH (2013) Th9 cell development requires a BATF-regulated transcriptional network. J Clin Invest 123(11):4641–4653. doi:10.1172/JCI69489
38. Chang HC, Sehra S, Goswami R, Yao W, Yu Q, Stritesky GL, Jabeen R, McKinley C, Ahyi AN, Han L, Nguyen ET, Robertson MJ, Perumal NB, Tepper RS, Nutt SL, Kaplan MH (2010) The transcription factor PU.1 is required for the development of IL-9-producing T cells and allergic inflammation. Nat Immunol 11(6):527–534. doi:10.1038/ni.1867
39. Migone TS, Zhang J, Luo X, Zhuang L, Chen C, Hu B, Hong JS, Perry JW, Chen SF, Zhou JX, Cho YH, Ullrich S, Kanakaraj P, Carrell J, Boyd E, Olsen HS, Hu G, Pukac L, Liu D, Ni J, Kim S, Gentz R, Feng P, Moore PA, Ruben SM, Wei P (2002) TL1A is a TNF-like ligand for DR3 and TR6/DcR3 and functions as a T cell costimulator. Immunity 16(3):479–492
40. Meylan F, Davidson TS, Kahle E, Kinder M, Acharya K, Jankovic D, Bundoc V, Hodges M, Shevach EM, Keane-Myers A, Wang EC, Siegel RM (2008) The TNF-family receptor DR3 is essential for diverse T cell-mediated inflammatory diseases. Immunity 29(1):79–89. doi:10.1016/j.immuni.2008.04.021
41. Pappu BP, Borodovsky A, Zheng TS, Yang X, Wu P, Dong X, Weng S, Browning B, Scott ML, Ma L, Su L, Tian Q, Schneider P, Flavell RA, Dong C, Burkly LC (2008) TL1A-DR3 interaction regulates Th17 cell function and Th17-mediated autoimmune disease. J Exp Med 205(5):1049–1062. doi:10.1084/jem.20071364
42. Jones GW, Stumhofer JS, Foster T, Twohig JP, Hertzog P, Topley N, Williams AS, Hunter CA, Jenkins BJ, Wang EC, Jones SA (2011) Naive and activated T cells display differential responsiveness to TL1A that affects Th17 generation, maintenance, and proliferation. FASEB J 25(1):409–419. doi:10.1096/fj.10-166843
43. Berg LJ, Finkelstein LD, Lucas JA, Schwartzberg PL (2005) Tec family kinases in T lymphocyte development and function. Annu Rev Immunol 23:549–600. doi:10.1146/annurev.immunol.22.012703.104743

Chapter 6

Polarizing Cytokines for Human Th9 Cell Differentiation

Prabhakar Putheti

Abstract

CD4+ T helper (Th) cell subset generation in vivo requires T cell receptor activation and surface CD28 co-stimulation in the presence of one or more cytokines. Similarly, Th cells can be generated in vitro by activating naïve CD4+CD25− T cells with plate bound-anti-CD3 monoclonal antibody (mAb) (pbCD3) and soluble-anti-CD28 mAb (sCD28) in the presence of polarizing recombinant (r) cytokines and anti-cytokine mAbs. In comparison to in vitro CD4+CD25− T cells, memory CD4+CD25−CD45RO+ T cells have been shown to convert to Th9 cells more efficiently. Here, protocol for in vitro generation of human Th9 cells by activating CD4+CD25−CD45RO+ memory T cells with pbCD3 and sCD28 in the presence of polarizing recombinant interleukin-4 (rIL-4) and transforming growth factor (rTGF-β) is described.

Key words T Helper cells, Human Th9 cells, IL-4, IL-9, TGF-β, Peripheral blood

1 Introduction

When the immune system is exposed to allergens, pathogens, or alloantigens, naive $CD4^+$ T cells get activated and differentiated producing pro- or anti-inflammatory cytokines. In 1978, Tada et al. showed that there are two types of T cells: (1) Th1 cells are nylon nonadherent and do not express Ia antigen; and (2) Th2 cells are nylon adherent and express Ia antigen [1]. In 1986, Coffman and Mosmann showed that in response to Con-A or antigen stimulation, Th1 cell lines produce IL-2 and interferon-γ (IFN-γ), whereas Th2 cell lines produce B cell stimulatory factor 1 [2]. Thereafter, Th cell subsets were extensively studied and more Th cell subsets were identified based on their cytokine profile and/or specific transcription factors. Th1 cells express transcription factor T-bet, produce pro-inflammatory cytokines—IL-2, IL-12, IFN-γ, and TNFα—and induce macrophage activation and clearance of intracellular pathogens [3]. Th2 cells express transcription factor GATA-3, produce anti-inflammatory cytokines, IL-4 and IL-10—and induce IgE production, eosinophil recruitment, and clearance of extracellular parasites [4]. Th17 cells

Ritobrata Goswami (ed.), *Th9 Cells: Methods and Protocols*, Methods in Molecular Biology, vol. 1585,
DOI 10.1007/978-1-4939-6877-0_6, © Springer Science+Business Media LLC 2017

express transcription factor RORγt, produce pro-inflammatory cytokines—IL-17A, IL-17F, IL-21, IL-22, and IL-23—and induce immune responses against extracellular bacteria and fungi [5–7]. T regulatory cells (Treg) express transcription factor Foxp3, produce TGF-β, and help in tolerance [8, 9]. Th9 cells express transcription factors PU.1 and BATF, and produce pro-inflammatory cytokine IL-9 [10–13].

Unlike human Th1, Th2, Th17, and Treg cells that are generated from naive CD4+CD25−T cells, human Th9 cells are generated from memory CD4+CD25−CD45RO+ T cells [12]. In vitro, IL-4 induces Th2 cell generation and TGF-β induces Treg cell generation, but IL-4 in combination with TGF-β induces Th9 cell generation [12].

For in vitro human peripheral blood Th9 cell generation, memory CD4+CD25−CD45RO+ T cells should be isolated from human peripheral blood in the sequence described below: (1) human peripheral blood mononuclear cells (PBMC) should be isolated from healthy adult peripheral blood; (2) CD4+ T cells should preferably be isolated from PBMC by *RosetteSep™ Human CD4+ T cell enrichment cocktail*; (3) CD4+CD25− T cells should preferably be isolated from CD4+ T cells by *EasySep™ Human PE Positive Selection Kit*; and (4) finally, memory CD4+CD25−CD45RO+ T cells should preferably be isolated from CD4+CD25− T cells by *EasySep™ Human Biotin Positive Selection Kit* (*RosetteSep™ and EasySep™* are from Stemcell Technologies). Then the memory CD4+CD25−CD45RO+ T cells should be cultured in vitro for 4 days in the presence of pbCD3 and sCD28 activation and rIL-4 and rTGF-β.

2 Materials

2.1 Isolation of CD4+CD25−CD45RO+ Memory T Cells

1. Peripheral Blood: Obtain 50 mL of heparinized peripheral blood from adult healthy volunteer (*see* **Note 1**).
2. RPMI 1640 media with L-glutamine.
3. Centrifuge tubes: 5 mL round-bottom tubes, 15 and 50 mL conical bottom tubes, and 2 mL microcentrifuge tubes.
4. RosetteSep™ and EasySep™ cell isolation kits: RosetteSep™ Human CD4+ T cell enrichment cocktail, EasySep™ Human PE Positive Selection Kit, EasySep™ Human Biotin Positive Selection Kit, and EasySep™ magnet (RosetteSep™ and EasySep™ are Stemcell Technologies Inc. kits; selection kits consist of selection cocktail that binds to appropriate fluorochrome or biotin and magnetic nanoparticles for attaching the fluorochrome labeled cells to the magnet). For isolating CD4+ T cells, CD4+CD25− T cells, and CD45RO+ T cells by other manufacturer's kits, please follow the manufacturer's instructions.

5. Lymphoprep™ (alternatively Ficoll-Paque can be used).
6. Phosphate buffered saline (PBS).
7. Wash buffer: PBS+ 2% (v/v) Fetal Calf Serum (FCS). Add 4 mL of filtered FCS to 196 mL of sterile PBS.
8. Phycoerythrin (PE, orange fluorescence)-labeled mouse anti-human CD25 mAb (Clone: M-A251) and biotin labeled mouse anti-human-CD45RO mAb (Clone: UCHL1).
9. 5 and 10 mL pipettes.
10. 1, 10, 200, and 1000 μL tips and pipettes.

2.2 Cell Culture Reagents

1. Culture Media (100 mL): RPMI 1640 media with L-glutamine, 10% FCS (v/v), 1% non-essential amino acids (v/v), and 50 U/mL penicillin and 50 μg/mL streptomycin. Take 89 mL of RPMI 1640 with L-glutamine in a sterile glass beaker. Add 5 μL of 10,000 U/ml penicillin and 10,000 μg/mL of streptomycin. Add 1 mL of 100× non-essential amino acids and 10 mL of FCS. Mix the solution gently and filter it through sterile 0.22 μm filtration systems.
2. Monoclonal Antibodies: anti-human CD3 (Clone: UCHT1) and anti-human CD28 (Clone: CD28.2) with no azide/low endotoxin, anti-human CD4 (Clone: L200), anti-human IL-9 (Clone: MH9A4), and isotype controls.
3. Recombinant Cytokines: rIL-2, rIL-4, and rTGF-β.
4. Sterile 24-well polystyrene plates.

3 Methods

Perform all the steps in sterile conditions and in laminar flow hood.

3.1 Isolation of CD4+ T Cells

1. Take 25 mL of peripheral blood per 50 mL centrifuge tube (*see* **Note 1**).
2. Add Human CD4+ T cell enrichment cocktail at 50 μL/mL peripheral blood and mix well (*see* **Note 1**). Incubate for 20 min at room temperature.
3. Dilute sample with an equal volume of PBS. Layer the diluted sample on lymphoprep in a 50 mL centrifuge tube and centrifuge at 750 × *g* for 20 min at room temperature with no brakes.
4. Collect white layer (enriched CD4+ T cells) in the middle and transfer to fresh 50 mL centrifuge tubes.
5. Centrifuge at 750 × *g* for 20 min at room temperature with full brakes. Discard the supernatant and transfer the cell pellet to 15 mL centrifuge tubes, fill the tube up to 15 mL mark with PBS. Centrifuge at 750 × *g* for 20 min at room temperature

with full brakes. Discard the supernatant and resuspend cells in 1 mL of wash buffer.

6. Count the cells (*see* **Note 2**) and proceed with CD4+CD25− T cell isolation (*see* **Note 3**).

3.2 Isolation of CD4+CD25− T Cells

1. Stain the cells with mouse anti-human-CD25 PE (1 μg/250 × 10^6 cells/mL) for 15 min at 4 °C in the dark (*see* **Notes 4–6**).
2. Add 100 μL/mL PE selection cocktail and incubate for 15 min at room temperature.
3. Add nanoparticles at 50 μL/mL and incubate for 15 min at 20 °C.
4. Place the tube (without cap) into the EasySep magnet. Incubate at room temperature for 5 min.
5. Invert the magnet with tube and collect the buffer with CD4+CD25− T cells into a 15 mL tube. Discard the tube with CD25+ T cells (*see* **Note 6**).
6. Count CD4+CD25− T cells and resuspend them at 1 × 10^6 cells per mL in complete media (*see* **Note** 7).

3.3 Isolation of CD4+CD25− CD45RO+ Memory T Cells from Human PBMCs

1. Add biotin labeled mouse anti-human CD45RO mAb to the CD4+CD25− T cells at the concentration of 1 μg/250 × 10^6 cells/mL, mix well, and incubate for 15 min at 4 °C in the dark (*see* **Note 8**).
2. Add 100 μL/mL of biotin positive selection cocktail (EasySep™ Human biotin Positive Selection Kit), mix, and incubate for 15 min at room temperature.
3. Add anti-biotin nanoparticles at 50 μL/mL and incubate for 15 min at room temperature.
4. Place the tube (without cap) into the EasySep magnet. Incubate at room temperature for 5 min.
5. Invert the magnet with tube into a waste container and discard the liquid containing CD45RO− cells (termed "pour-off CD45RO− T cells" here-after).
6. Gently add 4 mL of wash buffer into the tube still placed in magnet and invert the magnet with tube into a waste container to remove pour-off CD45RO− T cells (*see* **Note 8**).
7. Take the tube off the magnet and pipette 2 mL of wash buffer onto the wall of the tube to dislodge the cells from the wall.
8. Repeat the step and gently vortex the tube.
9. Centrifuge the cells at 525 × *g* for 5 min at room temperature with full brakes.
10. Discard supernatant and resuspend the cells in 3 mL of wash buffer. Centrifuge at 525 × *g* for 5 min at room temperature with full brakes.

11. Discard supernatant and resuspend the cells in 1 mL of complete media.
12. Count CD4+CD25−CD45RO+ T cells and adjust the concentration to 2 × 10^6 cell/mL in complete media (*see* **Notes 8** and **9**).

3.4 In Vitro Generation of Polarized Human Th9 Cells

1. Prepare 3 mL of 5 μg/mL mouse anti-human-CD3 in sterile wash buffer.
2. Take a 24-well plate and pipette 1 mL of mouse anti-human CD3 into any two central wells of the plate (*see* **Note 10**).
3. Place the lid back and wrap the plate with Saran wrap.
4. Wrap the plate with aluminum foil, and incubate it overnight at 4 °C (*see* **Note 11**). Leave the four corner wells empty.
5. Before plating the cells, unwrap the plate, and aspirate out the PBS from the anti-human CD3 coated plate gently without touching or scratching the bottom of the wells.
6. Dispense 500 μL of PBS gently to wash off unbound anti-human CD3, and aspirate out the PBS. Repeat this step one more time (*see* **Note 12**).
7. Immediately add 400 μL of complete media into both wells (*see* **Note 13**). On the lid, mark one well as Th0 and the other well as Th9.
8. Add 1 μg mouse anti-human-CD28 to Th0 and Th9 wells. Add 1.5 ng of IL-2 to Th0 well. Add 20 ng of rIL-4 and 1 ng of rTGF-β to the Th9 well.
9. Make up the volume in each well to 500 μL with complete media and mix well with pipette (*see* **Note 14**). Add 1 mL PBS to corner four wells (*see* **Note 10**).
10. Add 500 μL of cell suspension to each well and mix it gently. [Final concentration now is as follows: (a) Th0 well: 1 × 10^6 cells/mL, 5 μg/mL of anti-human CD3, 1 μg/mL of mouse anti-human CD28 mAb, 1.5 ng/mL of IL-2; and (b) Th9 well: 1 × 10^6 cells/mL, 5 μg/mL of anti-human CD3, 1 μg/mL of mouse anti-human CD28 mAb, 20 ng/mL of rIL-4 and 1 ng/mL of rTGF-β].
11. Add 1 mL of complete media to any empty well in the plate, and label it on the lid as blank well (*see* **Note 15**).
12. Place the lid back and place it for incubation at 37 °C/5% CO_2 for 96 h (*see* **Note 15**).
13. At 96 h, using the pipette collect the cells from the wells by thoroughly flushing the corners of the well and by gently scratching the bottom of the well (*see* **Note 16**).
14. Transfer the cell suspension to 2 mL of microcentrifuge tube.
15. Add 1 mL of complete media to well, pipette thoroughly, and transfer the cell suspension to the same 2 mL of microcentrifuge tube.

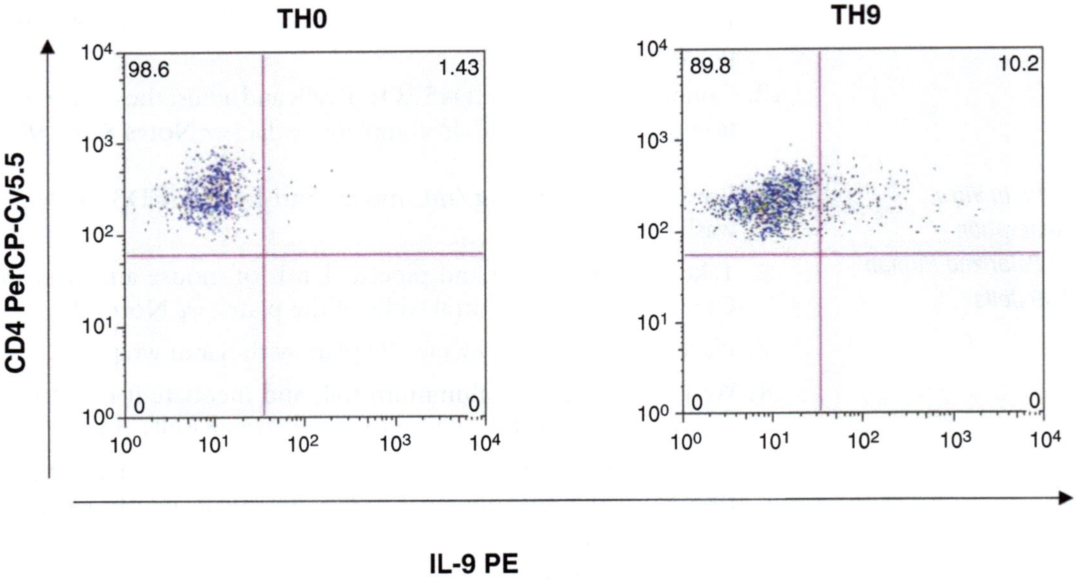

Fig. 1 IL-4+TGF-β in the presence of pbCD3/sCD28 activation induces generation of CD4+IL-9+ T cells. CD4+CD25-CD45RO+ T cells (1×10^6 cells/mL in 24-well plates) were activated with plate bound-anti-CD3 mAb/soluble-anti-CD28 mAb in the presence of rIL-2 (Th0) or of rIL-4+ rTGF-β for 96 h (Th9) and analyzed by flow cytometry for IL-9 expression. rIL-4+ rTGF-β in combination induces significantly higher percentage of CD4+ T cells positive for IL-9 as compared with rIL-2

16. Quality of Th9 cell generation may be checked by performing surface staining for CD4 and intracellular staining for IL-9 and acquiring in a flow cytometer (*see* **Notes 5** and **17–19**; Fig. 1). Use Th0 cells as control. Discard blank complete media well.

4 Notes

1. Leukofilters can also be used for PBMC isolation. In case of leukofilter, take a 50 mL syringe and back-wash cells from leukofilter with 50–100 mL of RPMI media. Collect cell suspension in 50 mL centrifuge tubes and centrifuge at 750 × *g* for 20 min at room temperature with full brakes. Discard supernatant and resuspend cell pellet in 10 mL of PBS. Count the cells. Cells should be 30–50 RBCs to one leukocyte. Adjust cell concentration to 50×10^6 leukocytes/mL (*see* **Note 2**). Add RosetteSep Human CD4+ T cell enrichment cocktail at 50 μL/mL of cell suspension.
2. Cell count varies from operator to operator, and this can affect the efficiency of in vitro generation of Th9 cells. In a given project, it is important to designate one operator to count the cells. Alternatively, automated cell counter may be used.

3. If there are any trace red blood cells (RBC) in the PBMC, they will be removed when memory CD4+CD25−CD45RO+ T cells are positively selected in the later steps. Do not perform hemolysis to remove trace of RBC.
4. Different or more advanced kits from StemCell technologies or other manufacturers for human CD4+ T cell isolation may or may not give similar Th9 cell generation results. Always perform enough parallel testing with the protocol described in this chapter to make sure the Th9 cell generation is not affected.
5. Phycoerythrin is a 240-kD protein with 23 phycoerythrobilin chromophores per molecule in vitro. PE is a bright fluorochrome used specifically for identifying molecules expressed in low levels on the cell surface or intracellular by flow cytometry. As average expression of surface CD25 is 7% of peripheral blood CD4+ T cells in healthy individuals and intracellular IL-9 expression varies from 4 to 15% in in vitro generated Th9 cells, PE makes a good choice for staining these molecules [12, 14]. However, PE is also a large molecule and may not pass through the permeabilized cells as efficiently as other fluorescent molecules. Alternatively, APC tagged antibodies may be used for intracellular staining. As PE is sensitive to photobleaching, while working with PE always work in dim light and store the cells in the dark by wrapping the tubes and containers in aluminum foil.
6. Keep an aliquot of CD4+ T cells stained with CD25PE and of CD4+CD25− T cells to check the purity of CD4+CD25− T cell isolation. Perform cell surface staining with anti-human CD4 PerCP Cy5.5.
7. The fraction of CD25+ T cells (from **step 1** of Subheading 3.2) can be used for Treg cell isolation for other studies (from **step 5** of Subheading 3.2).
8. Although not as efficient as memory CD4+CD25−CD45RO+ T cells, naive CD4+CD25−CD45RA+ T cells can also be converted in vitro to Th9 cells [12]. Upon CD4+CD25−CD45RO+ T cell isolation, presence of residual CD4+CD25−CD45RO− T cells (mostly naive CD4+CD25−CD45RA+ T cells) does not affect Th9 cell generation. Hence, purity check of memory CD4+CD25−CD45RO+ T cells is not required at this step. In case purity of memory CD4+CD25−CD45RO+ T cells isolation is required to be checked, collect an aliquot of CD4+CD25− T cells before adding biotin labeled mouse anti-human-CD45RO and an aliquot of *pour-off CD45RO− T cells*, and perform surface staining with appropriate fluorescent labeled mouse anti-human-CD4, CD25, and CD45RO. Purity of CD45RO+ T cells can only be indirectly interpreted by the

absence of CD45RO+ T cells in the aliquot of *pour-off* CD45RO− T cells.

9. This is an ideal procedure where clean CD45RO+ T cells are isolated and CD45RO− T cells (or CD45RA+ T cells) can be used for other experiments such as in vitro Th1, Th2, Th17, or Treg generation. Otherwise, anti-CD25-PE and anti-CD45RA-PE can be added in Subheading 3.2, and unbound CD45RO+ T cells may be isolated in Subheading 3.2 (without performing Subheading 3.3). If this modification is pursued, perform enough parallel testing with the protocol described in this chapter to make sure the Th9 cell generation is not affected.
10. Evaporation of media from cell cultures is a common problem that disturbs the concentration of cytokines in cultures. In addition, the evaporation is more in the peripheral wells of the plate as compared to central wells. So, central wells of the plate are preferred for cell cultures as compared to peripheral wells. Adding PBS to the wells in the four corners of the plate helps in providing added humidity to the cells in the plate and reduce evaporation of media from cell cultures.
11. Alternatively, the plate can be incubated with anti-human CD3 for 2 h at 37 °C. While performing a set of experiments in a study, it is important to keep the methodology consistent. Overnight incubation with anti-human CD3 at 4 °C for some experiments and 2 h incubation with anti-human CD3 at 37 °C for the rest can affect the Th9 cell generation and other downstream applications.
12. The idea behind washing the wells with PBS is only to remove unbound mouse anti-human CD3. Flushing hard with PBS can remove bound anti-human CD3, resulting in nonactivation of some T cells.
13. With the air-flow in laminar flow hood and with hydrophobic surface of polystyrene 24-well plate material, the wells dry fast. To prevent the wells from drying, immediately add 400 μL of complete media to the wells.
14. Follow manufacturer's instructions for reconstituting lyophilized recombinant cytokines and filtering. In order to avoid frequent freeze and thawing, prepare aliquots in volumes just sufficient for the experiments. Filtering removes any undissolved clumps that can kill the cells in cultures.
15. Complete media in the blank well helps identify any bacterial or fungal contamination of cultures. If there is contamination in the cell cultures only and not in the blank complete media well, it suggests the source of contamination is cell preparation. If there is contamination in cell cultures and in the blank wells, the source of contamination may be complete media

and/or incubator. Using center wells for cell cultures gives extra protection from contamination from adjacent plates in the incubator. After 48 h, measure the volume of complete media in the blank well to identify the evaporation volume. Accordingly, add fresh complete media to culture wells to make it 1 mL/well. For example, if the volume of complete media is 800 μL after 48 h, this would mean 200 μL of complete media has evaporated in the plate and that 200 μL of complete media should be added to the culture wells.

16. Supernatants of memory CD4+CD25−CD45RO+ T cells cultures may be collected for quantitation of IL-9 by ELISA.
17. At the end of 96 h of in vitro Th9 cell generation, percentage of CD4+IL-9+ Th9 cells varies from specimen to specimen. For example, one specimen may show 5% Th9 cell generation, while the other may show 12% Th9 cell generation. Normal ranges for in vitro of Th9 cell generation should be developed within the laboratory. Specimens with results out of the normal range should be checked if it is specific to any particular donor or if there is any problem with one or more of the reagents.
18. Cultured cells may be stained with fluorescence isothiocyanate (FITC, green fluorescence)-labeled-Annexin-V and Propidium Iodide to identify live Th9 cell percentage.
19. While performing Th9 cell surface and/or intracellular staining, some operators have the habit of marking the wells of the plate or the opening of 5 mL round-bottom polystyrene tubes used for flow cytometry. When the marker ink mixes with the sample during the washing steps, it gives false-positive results in the PE channel. When needed, make marking on the lid of the plate and on the sides of the tubes.

References

1. Tada T, Takemori T, Okumura K, Nonaka M, Tokuhisa T (1978) Two distinct types of helper T cells involved in the secondary antibody response: independent and synergistic effects of Ia− and Ia+ helper T cells. J Exp Med 147(2):446–458
2. Mosmann TR, Cherwinski H, Bond MW, Giedlin MA, Coffman RL (1986) Two types of murine helper T cell clone I. Definition according to profiles of lymphokine activities and secreted proteins. J Immunol 136(7): 2348–2357
3. Szabo SJ, Sullivan BM, Peng SL, Glimcher LH (2003) Molecular mechanisms regulating Th1 immune responses. Annu Rev Immunol 21: 713–758
4. Ansel KM, Djuretic I, Tanasa B, Rao A (2006) Regulation of Th2 differentiation and Il4 l0Cus accessibility. Annu Rev Immunol 24: 607–656
5. Harrington LE, Hatton RD, Mangan PR, Turner H, Murphy TL, Murphy KM, Weaver CT (2005) Interleukin 17-producing CD4+ effector T cells develop via a lineage distinct from the T helper type 1 and 2 lineages. Nat Immunol 6(11):1123–1132
6. Park H, Li Z, Yang XO, Chang SH, Nurieva R, Wang YH, Wang Y, Hood L, Zhu Z, Tian Q, Dong C (2005) A distinct lineage of CD4 T cells regulates tissue inflammation by producing interleukin 17. Nat Immunol 6(11): 1133–1141

7. Weaver CT, Hatton RD, Mangan PR, Harrington LE (2007) IL-17 family cytokines and the expanding diversity of effector T cell lineages. Annu Rev Immunol 25:821–852
8. Sakaguchi S (2004) Naturally arising CD4+ regulatory t cells for immunologic self-tolerance and negative control of immune responses. Annu Rev Immunol 22:531–562
9. Chen W, Jin W, Hardegen N, Lei KJ, Li L, Marinos N, McGrady G, Wahl SM (2003) Conversion of peripheral CD4+CD25− naive T cells to CD4+CD25+ regulatory T cells by TGF-beta induction of transcription factor Foxp3. J Exp Med 198(12):1875–1886
10. Jabeen R, Goswami R, Awe O, Kulkarni A, Nguyen ET, Attenasio A, Walsh D, Olson MR, Kim MH, Tepper RS, Sun J, Kim CH, Taparowsky EJ, Zhou B, Kaplan MH (2013) Th9 cell development requires a BATF-regulated transcriptional network. J Clin Invest 123(11):4641–4653
11. Chang HC, Han L, Jabeen R, Carotta S, Nutt SL, Kaplan MH (2009) PU.1 regulates TCR expression by modulating GATA-3 activity. J Immunol 183(8):4887–4894
12. Putheti P, Awasthi A, Popoola J, Gao W, Strom TB (2010) Human CD4 memory T cells can become CD4+IL-9+ T cells. PLoS One 5(1): e8706
13. Dardalhon V, Korn T, Kuchroo VK, Anderson AC (2008) Role of Thl and Th17 cells in organ-specific autoimmunity. J Autoimmun 31(3):252–256
14. Putheti P, Pettersson A, Soderstrom M, Link H, Huang YM (2004) Circulating CD4+CD25+ T regulatory cells are not altered in multiple sclerosis and unaffected by disease-modulating drugs. J Clin Immunol 24(2):155–161

Chapter 7

Determining the Frequencies of Th9 Cells from Whole Blood

Anuradha Rajamanickam and Subash Babu

Abstract

Th9 cells are a subset of CD4+ T cells producing the cytokine, IL-9. Th9 cells are increasingly recognized as being important player in allergy, autoimmunity, and antitumor responses. The polarization and expansion of Th9 cells requires the cytokines IL-4, TGF-β. In this chapter, we described the protocol for measuring Th9 frequencies using whole blood.

Key words Th9, IL-9 cytokine, Whole blood immunophenotyping, Intracellular cytokine assay, Flow cytometry

1 Introduction

CD4+ T helper cells differentiate into effector subsets in response to environmental cues including cytokines and ligand–receptor interactions from cell-to-cell contact. These environmental signals promote the development of distinct T helper (Th) subsets that secrete specific cytokines and perform distinct functions in regulating immunity and inflammation [1–4]. Schmitt et al. first described IL-9 production by activated mouse T cells and consequently defined the cytokines that promote the differentiation of IL-9-producing Th cells in culture [5, 6]. Traditionally associated with the Th2 response, IL-9 is a member of the common cytokine receptor γ chain family that use the common IL-2Rγc for signal transduction, and similar to other family members exerts broad effects on many cell types including mast cells, eosinophils, T cells and epithelial cells [7–10]. Th9 cells are a subset of CD4+ T cells producing copious amount of IL-9 and variable amount of IL-10 depending on the species. This CD4+ T cell subset is developed in the presence of IL-4 and TGF-β producing IL-9, either by itself or in conjunction with IL-10; however, these cells fail to produce IL-4 [11, 12]. Th9 cells in humans were initially described as $CD4^+$ T cells, IL-9+ and IL-17+; however, IL-9 producing CD4+ T cells

Ritobrata Goswami (ed.), *Th9 Cells: Methods and Protocols*, Methods in Molecular Biology, vol. 1585,
DOI 10.1007/978-1-4939-6877-0_7, © Springer Science+Business Media LLC 2017

that are distinct from Th1, Th2, and Th17 cells have been described elsewhere [13–15]. Th9 cells, in humans, can play protective (tumors) as well as pathogenic (allergy, atopy, asthma, and autoimmunity) role in differing disease states [10, 16–18]. While Th9 cells have been shown to act as mediators of inflammation in experimental disease models, such as colitis, peripheral neuritis, and experimental autoimmune encephalitis, IL-9 has also been shown to participate in peripheral tolerance by increasing the survival and function of regulatory T cells [19–22]. In this chapter, we describe how to quantify IL-9+ CD4+ T cells from human blood.

2 Materials

2.1 *In Vitro Culture*

1. Human Blood.
2. Complete RPMI Medium: RPMI 1640 medium, penicillin (100 U/mL) /streptomycin (100 mg/mL), L-glutamine (2 mM), HEPES (10 mM).
3. Brugia malayi Antigen (BmA), Microfilariae Antigen (Mf), Mycobacterial Purified Protein Derivative (PPD) Antigen.
4. Phorbol 12-myristate 13 acetate (PMA).
5. Ionomycin.
6. Co-stimulatory reagents: CD49d/CD28.
7. Fast Immune Brefeldin A Solution.
8. Phosphate buffered saline (PBS), without Ca++ or Mg++.
9. FACS lysing solution (BD Biosciences).
10. Cytofix/cytoperm buffer (BD Biosciences).
11. Albumin from bovine serum (BSA).
12. Dimethyl sulfoxide (DMSO).
13. Perm/Wash™ buffer (BD Biosciences).

2.2 *Intracellular Staining*

1. Antibodies: Anti-human CD3, CD4, CD8, IL-4, IL-10, and IL-9 that are fluorescently labeled.
2. FACS Canto II flow cytometer.
3. FACS Diva software v.6 (Becton Dickinson).

2.3 *ELISA*

1. IL-9 ELISA kit.
2. ELISA plate washer.
3. ELISA plate reader.
4. Microtubes.

2.4 *Plasticware and Other Instruments*

1. 50 mL tubes.
2. 2 mL centrifuge tubes.
3. 12-well tissue culture plates

4. Cell scraper.
5. CO_2 incubator.
6. Serological pipettes.
7. Water bath.
8. Aluminum foil.
9. Heparin tubes.

3 Methods

3.1 *In Vitro* Culture of Th9 Cells from Human Whole Blood

1. Collect 10 mL of whole blood in the sodium heparin tubes (*see* **Note 1**).
2. Spin the tubes at 1600 × *g* for 10 min at 4 °C.
3. Separate the plasma (fluid portion) and store at 4 °C in 2 mL microtubes.
4. Dilute whole blood at 1:1 ratio with complete RPMI 1640 medium.
5. In a 12-well tissue culture plate add 2 mL of the blood per well, followed by addition of BmA (10 μg/mL), Mf (10 μg/mL), PPD (10 μg/mL), and PMA (12.5 ng/mL) (*see* **Note 2**), and ionomycin (125 ng/mL) (*see* **Note 3**).
6. Add CD28/CD49d/blood sample (10 μL/mL) co-stimulatory reagent (*see* **Note 4**) to all the labeled wells and gently mix for uniform distribution.
7. Incubate the plates in CO_2 incubator at 37 °C for 2 h.
8. After 2 h, add Brefeldin A (10 μg/mL) (*see* **Note 5**) to each well in the culture plates and incubate it in CO_2 incubator at 37 °C for 4–6 h.
9. Post incubation, transfer the cells to 50 mL falcons and spin at 1600 × *g* for 10 min at 4 °C.
10. Carefully pipette out the supernatants and store them in labeled 2 mL microtubes at 4 °C.
11. Wash and scrape the wells with PBS using the cell scraper and add it to the pellets in the respective falcon tubes.
12. Make it up to 15 mL with PBS.
13. Vortex and spin at 300 × *g* for 10 min at 4 °C and then discard the supernatant by using the serological pipette.
14. Resuspend the cell pellet in 10 mL of the prepared BD lysing solution (BD lysing solution (10×): Distilled water (1:10) and vortex until the pellet dissolves in the solution completely.
15. Leave the tubes closed with cap at room temperature for 10 min.
16. Spin at 300 × *g* for 10 min at 4 °C.

17. Discard the supernatant and resuspend the pellet in 10 mL of PBS and vortex well.
18. Spin at 300 × *g* for 10 min at 4 °C.
19. Discard the supernatant, add 200 μL of cytofix/cytoperm to the pellet, vortex the tubes and keep them at 4 °C for 20 min.
20. Vortex and then add 12 mL of cold 1% PBS/BSA (v/v) solution and invert to mix (Do not vortex the tubes after adding PBS/BSA).
21. Spin at 1600 × *g* for 10 min at 4 °C.
22. Discard the supernatants and resuspend the cells (pellet) of each tube in 1 mL of cold PBS/DMSO (*see* **Note 6**) and aliquot into two separate tubes for each culture condition (Unstimulated, BmA, Mf, PPD, and P/I).
23. Store at −80 °C until further use.

3.2 Intra Cellular Cytokine Staining (ICS)

1. Thaw the cryopreserved fixed cells at 37 °C using a water bath (*see* **Note 7**).
2. Transfer the thawed cells to 50 mL falcon tubes. Wash the insides of the vials well with PBS.
3. Add 10 mL of cold PBS to the cells, vortex and spin at 1600 × *g* for 10 min at 4 °C.
4. Discard the supernatant and wash pellet with 12 mL of cold 1% PBS/BSA (v/v) to the cells, vortex and spin at 1600 × *g* for 10 min at 4 °C.
5. Discard the supernatant and add desired volume of 1% PBS/BSA (v/v) to the pellet.
6. Aliquot 500 μL of cells to each FACS tube.
7. Stain the cells with corresponding surface monoclonal antibodies and vortex.
8. Incubate the tubes at 4 °C for 30–60 min. Cover the tubes with aluminum foil to maintain the dark environment.
9. Add 1.5 mL of perm buffer to each tube and incubate at 4 °C for 15 min.
10. Post incubation, vortex the tubes and spin at 1600 × *g* for 10 min at 4 °C.
11. Discard the supernatant.
12. Resuspend the pellet in the unstained tube with 200 μL of cold PBS and add 150 μL to the FMO tubes. Vortex the tubes.
13. To the remaining tubes, add 500 μL of the perm buffer and stain the cells with intracellular monoclonal antibodies (*see* **Note 8**).
14. Incubate the tubes for 30 min at 4 °C.
15. Add 1 mL of perm buffer to each of the ICS tube.

16. Vortex and spin at 1600 × *g* for 10 min at 4 °C.
17. Discard the supernatant and resuspend the cells/pellet in 200 μL of PBS.
18. Flow cytometry was performed on a FACS Canto II flow cytometer with FACS Diva software v.6 (Becton Dickinson).
19. Lymphocytes were identified by typical forward (FCS) versus side scatter (SSC), and cell doublets were excluded using forward scatter area-SSC A versus height SSC H (*see* **Note 9**).
20. The lymphocyte gating was set by forward and side scatter and 100,000 gated lymphocyte events were acquired (*see* **Note 10**).
21. Data are collected and analyzed using FlowJo software (*see* **Note 9**).

3.3 ELISA for IL-9 Detection Using Culture Supernatants

1. Coat ELISA plate with 100 μL/well of capture antibody in coating buffer (dilute according to manufacturer's instruction), which is included with the reagent set. Seal the plate and incubate overnight at 4 °C (*see* **Note 10**).
2. Aspirate wells and wash three times with >250 μL/well Wash Buffer. Allowing time for soaking (~ 1 min) during each wash step increases the effectiveness of the washes. Blot plate on absorbent paper to remove any residual buffer.
3. Dilute 1 part 5× ELISA diluent with 4 parts DI water. Block wells with 200 μL/well of 1× ELISA diluent. Incubate at room temperature for 1 h.
4. Aspirate and wash at least once with Wash Buffer.
5. Using 1× ELISA diluent, dilute standards as noted on the C of A to prepare the top concentration of the standard. Add 100 μL/well of top standard concentration to the appropriate wells.
6. Perform two-fold serial dilutions of the top standards to make the standard curve for a total of eight points.
7. Add 100 μL/well of your samples to the appropriate wells. Seal the plate and incubate at room temperature for 2 h (or overnight at 4 °C for maximal sensitivity).
8. Aspirate/wash as in **step 2**. Repeat for a total of 3–5 washes.
9. Add 100 μL/well of detection antibody diluted in 1× ELISA diluent (dilute according to manufacturer's instruction). Seal the plate and incubate at room temperature for 1 h.
10. Aspirate/wash as in **step 2**. Repeat for a total of 3–5 washes.
11. Add 100 μL/well of Avidin-HRP diluted in 1× ELISA diluent (dilute according to manufacturer's instruction). Seal the plate and incubate at room temperature for 30 min.
12. Aspirate/wash as in **step 2**. In this wash step, soak wells in Wash Buffer for 1–2 min prior to aspiration. Repeat for a total of 5–7 washes.

13. Add 100 μL/well of 1× TMB Solution to each well. Incubate plate at room temperature for 15 min.
14. Add 50 μL of Stop Solution to each well.
15. Read plate at 450 nm. If wavelength subtraction is available, subtract the values of 570 nm from those of 450 nm and analyze the data.

4 Notes

1. Whole blood—collected into sodium heparin VACUTAINER tubes. For best results, assay blood within 8 h of collection since a minor loss of activity can be expected beyond 8 h; typically, the percent of cytokine-positive cells is reduced by approximately 5%. If blood cannot be used within 8 h, store VACUTAINER tubes horizontally at room temperature.
2. Stimulation with BmA antigen is required as it helps to identify antigen specific expression. PPD is non-helminth control antigen and P/I is used as a positive control antigen. Phorbol 12-myristate 13 acetate (PMA) can be used as positive control because PMA activates all the T cell (unspecific) as maximum detection of intracellular cytokines. PMA needs to be reconstituted in DMSO at 0.1 mg/mL and stored in small aliquots (20 μL) at −20 °C. PMA is used by diluting the stock 1:100 in sterile PBS for each experiment at final concentration of 10 ng/mL of cell suspension.
3. Ionomycin is another reagent that needs to be used together with PMA as a positive control in experiments. This is reconstituted in ethanol at 0.5 mg/mL, stored at −20 °C, and used by diluting the stock 1:10 in sterile PBS for experiment, at a final concentration of 1 μg/mL of cell suspension.
4. Co-stimulatory reagent CD49d/CD28 also can be used to activate T cells; antibodies to CD28 and CD49d can increase the cytokine response to protein antigens, peptides and BmA by amplifying the signal for low-affinity T cells [24]. Experiments comparing the apoptosis inducing effects of anti-CD28 found that plate-bound anti-CD28 resulted in a greater induction of IL-9 production than cultures stimulated with soluble anti-CD28 [25].
5. Brefeldin-A (BFA) is a protein transport inhibitor commonly used to enhance intracellular cytokine staining signals by blocking transport processes during cell activation. Especially useful for the intracellular staining of cytokines, BFA leads to the accumulation of most cytokines at the Golgi complex/endoplasmic reticulum [24]. Typically, protein transport inhibitors are included during in vitro cell activation cultures for 4–24 h prior to harvest is used by at 10 μg/mL of cell suspension for

the last 4–5 h of activation. Extensive incubation with BFA can reduce cell viability.

6. DMSO is extremely harmful to cells when not used in cold conditions. Take utmost care to maintain the temperature while adding PBS/DMSO to the cells. Transfer the cells to −80 °C freezer immediately after aliquoting.
7. Intracellular cytokine staining (ICS) assays measure a functional readout (cytokine production) as opposed to antigenic stimulation, which measure antigen specificity without regard to function. Those assays can provide quantitative readouts since they enumerate antigen-specific cells without lengthy in vitro restimulation, necessary for proliferation or apoptosis or both. A major advantage of intracellular cytokine staining is the ability to analyze multiple parameters per cell. ICS flow cytometry, especially when combined with cell surface staining, is a powerful tool for the detection of cytokines from both homogenous and mixed cell populations. For intracellular protein detection, cells must be fixed and permeabilized to allow a fluorescent antibody to enter and detect the target protein of interest. Different antigens have different sensitivities to and requirements for fixation and permeabilization, requiring additional optimization of protocols. To detect cytokines, which are secreted proteins, protein transport inhibitors such as brefeldin A are used to trap proteins inside the cell.
8. Monoclonal antibodies are light and temperature sensitive in nature. Take care to maintain them in appropriate dark and cold conditions.
9. To set the gates "Fluorescence Minus One" controls during the analysis of the data, it is very important to distinguish the presence or not of the antigen of interest. This is very simple when investigating antigens whose positive expression is clearly distinguishable from the background fluorescence, CD3, CD4, CD8, etc. However, in most cases, antigens with unknown patterns of expression are investigated, and it is difficult to set the threshold of positivity. Unstained samples, or those incubated with irrelevant monoclonal antibodies of the same isotype and color, are not reliable for this purpose. "Fluorescence minus one" (FMO) controls must be utilized in those cases [23]. FMO controls are samples that have been stained with all antibodies except the one of interest. When performing multicolor experiments, in addition to single stained samples, prepare FMO controls for all the antigens under investigation, and then set the gates to distinguish positive and negative expression.
10. All data are depicted as frequency of CD4+ T cells expressing cytokine(s) (Fig. 1) or as the mean fluorescence intensity (MFI) of cytokine expression within a particular subset.

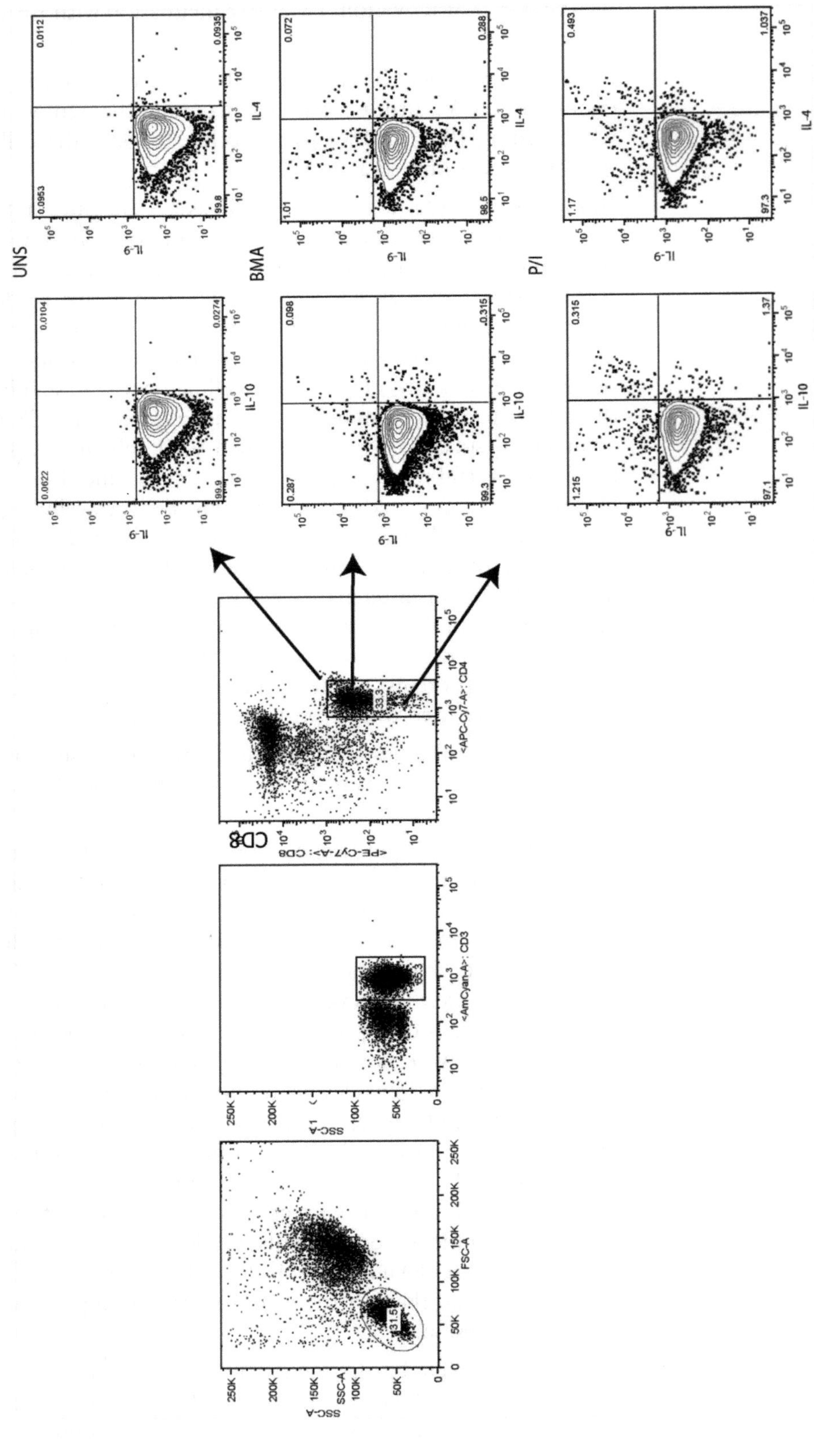

Fig. 1 Representative flow plot depicting the gating strategy of CD4+ T cells expressing Th9 (IL-9 and IL-10) cells of Unstimulated, BmA and P/I

Values following media stimulation are depicted as baseline frequency while frequencies following stimulation with antigens or PMA/ionomycin are depicted as net frequencies (with baseline values subtracted). The enzyme linked immunosorbent assay (ELISA) is used for the detection and quantification of proteins typically secreted or released from cells. Immobilizing a target-specific capture antibody onto a high protein binding capacity ELISA plate enables capture of target protein. A protein-specific biotinylated antibody then detects the captured protein. The target protein is quantified using a colorimetric reaction based on activity of avidin–horseradish peroxidase (bound to the biotinylated detection antibody) on a specific substrate (e.g., ABTS, Super AquaBlue, or TMB). The optical density of the end product is measured using a spectrophotometer.

References

1. Chang HC, Sehra S, Goswami R, Yao W, Yu Q, Stritesky GL, Jabeen R, McKinley C, Ahyi AN, Han L, Nguyen ET, Robertson MJ, Perumal NB, Tepper RS, Nutt SL, Kaplan MH (2010) The transcription factor PU.1 is required for the development of IL-9-producing T cells and allergic inflammation. Nat Immunol 11(6):527–534. doi:10.1038/ni.1867
2. Dardalhon V, Awasthi A, Kwon H, Galileos G, Gao W, Sobel RA, Mitsdoerffer M, Strom TB, Elyaman W, Ho IC, Khy S, Oukka M, Kuchroo VK (2008) IL-4 inhibits TGF-beta-induced Foxp3+ T cells and, together with TGF-beta, generates IL-9+ IL-10+ Foxp3(−) effector T cells. Nat Immunol (12):1347–1355. doi:10.1038/ni.1677
3. Veldhoen M, Uyttenhove C, van Snick J, Helmby H, Westendorf A, Buer J, Martin B, Wilhelm C, Stockinger B (2008) Transforming growth factor-beta 'reprograms' the differentiation of T helper 2 cells and promotes an interleukin 9-producing subset. Nat Immunol 9(12):1341–1346. doi:10.1038/ni.1659
4. Kaplan MH, Glosson NL, Stritesky GL, Yeh N, Kinzfogl J, Rohrabaugh SL, Goswami R, Pham D, Levy DE, Brutkiewicz RR, Blum JS, Cooper S, Hangoc G, Broxmeyer HE (2011) STAT3-dependent IL-21 production from T helper cells regulates hematopoietic progenitor cell homeostasis. Blood 117(23):6198–6201. doi:10.1182/blood-2011-02-334367
5. Schmitt E, Germann T, Goedert S, Hoehn P, Huels C, Koelsch S, Kuhn R, Muller W, Palm N, Rude E (1994) IL-9 production of naive CD4+ T cells depends on IL-2, is synergistically enhanced by a combination of TGF-beta and IL-4, and is inhibited by IFN-gamma. J Immunol 153(9):3989–3996
6. Schmitt E, Van Brandwijk R, Van Snick J, Siebold B, Rude E (1989) TCGF III/P40 is produced by naive murine CD4+ T cells but is not a general T cell growth factor. Eur J Immunol 19(11):2167–2170. doi:10.1002/eji.1830191130
7. O'Shea JJ, Paul WE (2010) Mechanisms underlying lineage commitment and plasticity of helper CD4+ T cells. Science 327(5969):1098–1102. doi:10.1126/science.1178334
8. Li XC, Schachter AD, Zand MS, Li Y, Zheng XX, Harmon WE, Strom TB (1998) Differential expression of T-cell growth factors in rejecting murine islet and human renal allografts: conspicuous absence of interleukin (IL)-9 despite expression of IL-2, IL-4, IL-7, and IL-15. Transplantation 66(2):265–268
9. Noelle RJ, Nowak EC (2010) Cellular sources and immune functions of interleukin-9. Nat Rev Immunol 10(10):683–687. doi:10.1038/nri2848
10. Wilhelm C, Turner JE, Van Snick J, Stockinger B (2012) The many lives of IL-9: a question of survival? Nat Immunol 13(7):637–641. doi:10.1038/ni.2303
11. Kaplan MH (2013) Th9 cells: differentiation and disease. Immunol Rev 252(1):104–115. doi:10.1111/imr.12028
12. Kaplan MH, Hufford MM, Olson MR (2015) The development and in vivo function of T helper 9 cells. Nat Rev Immunol 15(5):295–307. doi:10.1038/nri3824

13. Beriou G, Bradshaw EM, Lozano E, Costantino CM, Hastings WD, Orban T, Elyaman W, Khy SJ, Kuchroo VK, Baecher-Allan C, Hafler DA (2010) TGF-beta induces IL-9 production from human Th17 cells. J Immunol 185(1):46–54. doi:10.4049/jimmunol.1000356
14. Jones CP, Gregory LG, Causton B, Campbell GA, Lloyd CM (2012) Activin A and TGF-beta promote T(H)9 cell-mediated pulmonary allergic pathology. J Allergy Clin Immunol 129(4):1000–1010. doi:10.1016/j.jaci.2011.12.965e1003
15. Yao W, Zhang Y, Jabeen R, Nguyen ET, Wilkes DS, Tepper RS, Kaplan MH, Zhou B (2013) Interleukin-9 is required for allergic airway inflammation mediated by the cytokine TSLP. Immunity 38(2):360–372. doi:10.1016/j.immuni.2013.01.007
16. Soroosh P, Doherty TA (2009) Th9 and allergic disease. Immunology 127(4):450–458. doi:10.1111/j.1365-2567.2009.03114.x
17. Yao W, Tepper RS, Kaplan MH (2011) Predisposition to the development of IL-9-secreting T cells in atopic infants. J Allergy Clin Immunol 128(6):1357–1360 . doi:10.1016/j.jaci.2011.06.019e1355
18. Pan HF, Leng RX, Li XP, Zheng SG, Ye DQ (2013) Targeting T-helper 9 cells and interleukin-9 in autoimmune diseases. Cytokine Growth Factor Rev 24(6):515–522
19. Porto AF, Neva FA, Bittencourt H, Lisboa W, Thompson R, Alcantara L, Carvalho EM (2001) HTLV-1 decreases Th2 type of immune response in patients with strongyloidiasis. Parasite Immunol 23(9):503–507
20. Rotman HL, Schnyder-Candrian S, Scott P, Nolan TJ, Schad GA, Abraham D (1997) IL-12 eliminates the Th-2 dependent protective immune response of mice to larval Strongyloides stercoralis. Parasite Immunol 19(1):29–39
21. Gordy C, Pua H, Sempowski GD, He YW (2011) Regulation of steady-state neutrophil homeostasis by macrophages. Blood 117(2):618–629. doi:10.1182/blood-2010-01-265959
22. Jia W, Pua HH, Li QJ, He YW (2011) Autophagy regulates endoplasmic reticulum homeostasis and calcium mobilization in T lymphocytes. J Immunol 186(3):1564–1574. doi:10.4049/jimmunol.1001822
23. Perfetto SP, Chattopadhyay PK, Roederer M (2004) Seventeen-colour flow cytometry: unravelling the immune system. Nat Rev Immunol 4(8):648–655. doi:10.1038/nri1416
24. Takami M, Love RB, Iwashima M (2012) TGF-beta converts apoptotic stimuli into the signal for Th9 differentiation. J Immunol 188(9):4369–4375. doi:10.4049/jimmunol.1102698
25. Waldrop SL, Davis KA, Maino VC, Picker LJ (1998) Normal human CD4+ memory T cells display broad heterogeneity in their activation threshold for cytokine synthesis. J Immunol 161(10):5284–5295

Chapter 8

IL-9 Production by Nonconventional T helper Cells

Silvia C.P. Almeida and Luis Graca

Abstract

IL-9 is a pro-inflammatory cytokine implicated in certain immune-mediated diseases where chronic or acute inflammation of the mucosa plays an important role. Although initially described as being produced by what was then thought to be Th2 cells, it was later described that specialized lymphocyte populations are involved in IL-9 production. In addition to the classical Th9 effector (subset of CD4+ T cells), IL-9 is also produced by nonconventional lymphocytes, namely invariant natural killer T (iNKT) cells and innate lymphoid cells (ILCs). The identification of IL-9-producing cells by flow cytometry and cytokine measurements are pivotal for assigning and defining functional cellular phenotypes. In this chapter we provide methods for the in vitro polarization of IL-9-producing nonconventional lymphocytes and the best conditions for the detection of IL-9 production by intracellular staining.

Key words IL-9, Th9, Nonconventional T cells, iNKT cells, ILCs, Intracellular cytokine staining, Cytokine, Flow cytometry

1 Introduction

Interleukin-9 (IL-9) was discovered in 1988 by Uyttenhove and colleagues as a protein produced by mouse helper T cell clones following in vitro stimulation with concanavalin A, long before Th9 cells had been formally described [1]. This glycoprotein was attributed a supportive role for the growth of helper T cells when IL-2, IL-4, and antigen were absent [2] and was also shown to be produced by naïve murine CD4⁺ T cells stimulated with IL-2, IL-4, and TGF-β [3]. Importantly, a human homologue of the murine IL-9 was identified as being produced by activated T cells [4]. It also became known that the IL-9 receptor is expressed on a multitude of both immune and nonimmune cells, such as dendritic cells, macrophages, mast cells, epithelial cells, and lymphocytes [5–8].

When initially identified, IL-9 was associated to Th2 like responses, although it rapidly became evident that IL-9 production by $CD4^+$ T cells cannot be triggered by IL-4 stimulation alone, but does, indeed, rely on the simultaneous presence of TGF-β and IL-4 [8, 9]. In fact, while T cell activation in presence

Ritobrata Goswami (ed.), *Th9 Cells: Methods and Protocols*, Methods in Molecular Biology, vol. 1585,
DOI 10.1007/978-1-4939-6877-0_8, © Springer Science+Business Media LLC 2017

of TGF-β and IL-2 favors acquisition of Foxp3 expression leading to induction of regulatory T cells, the presence of IL-4 favors IL-9 production by T cells while preventing Foxp3 expression [10]. This requirement for stimulating conditions, distinct from that of Th2 cells, led to the identification of a new functional subset of T cells that was named Th9 cells [11, 12].

IL-9 production is under the positive direct control of the pleiotropic transcription factors NF-κB and NFAT [13–15], as well as of STAT5 [16, 17] and STAT6, that is engaged by IL-4 and suppresses the TGF-β-induced Foxp3 and T-bet induction, simultaneously blocking Treg and Th1 differentiation [8, 10, 18]. The transcription factors IRF4, BATF, and PU.1 were found to be crucial regulators of IL-9 production through direct binding to the *Il9* promoter region [19–21].

1.1 Nonconventional T Helper Cells

Innate lymphoid cells (ILCs) are immune cells that belong to the lymphoid lineage, yet they do not express antigen-specific receptors. They constitute the most recently described cells of the innate immune system. ILCs characteristically do not express lymphoid lineage markers (Lin-) but express CD161 and CD127 (IL-7 receptor alpha) [22, 23]. They are present mostly in mucosal sites, both in mice and humans, and despite their existence as a minor population, they compensate low numbers with an ability to quickly respond to stimulation producing copious amounts of cytokines [23–28]. Several ILC subpopulations have been described and among those ILC2s are characterized by production of IL-4, IL-5 IL-9, and IL-13 [22, 29]. ILC2 development is dependent on the transcription factors RORα, GATA-3, and Tcf [30–33], and these cells have been associated with the development of asthma and allergic diseases [22].

Invariant type 1 natural NKT (iNKT) cells constitute a minor population of lymphocytes characterized by co-expression of some NK receptors, like NK1.1 and NKG2D, and of semi-invariant T-cell receptors [34]. iNKT cells recognize glycolipid antigens presented by APCs on the surface of CD1d, a major histocompatibility complex (MHC) class I-like protein [34]. iNKT cells can acquire functional characteristics in the thymus, that have been associated with subsets that mirror Th1, Th2, and Th17 functions—being therefore named as iNKT1, iNKT2, and iNKT17 cell types [35]. However, iNKT cells retain plasticity to be polarized in the periphery in response to local cytokines, namely with the acquisition of Foxp3 expression and regulatory function [36] or a Th17-like phenotype [37]. More recently, it has been demonstrated that iNKT cells can also produce IL-9 and protect against dextran sulfate sodium (DSS)-induced colitis in a manner that is dependent on IL-4 [38]. In addition, IL-9-producing iNKT cells do not acquire their function during their thymic development, as iNKTs from naïve mice do not express IL-9, but rather do so following their peripheral activation in the right inflammatory context [39]. The authors have shown that IL-9 production by iNKT cells can be induced in different peripheral iNKT subsets defined by CD4, NK1.1 and Neuropilin-1, and is highly impaired by the deficiency of IRF4 and BATF, similar to what has been described for Th9 cells [8, 10, 39, 40].

1.2 IL-9 in Human Disease

IL-9 is a pro-inflammatory cytokine and a pivotal driver of chronic inflammation and autoimmune diseases in mucosal immunity in the gastrointestinal tract and the lung.

In humans, several genetic studies have established a link between asthma and bronchial hyperresponsiveness to the chromosome 5q31-q33, where the IL-9 gene is located, and also an allelic association between IL-9 and the serum levels of total IgE [41]. IL-9 has also been shown to be a trigger, responsible for severe asthma exacerbation in children with dust mite allergy [42].

In some atopic asthma patients, IL-9 production as well as IL-9 producing cells have been shown to be increased [42]. A similar trend has been described in patients with allergic rhinitis, especially during periods of higher exposure to the triggering allergen (e.g., during the pollen season) [43].

In the gastrointestinal tract, IL-9 is one of the most differentially expressed and upregulated genes in a cohort of children suffering from peanut allergy due to the presence of an elevated number of Th9 cells [44]. In atopic patients with food allergy, an association was established between the allergic response and IL-9 production by mast cells [45]. Interestingly, the role of IL-9 in allergy may extend beyond the participation in mucosal immunity as an enhanced production of IL-9 has been recently described in allergic contact dermatitis [46].

Several recent studies have tagged IL-9 as an important participant in processes of chronic intestinal inflammation namely in patients with ulcerative colitis [47], while others have detected elevated levels of IL-9 in a significant percentage of inflammatory bowel disease patients with severe prognosis [48].

Interestingly, IL-9 has been implicated in cancer, namely in human Hodgkin's lymphoma [49] and in melanoma where Th9 cells and IL-9 seem to promote an antitumor effect by favoring recruitment mostly of innate immunity cells (e.g., mast cells) [50].

A role in rheumatic diseases, like rheumatoid arthritis, lupus systemic erythematosus, and systemic sclerosis, has been described for IL-9, and it has been suggested that this effect may not be just due to direct production of IL-9 by Th9 cells but also from the cumulative pan-inflammatory environment generated [51, 52].

ELISA and, more recently, cytometric bead assays (CBA) allow the detection of cytokines with high sensitivity. However, they are of limited utility when studying phenotypically heterogeneous cell samples. Detection of expression of cytokines and its unequivocal attribution to a given cell is of crucial importance and relies on techniques that work towards the definition of a single-cell phenotype.

For this, intracellular cytokine staining (ICS) making use of fluorescently labeled monoclonal antibodies (MAbs) detected by flow cytometry became the technique of election for assessment of cytokine expression at the single-cell level. Due to recent advances in multiparametric flow cytometry, as many as 19 cellular features can be simultaneously investigated, from cell surface markers, cytokines, transcription factors, to proliferation dilution and cell viability [53]. This characteristic makes ICS most advantageous and the technique of choice over, for example, ELISPOT.

This chapter describes a detailed protocol to polarize and/or to stimulate in vitro/ex vivo iNKT cells toward IL-9 production, and the subsequent identification of IL-9 producing cells.

2 Materials

Solutions for cell culture should be prepared with apyrogenic water (endotoxin free) and with cell culture grade reagents. Use freshly made solutions. For intracellular cytokine staining, all solutions should be prepared using deionized water and analytical grade reagents. Preferably, prepare and use diluted solutions fresh.

In some experiments, in particular following in vitro assays, it is highly recommended to use a cell viability assay, using a fluorescent reactive dye that allows discrimination of live and dead cells (live–dead staining).

2.1 Plastic Wares

1. U-bottom and flat bottom 96-well plates, tissue culture treated.
2. U-bottom 96-well plates, polystyrene, clear (to perform staining).
3. Hemocytometer (cell counting chamber).

2.2 Solutions

1. H_2O: apyrogenic for cell culture and deionized for preparation of solutions for staining.
2. Phosphate-buffered saline (PBS): 137 mM NaCl, 2.7 M KCl, 10 mM Na_2HPO_4, 1.8 mM KH_2PO_4, pH 7.2. Store at room temperature.
3. FCS: fetal calf serum, heat inactivated, store at −20 °C.
4. PBS-2% FCS: PBS with 2% FCS (v/v).
5. Trypan Blue solution: 0.4% (w/v), sterile-filtered.
6. Viability Dye: Live/Dead Fixable Stain Kit (Aqua fluorescent reactive dyes), prepare stock solution in DMSO adding 50 μL per tube of reactive dye as supplied, working concentration is 1000 fold dilution in PBS.
7. Staining buffer: PBS with 2% FCS (v/v). Store at 4 °C.

8. Fc Block: anti-mouse CD16/CD32 (clone 2.4G2), 0.5 mg/mL ready-made stock solution, working concentration 1.25 μg/mL in the appropriate buffer (*see* **Note 1**). Store at 4 °C.
9. Normal rat/mouse serum.
10. Surface mouse antibodies: anti-mouse CD19 (clone MB19-1), anti-mouse CD25 (clone PC61.5), anti-mouse TCRβ (H57-597). Store at 4 °C.

    Surface human antibodies: anti-human CD19 (clone HIB19), anti-human CD14 (clone 61D3), anti-human CD123 (clone 6H6), anti-human CD3 (clone OKT3), anti-human CD25 (clone BC96). Store at 4 °C.

    Tetramers for iNKT cell identification: mouse and human CD1d tetramers coupled with PE and loaded with PBS57 (an iNKT cell ligand analog of α-Galactosyl Ceramide) are supplied by the National Institutes of Health Tetramer Core Facility, USA.
11. Magnetic separation: Anti-PE MicroBeads and Anti-Biotin MicroBeads, LS Columns and MACS® Manual Separators (Miltenyi Biotec).
12. Purification of PBMCs from human peripheral blood: SepMate™-50 (STEMCELL Technologies) and Histopaque®-1077.
13. Intracellular mouse antibodies: anti-mouse/rat Foxp3 (clone FKJ-16 s), anti-mouse IL-9 (clone RM9A4). Store at 4 °C.

    Intracellular human antibodies: anti-human Foxp3 (clone PCH 101), anti-human IL-9 (clone MH9A4, Mouse IgG2b, κ chain), isotype control for human IL-9: Mouse IgG2b (κ chain). Store at 4 °C.
14. Fix/Perm Buffer: add 1 part Fixation/Permeabilization Concentrate (eBioscience, Ref 00-5123) to 3 parts of Fixation/Permeabilization Diluent (eBioscience, Ref 00-5223). Store at 4 °C.
15. Permeabilization Buffer: 10× Perm buffer (eBioscience, Ref 00–8333), use 1× diluted in water. Store at 4 °C.
16. FACS Buffer: PBS with 2% FCS (v/v), 0.02% NaN_3 (w/v). Store at 4 °C.

2.3 Cell Culture

1. Complete RPMI for mouse cells: RPMI 1640 medium with GlutaMAX, with FCS 10% (v/v), HEPES 10 mM, 100 Units/mL penicillin, 100 μg/mL streptomycin, 1 mM sodium pyruvate, and 50 μM β-mercaptoethanol.
2. Complete RPMI for human cells: RPMI 1640 medium with GlutaMAX, with FCS 10% (v/v), HEPES 10 mM, 100 Units/mL penicillin, 100 μg/mL streptomycin, 25 μg/mL gentamicin.

3. Anti-mouse CD3 (clone 145-2C11): 1 mg/mL of ready-made stock solution, working concentration 3 μg/mL in PBS. Store at 4 °C.
4. Anti-human CD3 (clone OKT3): 0.5 mg/mL of ready-made stock solution, working concentration 1 μg/mL in PBS. Store at 4 °C.
5. Anti-mouse CD28 (clone 37.51): 0.5 mg/mL of ready-made stock solution, working concentration 2 μg/mL in complete RPMI medium. Store at 4 °C.
6. Anti-human CD28 (clone CD28.2): 1 mg/mL of ready-made stock solution, working concentration 2 μg/mL in complete RPMI medium. Store at 4 °C.
7. TGF-β (recombinant human TGF-β, carrier free): 20 μg/mL stock solution in 4 mM HCl supplemented with 0.1% BSA (w/v) (store at −80 °C), working concentration 5 ng/mL in culture medium. Store at 4 °C.
8. Mouse IL-2: 100 μg/mL of ready-made stock solution (store at −80 °C), working concentration 5 ng/mL in culture medium.
9. Human IL-2: 100 μg/mL of stock solution (store at −80 °C), working concentration 10 ng/mL in culture medium.
10. Mouse IL-4: 100 μg/mL of ready-made stock solution (store at −80 °C), working concentration 20 ng/mL in complete RPMI medium.
11. Anti-mouse IL-4: in-house production, 1 mg/mL of stock solution (store at −20 °C), working concentration 100 μg/mL in complete RPMI medium. The monoclonal antibody was produced using Integra CL1000 flasks (IBS Integra Biosciences, Chur, Switzerland) and purified by 50% (w/v) ammonium sulfate precipitation. Following dialysis against PBS, purity was checked by native and SDS gel electrophoresis.
12. Human IL-4: 100 μg/mL of ready-made stock solution (store at −80 °C), working concentration 20 ng/mL in complete RPMI medium.
13. Anti-human IFN-γ (clone NIB42): 1 mg/mL of ready-made stock solution, working concentration 5 μg/mL in complete RPMI medium. Store at 4 °C.
14. Anti-human IL-12/IL-23 (clone C8.6): 1 mg/mL of ready-made stock solution, working concentration 5 μg/mL in complete RPMI medium. Store at 4 °C.
15. PMA (Phorbol myristate acetate): 5 mg/mL of stock solution in DMSO (store at −80 °C), working concentration 50 ng/mL in complete RPMI medium.
16. Ionomycin: 2.5 mg/mL of stock solutions in DMSO (store at 4 °C), working concentration 500 ng/mL in complete RPMI medium.

17. Brefeldin A: 5 mg/mL of stock solutions in DMSO (store at −80 °C), working concentration 5 μg/mL in complete RPMI medium.
18. GolgiStop: 1500× ready-made stock solution (store at 4 °C), working concentration 1× in complete RPMI medium.

3 Methods

The protocol indicated was specifically developed for iNKT cells. You may need to adapt to other cell types adjusting accordingly. Perform all procedures as indicated.

3.1 Cell Culture Under IL-9 Polarizing Conditions

Perform all procedures in a flow laminar hood to ensure sterility (*see* **Notes 2** and **3**).

1. At least 3 h ahead of plating the cells, coat U-bottom (mouse cells) or flat bottom (human cells) 96-well plates (tissue culture treated) with 50 μL per well of 3 μg/mL of anti-mouse CD3 or 1 μg/mL of anti-human CD3 in room temperature PBS. Place plates in an incubator at 37 °C, 5% CO_2.
2. Invariant NKT cells are obtained from mouse spleens or from human blood as described elsewhere and resuspended in PBS-2% FCS [36]. *See* Subheadings 3.1.1 and 3.1.2 for mouse and human iNKT cell enrichment.
3. Count the cell suspension with a hemocytometer by preparing appropriate dilutions of the cell suspension (*see* **Note 4**).
4. Immediately before plating, wash the plates with room temperature PBS, using 100 μL per well 1–2 times using a multichannel pipette. Be very careful in removing any remaining PBS and making sure the coated surface of the wells is not disturbed.
5. Plate 30,000 iNKT cells per well in the coated U-bottom 96-well plates for mouse cells and flat bottom 96-well plates for human cells. Centrifuge (300 × *g* for 5 min at 4 °C) and resuspend mouse cells in complete RPMI supplemented with anti-CD28 (2 μg/mL), TGF-β (5 ng/mL), IL-2 (5 ng/mL) and IL-4 (20 ng/mL). Centrifuge (300 × *g* for 5 min at 4 °C) and resuspend human cells in complete RPMI supplemented with anti-CD28 (1 μg/mL), TGF-β (5 ng/ml), IL-2 (10 ng/mL), IL-4 (20 ng/mL), anti-IFN-γ (5 μg/ml) and anti-human IL-12/IL-23 (5 μg/mL). Plate cells in 200 μL per well (*see* **Note 5**) to promote in vitro conversion into IL-9 production.
6. Place the plates in an incubator at 37 °C, 5% CO_2 for 3 and 5 days for mouse and human cells, respectively.

3.1.1 Mouse iNKT Cell Enrichment

1. Collect spleens from the appropriate mouse strain (consider using a reporter strain, in this case Foxp3-GFP knockin mice) and incubate cell suspensions (1×10^8 cells/mL) with 1.25 μg/mL of anti-CD16/32 Ab (to block nonspecific binding to FcR) for 15 min at 4 °C.
2. Without washing, add PE-conjugated CD1d-PBS57 tetramers at the optimal titrated working concentration and incubate for 30 min at 4 °C. Wash cells (300 × *g* for 5 min at 4 °C).
3. Resuspend cells in staining buffer (1×10^6 cells/μL), add magnetic beads (0.2 μL beads/1×10^6 cells) and incubate for 20 min at 4 °C. Ensure gentle cell suspension agitation during incubation to maximize binding to the beads.
4. Wash (300 × *g* for 5 min at 4 °C) and resuspend cell suspension in staining buffer (10^8 cells per 500 μL buffer).
5. The magnetically labeled fraction is isolated in a MACS Cell Separator as described by the manufacturer. Briefly, place the column and magnetic holder to magnetic stand and wash the column with 3 mL of staining buffer. Apply the cell suspension to the column. After flow through by gravity, discard the eluent.
6. Following three consecutive washes (3 mL each of staining buffer), displace the column from the magnetic stand, and recover the iNKT cell enriched population from the column using 5 mL of staining buffer and an adapted plunger. Repeat this recovery step with 5 mL of staining buffer to improve recovery.
7. Wash cells (300 × *g* for 5 min at 4 °C) and resuspend cells in the following mix of titrated working antibodies in staining buffer: anti-mouse TCRβ at 1 μg/mL (diluted 1:200 with respect to the stock), anti-mouse CD25 at 0.5 μg/mL (diluted 1:400 with respect to the stock), anti-mouse CD19 at 1 μg/mL (diluted 1:200 with respect to the stock). Foxp3 expression is detected through GFP fluorescence. Incubate for 30 min at 4 °C. Wash cells (300 × *g* for 5 min at 4 °C). Proceed to the cell sorter.
8. Sorting strategy is (a) selection of lymphocytes, (b) selection of singlets, (c) selection of $CD19^-$ cells, (d) selection of $Foxp3GFP^-CD25^-$ cells, and (e) selection of iNKT cells as $CD1d/PBS57^+TCRb^{int}$.

3.1.2 Human iNKT Cell Enrichment

1. For human iNKT cell enrichment, peripheral blood lymphocytes are purified from peripheral blood using SepMate™-50 as described by the manufacturer. Briefly, the bottom chamber of the 50 mL tube is loaded with 15 mL of Histopaque®-1077. Apply the blood diluted in an equal volume of PBS+2% FCS and centrifuge at 1200 × *g* for 10 min at room temperature with the brake on.

2. Collect the supernatant and wash at 300 × *g* for 8 min at room temperature.
3. Suspend the cells in staining buffer (1 × 10^8 cells/mL) and label with biotinylated antibodies against human CD14, human CD19, and human CD123 all at 1 μg/mL (diluted between 1:100 and 1:200 with respect to the stock) by incubating for 30 min at 4 °C.
4. Wash (300 × *g* for 5 min) cells and resuspend cells in staining buffer (1 × 10^6 cells/μL), add anti-biotin magnetic beads (0.2 μL beads/10^6 cells) and incubate for 20 min at 4 °C. Ensure gentle cell suspension agitation during incubation to maximize binding to the beads.
5. The unlabeled fraction, enriched for the iNKT cells, is isolated in a MACS Cell Separator as described by the manufacturer. Briefly, place the column and magnetic holder to the magnetic stand and wash the column with 3 mL of staining buffer. Apply the cell suspension to the column and recover the unlabeled fraction by gravity. The magnetically labeled fraction that remains attached to the column should be discarded.
6. Wash cells (300 × *g* for 5 min) and resuspend cells in the following mix of titrated working antibodies in staining buffer: anti-human CD25 and anti-human CD3 at 1 μg/mL (diluted 1:200 with respect to the stock) and anti-human CD1d/PBS57$^+$ (diluted 1:200 with respect to the stock). Incubate for 30 min at 4 °C. Wash cells (300 × *g* for 5 min). Proceed to the cell sorter.
7. Sorting strategy is (a) selection of lymphocytes, (b) selection of singlets, (c) selection of CD25$^-$ cells, and (d) selection of iNKT cells as CD1d/PBS57$^+$CD3$^+$.

3.2 Stimulation with PMA/Ionomycin/Brefeldin A/Golgi Stop

Perform all procedures in a flow laminar hood to ensure sterility (*see* **Note 6**).

1. Following the in vitro polarization, centrifuge the plates at 300 × *g* for 3 min at room temperature, and remove the medium carefully without disturbing the cell pellet.
2. Prepare a mix of complete RPMI medium supplemented with 50 ng/mL PMA, 500 ng/mL ionomycin, 5 μg/mL Brefeldin A, and GolgiStop, plating 200 μL per well. Gently resuspend the cells with a multichannel pipette to ensure even distribution of the cells in the suspension.
3. Incubate the plates in an incubator at 37 °C, 5% CO_2 for 2 h and 3 h for mouse and human cells, respectively.

3.3 Surface Staining

1. Live/dead staining (*see* **Notes 7–9**): Transfer the cells to a polystyrene U-bottom 96-well plate, pellet the cells (300 × *g* for 3 min) and resuspend in 50–100 μL per well of PBS with the Live/dead reagent diluted 1000 times in relation to the stock. Incubate at room temperature or on ice for 30 min pro-

tected from light. Wash with 150 μL of PBS (300 × *g* for 3 min at 4 °C) and flick out the buffer.

2. Perform blocking of Fc receptors by adding 10 μL of 1.25 μg/mL Fc Block diluted in PBS-2% FCS. Incubate at 4 °C for 15 min.
3. Prepare a solution with the appropriate combination of antibodies and add 15 μL per well, without removing the Fc Block. In this case, add anti-mouse CD25 at 0.5 μg/mL (diluted 1:400 with respect to the stock) or anti-human CD25 at 0.5 μg/mL (diluted 1:100 with respect to the stock). Incubate at 4 °C for 30 min protected from light (*see* **Notes 1** and **10**).
4. Wash with 200 μL of PBS-2% FCS (300 × *g* for 3 min at 4 °C) and flick out the buffer.

3.4 Intracellular Staining (IL-9 and Foxp3)

(*See* **Notes 10** and **11**)

1. Add 100 μL per well of Fix/Perm buffer diluting 25 μL of buffer concentrate in 75 μL of buffer diluent. Incubate at room temperature for 30 min protected from light.
2. Pellet cells (300 × *g* for 3 min) and flick out the buffer.
3. Add 100 μL per well of 1× perm buffer diluting 10 μL of 10× Perm Buffer concentrate in 90 μL of deionized water.
4. Pellet cells (300 × *g* for 3 min) and flick out the buffer.
5. Add 20 μL of perm buffer containing 1.5 μg/mL of Fc Block and 2% (v/v) normal rat/mouse serum and incubate at room temperature for 15 min protected from light.
6. Without washing, add 5 μL of the antibody mix diluted in Perm buffer (IL-9 and Foxp3 antibodies diluted 1:20) per well. Incubate at room temperature for 30–45 min at dark.
7. Add 100 μL of perm Buffer, pellet cells (300 × *g* for 3 min) and flick out the buffer.
8. Repeat **step** 7 one more time.
9. Resuspend cells in 200 μL of FACS Buffer. Transfer cell suspensions to FACS tubes and proceed for acquisition in the flow cytometer (Figs. 1 and 2).

Figures 1 and 2 illustrate the production of IL-9 by mouse and human iNKT cells, respectively, detected by intracellular staining as described in Subheading 3.4.

4 Notes

1. The 25 μL volume indicated for the staining is suitable to stain up to 2.5 × 10^6 cells/well. When staining higher numbers of cells, upscale the volume proportionally. When staining for a

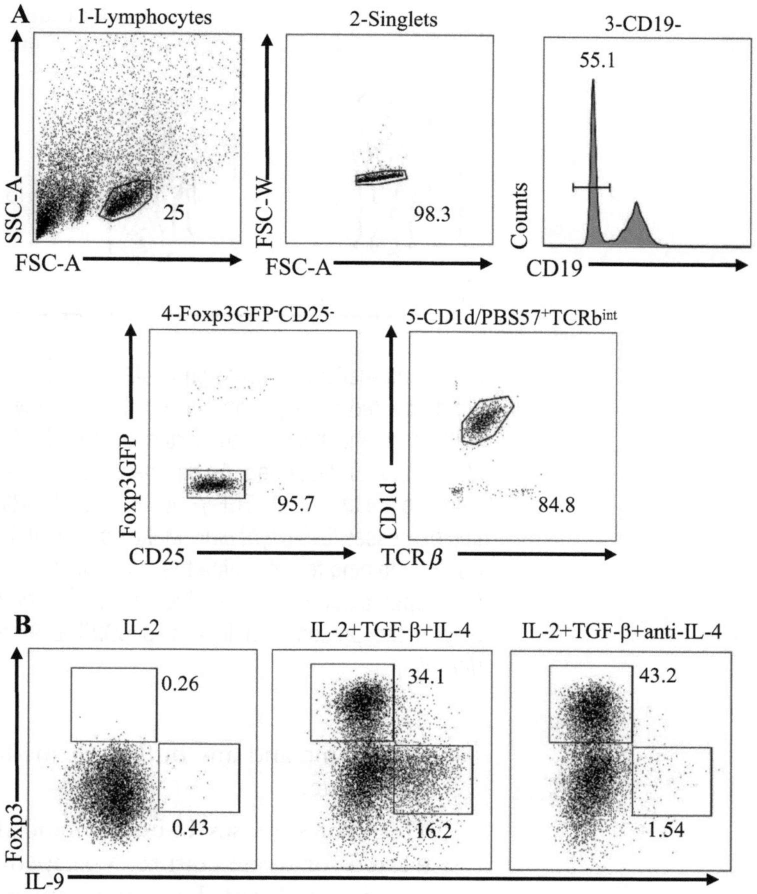

Fig. 1 Intracellular expression of IL-9 by mouse iNKT cells. Mouse iNKT cells were isolated from spleens of reporter Foxp3GFP reporter mice as described. (**a**) Sorting staining: briefly, following red blood cell lysis, total spleen suspensions were incubated with CD1d/PBS57-PE and then with anti-PE beads, according to the manufacturer's instructions. Cells were surface-stained with the following mix of antibodies: anti-mouse CD19, anti-mouse CD25 and anti-mouse TCRβ. Sorting strategy: (*1*) selection of lymphocytes, (*2*) selection of singlets, (*3*) selection of $CD19^-$ cells, (*4*) $Foxp3GFP^-CD25^-$ cells, (*5*) selection of iNKT cells as CD1d/$PBS57^+TCRb^{int}$. (**b**) Purified mouse iNKT cells were cultured and stimulated as described. As can be seen, in the presence of the IL-9-inducing cytokine mixture (IL-2+TGF-β+IL-4), up to 16.2% and 34.1% of the iNKT cells express IL-9 and Foxp3, respectively (*middle plot*). Note the lack of IL-9 staining when the cells are stimulated in the presence of an antibody against IL-4, without any impact on the expression on Foxp3, demonstrating that IL-9 production by iNKT cells is dependent on presence of IL-4 (*right plot*). When the cells are cultured only in the presence of IL-2, this iNKT population expresses neither IL-9 nor TGF-β (*left plot*)

much larger amount (more than 5×10^6), plates are not adequate and thus the staining should be performed in tubes.

2. Solutions should always be prepared according to the manufacturer's instructions. Take special care when preparing solutions to be used in tissue culture as water should be

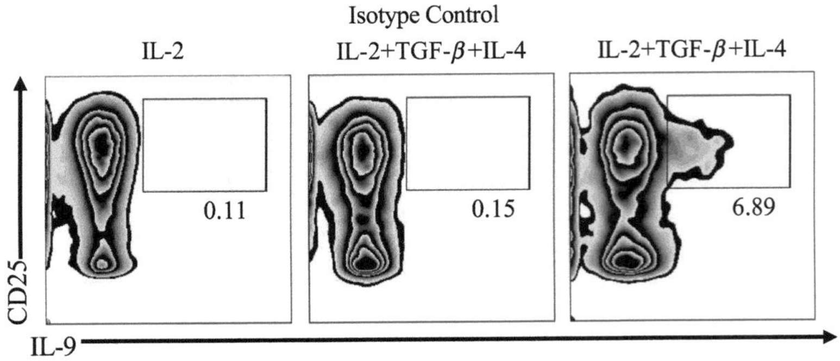

Fig. 2 Intracellular expression of IL-9 in human iNKT cells. Human iNKT cells were isolated from peripheral blood of healthy donors as described, enriched with magnetic beads and sorted as $CD25^{-}CD1d/PBS57^{+}CD3^{+}$, cultured and stimulated as described. As can be seen, in the presence of the IL-9-inducing cytokine mixture (IL-2+TGF-β+IL-4), up to 6.89% of the $CD25^{+}$ iNKT cell population expresses IL-9 (*right plot*). Make note of the importance of having the Isotype Control staining to determine the separation between the positive and the negative populations for IL-9 production (*middle plot*). When the cells are cultured only in the presence of IL-2, this $CD25^{+}IL\text{-}9^{+}$ population cannot be detected (*left plot*)

apyrogenic and any buffers or medium should be tissue culture grade.

3. This protocol uses U-bottom and flat-bottom 96-well plates treated for tissue culture. V-shaped plates are also an option. Bear in mind that the configuration of the well will determine the area available for cell growth and, as such, you should adjust the amount of cells you plate per well accordingly.
4. Do not forget that flow cytometry will only give relative proportions of the different populations, and the staining procedure will result in some cell loss. To calculate the cell number of each identified population it is pivotal that you know the starting number of cells. Therefore, one needs to count cells before beginning the staining. You can count manually using a hemocytometer. As an alternative, you can add counting beads (with a known concentration) to the stained cells allowing cell number quantification by flow cytometry.

   Always keep your cell suspension homogeneous at the time of cell counting and of preparation of the suspension to plate for cell culture and to perform the staining. Always use a viability dye like Trypan Blue to assess the percentage of dead cells in your sample.
5. Follow the instructions provided by the manufacturer. Should these not be informative enough, always prepare concentrated stock solutions and store single use aliquots at −80 °C. Do not freeze and thaw. Prepare diluted solutions immediately before

use. Remaining unused volumes should be disposed of at the end of the assay.

6. Incubation times with stimulating reagents may require optimization for your particular experimental setting. Please note that incubations with GolgiSTOP™ should not exceed 12 h, but shorter periods (typically up to 6 h) usually fit your assay.

   Phorbol 12-myristate 13-acetate (PMA) is a specific activator of Protein Kinase C (PKC) and Ionomycin is a calcium ionophore that promotes an influx of calcium, with a consequent activation of Protein Kinase C, leading to T cell activation.

   Brefeldin A and the BD GolgiStop™ are inhibitors of intracellular protein transport. Therefore, addition of Brefeldin A to cell cultures results in a blockade of protein transport of proteins in the endoplasmic reticulum, and GolgiStop™ blocks transport at the Golgi complex. The result is an intracellular accumulation of secreted proteins in the Golgi complex, leading to increased detection of cytokines by flow cytometry.

7. The staining and washing steps, as well as fixation and permeabilization, can be performed either in U-bottom 96-well plates or in eppendorf or bigger tubes, increase the volumes accordingly. Plates favor the utilization of smaller amounts of buffer and of antibodies, and also facilitate the simultaneous processing of a large number of samples.

8. Viability dyes like the Live/Dead Fixable Dead Cell staining reagent allows the discrimination between live and dead cells, particularly important when staining cells following in vitro assays.

   As the reagent reacts with amine groups (present both on the cell surface and inside the cells), the labeling reaction should be performed in PBS, Hanks' Balanced Salt Solution (HBSS), or Dulbecco's PBS (D-PBS) without the addition of protein (e.g., FCS, BSA) or sodium azide. The dye is recommended to be used 1000× diluted with respect to the stock solution prepared. You may need to concentrate to 500× dilution to better separate the positive and the negative peak. You can perform this labeling together with the cell surface staining, just make note that you should do it in PBS and not in PBS-2% FCS.

9. In every wash step be very careful to remove as much of the remaining buffer as possible. Particularly during staining spillover of buffers may compromise the integrity and stability of the cells and that of the staining.

10. All antibodies should be appropriately titrated in order to determine the best concentration for detection of the population of interest.

    For antibodies of lineage cellular markers (e.g., CD4, CD19, CD11c), working concentrations are usually lower.

However, when analyzing cellular markers whose expression varies depending on the state of activation of the cell, the working concentrations is typically more elevated. This is the case with cytokines and, often, the separation between the positive and the negative populations is more difficult, as the appearance is that of a smear. Isotype controls are absolutely critical to define this boundary. The isotype control should match the host species, the isotype and fluorochrome of the tested antibody and should be used at the same concentration of the tested antibody.

11. The abundance of protein one needs to detect may influence the decision of choosing a polyclonal antibody instead of a monoclonal one. Polyclonal antibodies recognize multiple epitopes on a given antigen and, as the serum obtained will contain a heterogenous complex mixture of antibodies of different affinity, this may be particularly useful when assessing cytokines, normally not very abundant. It is possible to gain amplification of signal from a target protein with low expression level, as the target protein may bind more than one antibody molecule on its multiple epitopes, allowing a more robust detection.

Acknowledgments

We thank Marta Monteiro for her contribution to the iNKT cell work in the lab, including establishing the protocols described in this paper. Work in the host laboratory is supported by Fundação para a Ciencia e Tecnologia (PTDC/SAU-TOX/11424/2009) and FAPESP (FAPESP/19906/2014).

References

1. Uyttenhove C, Simpson RJ, Van Snick J (1988) Functional and structural characterization of P40, a mouse glycoprotein with T-cell growth factor activity. Proc Natl Acad Sci U S A 85:6934–6938. doi:10.1073/pnas.85.18.6934
2. Simpson RJMR, Rubira MR, Gorman JJ, Van Snick J (1989) Complete amino acid sequence of a new murine T-cell growth factor P40. Eur J Biochem 183:715–722
3. Schmitt E, Van Brandwijk R, Van Snick J, Siebold B, Rüde E (1989) TCGF III/P40 is produced by naive murine CD4+ T cells but is not a general T cell growth factor. Eur J Immunol 19:2167–2170. doi:10.1002/eji.1830191130
4. Yang YCRS, Ciarletta A, Calvetti J, Kelleher K, Clark SC (1989) Expression cloning of cDNA encoding a novel human hematopoietic growth factor: human homologue of murine T-cell growth factor P40. Blood 74:1880–1884
5. Elyaman WBE, Uyttenhove C, Dardalhon V, Awasthi A, Imitola J, Bettelli E, Oukka M, van Snick J, Renauld JC, Kuchroo VK, Khoury SJ (2009) IL-9 induces differentiation of TH17 cells and enhances function of FoxP3+ natural regulatory T cells. Proc Natl Acad Sci U S A 106:12885–12890
6. Hültner LKS, Stassen M, Kaspers U, Kremer JP, Mailhammer R, Moeller J, Broszeit H, Schmitt E (2000) In activated mast cells, IL-1 up-regulates the production of several Th2-

related cytokines including IL-9. J Immunol 164:5556–5563
7. Mohamadzadeh MAK, Sugamura K, Bergstresser PR, Takashima A (1996) Expression of the common cytokine receptor gamma chain by murine dendritic cells including epidermal Langerhans cells. Eur J Immunol 26:156–160
8. Veldhoen MUC, van Snick J, Helmby H, Westendorf A, Buer J, Martin B, Wilhelm C, Stockinger B (2008) Transforming growth factor-beta 'reprograms' the differentiation of T helper 2 cells and promotes an interleukin 9-producing subset. Nat Immunol 9:1341–1346. doi:10.1038/ni.1659
9. Schmitt EGT, Goedert S, Hoehn P, Huels C, Koelsch S, Kühn R, Müller W, Palm N, Rüde E (1994) IL-9 production of naive CD4+ T cells depends on IL-2, is synergistically enhanced by a combination of TGF-beta and IL-4, and is inhibited by IFN-gamma. J Immunol 153:3989–3996
10. Dardalhon VAA, Kwon H, Galileos G, Gao W, Sobel RA, Mitsdoerffer M, Strom TB, Elyaman W, Ho IC, Khoury S, Oukka M, Kuchroo VK (2008) IL-4 inhibits TGF-beta-induced Foxp3+ T cells and, together with TGF-beta, generates IL-9+ IL-10+ Foxp3(−) effector T cells. Nat Immunol 9:1347–1355
11. Kaplan MHHM, Olson MR (2015) The development and in vivo function of T helper 9 cells. Nat Rev Immunol 15:295–307
12. Kaplan MH (2013) Th9 cells: differentiation and disease. Immunol Rev 252:104–115
13. Jash ASA, Kim GC, Chae CS, Hwang JS, Kim JE, Im SH (2012) Nuclear factor of activated T cells 1 (NFAT1)-induced permissive chromatin modification facilitates nuclear factor-κB (NF-κB)-mediated interleukin-9 (IL-9) transactivation. J Biol Chem 278:15445–15457
14. Xiao XBS, Liu W, Chu X, Wang H, Taparowsky EJ, Fu YX, Choi Y, Walsh MC, Li XC (2012) OX40 signaling favors the induction of T(H)9 cells and airway inflammation. Nat Immunol 13:981–990
15. Zhu YXKL, Luo W, Li CC, Yang L, Yang YC (1996) Multiple transcription factors are required for activation of human interleukin 9 gene in T cells. J Biol Chem 271: 15815–15822
16. Bassil ROW, Olah M, Kurdi AT, Frangieh M, Buttrick T, Khoury SJ, Elyaman W (2014) BCL6 controls Th9 cell development by repressing Il9 transcription. J Immunol 193:198–207
17. Liao WSR, Li P, Du N, West EE, Ren M, Mitra S, Leonard WJ (2014) Opposing actions of IL-2 and IL-21 on Th9 differentiation correlate with their differential regulation of BCL6 expression. Proc Natl Acad Sci U S A 111:3508–3513
18. Goswami R, Jabeen R, Yagi R, Pham D, Zhu J, Goenka S, Kaplan MH (2012) STAT6-dependent regulation of Th9 development. J Immunol 188:968–975. doi:10.4049/jimmunol.1102840
19. Chang H-C, Sehra S, Goswami R, Yao W, Yu Q, Stritesky GL, Jabeen R, McKinley C, Ahyi A-N, Han L, Nguyen ET, Robertson MJ, Perumal NB, Tepper RS, Nutt SL, Kaplan MH (2010) The transcription factor PU.1 is required for the development of IL-9-producing T cells and allergic inflammation. Nat Immunol 11:527–534. doi:10.1038/ni.1867
20. Jabeen R, Goswami R, Awe O, Kulkarni A, Nguyen ET, Attenasio A, Walsh D, Olson MR, Kim MH, Tepper RS, Sun J, Kim CH, Taparowsky EJ, Zhou B, Kaplan MH (2013) Th9 cell development requires a BATF-regulated transcriptional network. J Clin Invest 123:4641–4653. doi:10.1172/JCI69489
21. Staudt V, Bothur E, Klein M, Lingnau K, Reuter S, Grebe N, Gerlitzki B, Hoffmann M, Ulges A, Taube C, Dehzad N, Becker M, Stassen M, Steinborn A, Lohoff M, Schild H, Schmitt E, Bopp T (2010) Interferon-regulatory factor 4 is essential for the developmental program of T helper 9 cells. Immunity 33:192–202. doi:10.1016/j.immuni.2010.07.014
22. Barlow JL, McKenzie ANJ (2014) Type-2 innate lymphoid cells in human allergic disease. Curr Opin Allergy Clin Immunol 14:397–403. doi:10.1097/ACI.0000000000000090
23. Mjösberg JM, Trifari S, Crellin NK, Peters CP, van Drunen CM, Piet B, Fokkens WJ, Cupedo T, Spits H (2011) Human IL-25- and IL-33-responsive type 2 innate lymphoid cells are defined by expression of CRTH2 and CD161. Nat Immunol 12:1055–1062. doi:10.1038/ni.2104
24. Barlow JL, Bellosi A, Hardman CS, Drynan LF, Wong SH, Cruickshank JP, McKenzie ANJ (2012) Innate IL-13-producing nuocytes arise during allergic lung inflammation and contribute to airways hyperreactivity. J Allergy Clin Immunol 129:191–198.e191–194. doi:10.1016/j.jaci.2011.09.041
25. Bartemes KR, Iijima K, Kobayashi T, Kephart GM, McKenzie AN, Kita H (2012) IL-33-responsive lineage- CD25+ CD44(hi) lymphoid cells mediate innate type 2 immunity and allergic inflammation in the lungs. J Immunol 188:1503–1513. doi:10.4049/jimmunol.1102832

26. Gold MJ, Antignano F, Halim TYF, Hirota JA, Blanchet M-R, Zaph C, Takei F, McNagny KM (2014) Group 2 innate lymphoid cells facilitate sensitization to local, but not systemic, TH2-inducing allergen exposures. J Allergy Clin Immunol 133:1142–1148. doi:10.1016/j.jaci.2014.02.033
27. Halim TYF, Krauss RH, Sun AC, Takei F (2012) Lung natural helper cells are a critical source of Th2 cell-type cytokines in protease allergen-induced airway inflammation. Immunity 36:451–463. doi:10.1016/j.immuni.2011.12.020
28. Kim HY, Chang Y-J, Subramanian S, Lee H-H, Albacker LA, Matangkasombut P, Savage PB, McKenzie ANJ, Smith DE, Rottman JB, DeKruyff RH, Umetsu DT (2012) Innate lymphoid cells responding to IL-33 mediate airway hyperreactivity independently of adaptive immunity. J Allergy Clin Immunol 129:216–227.e211–216. doi:10.1016/j.jaci.2011.10.036
29. Artis D, Spits H (2015) The biology of innate lymphoid cells. Nature 517:293–301. doi:10.1038/nature14189
30. Hoyler T, Klose CSN, Souabni A, Turqueti-Neves A, Pfeifer D, Rawlins EL, Voehringer D, Busslinger M, Diefenbach A (2012) The transcription factor GATA-3 controls cell fate and maintenance of type 2 innate lymphoid cells. Immunity 37:634–648. doi:10.1016/j.immuni.2012.06.020
31. Liang H-E, Reinhardt RL, Bando JK, Sullivan BM, Ho I-C, Locksley RM (2012) Divergent expression patterns of IL-4 and IL-13 define unique functions in allergic immunity. Nat Immunol 13:58–66. doi:10.1038/ni.2182
32. Wong SH, Walker JA, Jolin HE, Drynan LF, Hams E, Camelo A, Barlow JL, Neill DR, Panova V, Koch U, Radtke F, Hardman CS, Hwang YY, Fallon PG, McKenzie ANJ (2012) Transcription factor RORα is critical for nuocyte development. Nat Immunol 13:229–236. doi:10.1038/ni.2208
33. Yang Q, Monticelli LA, Saenz SA, Chi AW-S, Sonnenberg GF, Tang J, De Obaldia ME, Bailis W, Bryson JL, Toscano K, Huang J, Haczku A, Pear WS, Artis D, Bhandoola A (2013) T Cell factor 1 is required for group 2 innate lymphoid cell generation. Immunity 38:694–704. doi:10.1016/j.immuni.2012.12.003
34. Bendelac A (1995) Positive selection of mouse NK1+ T cells by CD1-expressing cortical thymocytes. J Exp Med 182:2091–2096
35. Constantinides MG, Bendelac A (2013) Transcriptional regulation of the NKT cell lineage. Curr Opin Immunol 25:161–167. doi:10.1016/j.coi.2013.01.003
36. Monteiro M, Almeida CF, Caridade M, Ribot JC, Duarte J, Agua-Doce A, Wollenberg I, Silva-Santos B, Graca L (2010) Identification of regulatory Foxp3+ invariant NKT cells induced by TGF-beta. J Immunol 185:2157–2163. doi:10.4049/jimmunol.1000359
37. Monteiro M, Almeida CF, Agua-Doce A, Graca L (2013) Induced IL-17-producing invariant NKT cells require activation in presence of TGF-β and IL-1β. J Immunol 190:805–811. doi:10.4049/jimmunol.1201010
38. Kim HS, Chung DH (2013) IL-9-producing invariant NKT cells protect against DSS-induced colitis in an IL-4-dependent manner. Mucosal Immunol 6:347–357. doi:10.1038/mi.2012.77
39. Monteiro M, Agua-Doce A, Almeida CF, Fonseca-Pereira D, Veiga-Fernandes H, Graca L (2015) IL-9 expression by invariant NKT cells is not imprinted during thymic development. J Immunol 195:3463–3471. doi:10.4049/jimmunol.1403170
40. McNab FW, Pellicci DG, Field K, Besra G, Smyth MJ, Godfrey DI, Berzins SP (2007) Peripheral NK1.1 NKT cells are mature and functionally distinct from their thymic counterparts. J Immunol 179:6630–6637
41. Doull IJ, Lawrence S, Watson M, Begishvili T, Beasley RW, Lampe F, Holgate T, Morton NE (1996) Allelic association of gene markers on chromosomes 5q and 11q with atopy and bronchial hyperresponsiveness. Am J Respir Crit Care Med 153:1280–1284. doi:10.1164/ajrccm.153.4.8616554
42. Sordillo JE, Kelly R, Bunyavanich S, McGeachie M, Qiu W, Croteau-Chonka DC, Soto-Quiros M, Avila L, Celedón JC, Brehm JM, Weiss ST, Gold DR, Litonjua AA (2015) Genome-wide expression profiles identify potential targets for gene-environment interactions in asthma severity. J Allergy Clin Immunol 136:885–892.e882. doi:10.1016/j.jaci.2015.02.035
43. Nouri-Aria KT, Pilette C, Jacobson MR, Watanabe H, Durham SR (2005) IL-9 and c-Kit+ mast cells in allergic rhinitis during seasonal allergen exposure: effect of immunotherapy. J Allergy Clin Immunol 116:73–79. doi:10.1016/j.jaci.2005.03.011
44. Brough HA, Cousins DJ, Munteanu A, Wong YF, Sudra A, Makinson K, Stephens AC, Arno M, Ciortuz L, Lack G, Turcanu V (2014) IL-9 is a key component of memory TH cell peanut-specific responses from children with peanut

allergy. J Allergy Clin Immunol 134:1329–1338.e1310. doi:10.1016/j.jaci.2014.06.032

45. Chen C-Y, Lee J-B, Liu B, Ohta S, Wang P-Y, Kartashov AV, Mugge L, Abonia JP, Barski A, Izuhara K, Rothenberg ME, Finkelman FD, Hogan SP, Wang Y-H (2015) Induction of Interleukin-9-Producing Mucosal Mast Cells Promotes Susceptibility to IgE-Mediated Experimental Food Allergy. Immunity 43:788–802. doi:10.1016/j.immuni.2015.08.020
46. Liu J, Harberts E, Tammaro A, Girardi N, Filler RB, Fishelevich R, Temann A, Licona-Limón P, Girardi M, Flavell RA, Gaspari AA (2014) IL-9 regulates allergen-specific Th1 responses in allergic contact dermatitis. J Invest Dermatol 134:1903–1911. doi:10.1038/jid.2014.61
47. Gerlach K, Hwang Y, Nikolaev A, Atreya R, Dornhoff H, Steiner S, Lehr H-A, Wirtz S, Vieth M, Waisman A, Rosenbauer F, McKenzie ANJ, Weigmann B, Neurath MF (2014) TH9 cells that express the transcription factor PU.1 drive T cell-mediated colitis via IL-9 receptor signaling in intestinal epithelial cells. Nat Immunol 15:676–686. doi:10.1038/ni.2920
48. Defendenti C, Sarzi-Puttini P, Saibeni S, Bollani S, Bruno S, Almasio PL, Declich P, Atzeni F (2015) Significance of serum Il-9 levels in inflammatory bowel disease. Int J Immunopathol Pharmacol 28:569–575. doi:10.1177/0394632015600535
49. Knoops L, Renauld J-C (2004) IL-9 and its receptor: from signal transduction to tumorigenesis. Growth Factors (Chur) 22:207–215. doi:10.1080/08977190410001720879
50. Purwar R, Schlapbach C, Xiao S, Kang HS, Elyaman W, Jiang X, Jetten AM, Khoury SJ, Fuhlbrigge RC, Kuchroo VK, Clark RA, Kupper TS (2012) Robust tumor immunity to melanoma mediated by interleukin-9-producing T cells. Nat Med 18:1248–1253. doi:10.1038/nm.2856
51. Pan H-F, Leng R-X, Li X-P, Zheng SG, Ye D-Q (2013) Targeting T-helper 9 cells and interleukin-9 in autoimmune diseases. Cytokine Growth Factor Rev 24:515–522
52. Rojas-Zuleta WG, Vásquez G (2016) Th9 lymphocytes: A recent history from IL-9 to its potential role in rheumatic diseases. Autoimmun Rev. doi:10.1016/j.autrev.2016.02.020
53. Perfetto SP, Chattopadhyay PK, Roederer M (2004) Seventeen-colour flow cytometry: unravelling the immune system. Nat Rev Immunol 4:648–655. doi:10.1038/nri1416

Chapter 9

Prediction and Validation of Transcription Factors Binding Sites in the *Il9* Locus

William Orent and Wassim Elyaman

Abstract

Over the past decade, multiple effector T cell subsets have been identified with varying differentiation conditions in the milieu as well as a broad diversity of cytokine expression. Interleukin-9 (IL-9) secreting T helper 9 (Th9) cells are the newest member of this family. T helper cell differentiation including Th9 cells appears to be an epigenetic phenomenon requiring the coordination of a large variety of transcription factors to reshape the chromatin landscape and generate various T helper phenotypes. This chapter details methods for both predicting and validating potential transcription factor binding sites as well as their downstream epigenetic effect using a variety of in silico and in vitro methods in both primary Th9 cells and IL-9-producing T cell lines.

Key words T cells, Th9, Chromatin immunoprecipitation, Transcription factor

1 Introduction

Interleukin-9 (IL-9)-producing CD4+ T cells termed as T helper 9 (Th9) cells are the latest addition to the family of effector CD4+ T cells. Human and murine Th9 cells are differentiated with IL-4 plus transforming growth factor-beta (TGF-β) and produce large amounts of IL-9 [1, 2]. IL-9 can be also produced at moderate to high levels by mouse and human Th17 cells [2–4]. The Th9 developmental program has shown that this phenotype undergoes a complex differentiation process requiring the coordination of multiple transcription factors to induce, stabilize, and maintain the phenotype [5]. The relative contributions of these transcription factors and their impact upon effector phenotype have been shown to play a role in both the differentiation process as well as human and animal models of disease (Table 1) [6]. This chapter intends to discuss methods that have been successfully used for the prediction and validation of transcription factor binding sites in the *Il9* gene.

In mice and humans, induction of IL-9 production follows an upregulation of the transcription factor IRF4, which is

Ritobrata Goswami (ed.), *Th9 Cells: Methods and Protocols*, Methods in Molecular Biology, vol. 1585,
DOI 10.1007/978-1-4939-6877-0_9, © Springer Science+Business Media LLC 2017

Table 1
Transcription factors involved in positive or negative regulation of Il9 transcription

Transcription factor	Regulation
IRF4 [7]	+
PU.1 [8]	+
BATF [10]	+
SMAD2/3 [13, 18]	+
RBPJ [13]	+
BCL6 [16]	−
RORγt [19]	−
T-bet [9]	−
GATA3 [1, 9]	+/−
STAT1 [14]	−
STAT5 [16]	+
STAT6 [9]	+

+: Genes which are associated with upregulation of *Il9* expression
−: Genes which are associated with downregulation of *Il9* expression

required for IL-9 production in mouse T cells [7]. Chromatin immunoprecipitation (ChIP) studies have additionally shown that IRF4 binds to the *Il9* promoter in Th9 cells [7]. The same is true of the transcription factor PU.1, although it is currently not known if either transcription factor is sufficient for induction of IL-9 in the absence of the other. It is known that both transcription factors bind to different regions of the *Il9* promoter [8, 9]. Further complicating matters, the transcription factor BATF is enriched in Th9 cells and is necessary for IL-9 expression in both mice and humans [10]. Since both TGF-β and IL-4 are required for Th9 differentiation, it is likely that both cytokines activate pathways working synergistically to induce IL-9 expression. PU.1 is induced by TGF-β signaling and IRF4 by IL-4 [11, 12]. While it would be a simple conclusion that these transcription factors act in concert to induce IL-9 expression, further studies need to be performed to provide concrete evidence. In fact, another pathway downstream of TGF-β signaling has been implicated in IL-9 expression. We have shown that Notch co-stimulatory pathway induces Th9 cells in collaboration with TGF-β signaling [13]. We found that the transcription effector of the Notch pathway, RBPJ, cooperates with Smad3, a transcription factor downstream of TGF-β and together bind to the *Il9* promoter and promote its transactivation in T cells [13]. Multiple STAT transcription factors have been shown to play a role in the induction or repression of IL-9 expression in CD4+ T cells. STAT1 expression, induced by interferon (IFN)-β or IL-27, negatively regulates IL-9 expression [14]. Comparison of the *Il9* gene sequences among multiple species has identified the presence of multiple conserved noncoding sequences (CNS), which contain multiple transcription factor (TF) binding sites [15]. One of these, CNS0, includes a consensus STAT binding site where STAT1 could bind, although this has not been confirmed. STAT6, while not directly shown to bind in any of the CNS regions, is induced by IL-4 and appears to be responsible for repressing T-bet and Foxp3, both of which negatively regulate IL-9 production [9]. In our laboratory we were able to successfully predict STAT5 binding sites in the *Il9* promoter and confirmed STAT5 binding using ChIP [16, 17]. STAT5 appears to work in opposition to Bcl6 to induce IL-9 production by initiating epigenetic remodeling of the chromatin at the *Il9* promoter with more permissive histone marks [16]. Taken together, these data show the STAT family of proteins play a major role in the complex regulatory pathway of IL-9 expression. Transcription factors implicated in Th9 differentiation/IL-9 expression in CD4+ T cells and their influence on *Il9* gene transcription are shown below in Table **1**:

TRANSFAC (TRANScription FACtor database) is a unique knowledge-base containing published data on eukaryotic transcription factors and micro RNAs. TRANSFAC experimentally proven binding sites, and regulated genes, leverage compiled positional weight matrices (PWMs) to perform comparative promoter analyses of such high-throughput data, identifying the transcription factors most likely to have induced the experimentally observed pattern of gene expression. The origin of the database was an early data collection published in 1988 [20]. The content of the database is organized in a way that it is centered on the interaction between TFs and their DNA binding sites (TFBS). TFs are described with regard to their structural and functional features, extracted from the original scientific literature. They are classified to families, classes and super classes according to the features of their DNA binding domain [21, 22].

One of the principal methods for validating transcription factor binding to DNA sequences is ChIP. Using this technique, antibodies against a specific TF are employed to immunoprecipitate chromatin obtained from cell culture which has been cross-linked so that transcription factor–DNA interactions are maintained. Although any type of cell may be utilized for ChIP, it is important that a sufficient amount of starting material is available for the assay. In our laboratory, we have developed a ChIP assay using magnetic beads in a 96-well plate format. This greatly reduces both the amount of starting material necessary and the time required to immunoprecipitate multiple samples. Using this technique, researchers may test for multiple interactions in cell types which are not abundantly available (for instance primary human T cells). Sample preparation for ChIP involves a specific sequence of steps. The chromatin and DNA are first cross-linked by applying a formaldehyde containing solution directly to the cells in culture. This reaction is quenched to prevent the formaldehyde fixation from obscuring epitopes recognized by the antibodies to be used. The cells are then lysed and the chromatin is sheared to a specific size range. This ensures that the interactions identified by the ChIP experiment are specific. Following chromatin shearing, the samples are pre-cleared to prevent nonspecific binding to protein A. This step greatly decreases background. The samples are then bound with antibody to the relevant antigens. It is important that proper controls are utilized. A specific quantity of sample must be saved after the pre-clearing step to serve as the input. This allows for quantification of results (*see* **Note 1**). Additionally, a negative control such as IgG is included as a precipitation to account for background signal. When attempting to discover novel interactions, it is often difficult to include a separate antibody as a positive control and as a result, positive controls are often included downstream once purified DNA has been obtained. Once the chromatin has been bound by antibodies, it is immunoprecipitated using magnetic protein A beads. These chromatin-bound beads are washed to remove unbound chromatin and finally the remaining chromatin is eluted into solution. At this point the cross-linking is reversed to remove protein bound to the DNA and the DNA itself is purified to allow for various downstream applications, such as PCR or DNA sequencing. When planning an experiment, it is vital to have a grasp of a multitude of variables. First, the kinetics of the interaction one hopes to observe must be examined as TF binding can often be transient, especially with regard to differentiating T cells. Additionally, enough cells must be available to ensure a sufficient signal to noise ratio. The number of cells required depends upon the abundance of the interaction one hopes to observe and the quality of the antibodies available. While technical replicates are possible, in human studies it is often the case that biological replicates are both more easily added due to sample availability and of greater value statistically due to the heterogeneity of human samples. When selecting antibodies, care must be taken to use antibodies previously validated in ChIP experiments whenever possible. Both monoclonal and polyclonal antibodies have been used effectively for ChIP experiments. ChIP is a powerful technique for examining TF interactions with DNA in order to interrogate the induction of cytokine expression in T cells. Proper application of this assay can allow researchers to determine how various signaling pathways coalesce into various protein expression phenotypes, both directly and through chromatin remodeling to create more permissive or repressive environments.

2 Materials

2.1 Access Options to TRANSFAC

2.1.1 Online

An online subscription provides access to the TRANSFAC web interface. Online subscriptions are available at the academic/nonprofit laboratory, department, and institution-wide level. The academic/nonprofit laboratory subscription levels require IP plus username/password authentication. IP only authentication is optionally available for the institution-wide subscription level.

2.1.2 Download

A download subscription provides access to flat files containing data for factors, matrices, binding sites, genes, ChIP fragments, and other supporting information, as well as command line access to the Match tool. In this way, TRANSFAC data and positional weight matrices may be integrated into individual, or third party, analysis pipelines and tools.

2.2 Chromatin Immunoprecipitation (ChIP) PCR

2.2.1 Cross-Linking

1. Human effector Th9 cells (*see* **Note 2**).
2. 1 M NaCl.
3. 1 M HEPES.
4. 0.5 M EDTA.
5. DI water.
6. 37% formaldehyde (w/v) (*see* **Note 3**).
7. 1.4 M glycine.
8. Cross-Linking Solution: 100 mM NaCl, 50 mM HEPES, 1.5 mM EDTA, 11% formaldehyde.
9. 1.6 mL microcentrifuge tubes.
10. Ice bucket and ice.
11. Microcentrifuge.
12. Ice-cold phosphate buffered saline (PBS).
13. Protease inhibitor cocktail (100×): 100 mM AEBSF, 60 μM aprotinin, 5 mM bestatin, 1.5 mM E-64, 0.5 M EDTA, 2 mM leupeptin, 1 mM pepstatin A.
14. Pierce IP Buffer (Pierce).

2.2.2 Chromatin Shearing

1. Branson 250 Digital Sonifier with microtip and stand. While any sonicator suitable for chromatin shearing may be used, settings must be determined empirically.
2. Ice bucket and ice.
3. 5 M NaCl.
4. 20 mg/mL proteinase K.
5. Ultrapure agarose.

6. 10 mg/mL ethidium bromide.
7. Tris-acetate.
8. 0.5 M EDTA.
9. TAE Buffer (40×): 1.6 M Tris-acetate, 40 mM EDTA. One part 40× buffer diluted with 39 parts of double distilled H_2O.
10. 100 bp DNA ladder.
11. DI water.
12. DNA loading dye.
13. Gel electrophoresis equipment.
14. UV transilluminator.

2.2.3 Chromatin Immunoprecipitation

1. 96-well microtiter plate magnetic separation rack.
2. 96-well polystyrene flat bottom microwell plate.
3. 1 M Tris–HCl (pH 8).
4. 5 M NaCl.
5. 0.5 M EDTA.
6. Triton X-100.
7. 10% sodium dodecyl sulfate (SDS) (w/v).
8. 30% N-lauroylsarcosine sodium salt solution (w/v).
9. 1 M $NaHCO_3$.
10. Protein A magnetic beads.
11. Antibodies for transcription factors of interest.
12. Control IgG.
13. DI water.
14. IP Dilution Buffer: 20 mM Tris–HCl (pH 8.0), 150 mM NaCl, 1 mM EDTA, 1% Triton X-100 (v/v), 0.01% SDS (v/v).
15. IP Dialysis Buffer: 50 mM Tris–HCl (pH 8.0), 4 mM EDTA, 0.2% N-laurylsarcosine sodium salt solution (v/v).
16. Low-Salt Wash Buffer: 20 mM Tris–HCl (pH 8.0), 150 mM NaCl, 0.1% SDS (v/v), 0.1% Triton X-100 (v/v), 2 mM EDTA.
17. High-Salt Wash Buffer: 20 mM Tris–HCl (pH 8.0), 500 mM NaCl, 0.1% SDS (v/v), 0.1% Triton X-100 (v/v), 2 mM EDTA.
18. Elution Buffer: 0.1 M $NaHCO_3$, 1% SDS (v/v).
19. 20 mg/mL Proteinase K (Sigma).
20. Thermocycler.
21. 96-well microplate seal (suitable for PCR).
22. Microplate rotator (non-inverting).

23. Heated incubator with shaking platform (suitable for 96-well microplate).
24. 96-well polypropylene PCR plate.

2.2.4 DNA Purification

1. Whatman 96-well PCR Clean-up UNIFILTER filter plate.
2. Vacuum manifold system suitable for 96-well plate format OR alternatively, centrifuge capable of taking 96-well microplate.
3. Vacuum source (if using vacuum manifold system).
4. 96-well waste collection plate (*see* **Note 4**).
5. 8 M guanidine-HCl solution.
6. 5 M potassium acetate solution.
7. 1 M Tris–HCl (pH 8.0).
8. 5 M NaCl.
9. 0.5 M EDTA.
10. 200 proof ethanol.
11. DNA Binding Buffer: 4 M guanidine-HCl, 0.75 M potassium acetate, pH 4.6.
12. Wash Buffer (without ethanol): 100 mM Tris, 20 mM EDTA, 0.4 M NaCl, pH 7.5. Requires ethanol before use at ratio of 40% Wash Buffer and 60% ethanol (*see* **Note 5**).
13. TE Buffer: 10 mM Tris–HCl, 1 mM EDTA, pH 8.0.
14. 96-well polypropylene PCR plate.
15. Foil for 96-well plate seal suitable for −80 °C storage.

3 Methods

The TRANSFAC plugin can be used to search for putative transcription factor binding sites in DNA sequences of a specific gene. The binding site predictions are done by the Match tool, which uses the positional weight matrix library from TRANSFAC to analyze the sequences. For each of these matrices, Match contains optimized parameters, for minimization of error rates (false negative rate and false positive rate). As a special feature, Match provides the option to use profiles that are specific subsets of matrices with optimized cutoffs. These profiles allow users to adapt Match to their specific interests. For example, promoters of a certain tissue can be searched with tissue-specific profiles. In addition to several predefined profiles provided in the TRANSFAC plugin, one can create their own profiles in the online version of TRANSFAC. The matrices in the result list of the TRANSFAC plugin are hyperlinked to the respective Matrix Reports in the TRANSFAC online version. Once the plugin is installed and the preferences have been set, one can start searching for transcription factor binding sites in the promoter of a specific gene.

3.1 Searching for Predicted TF Binding Sites in the Murine and Human Il9 Gene Loci

1. In the BIOBASE search box, type the gene of interest.
2. Select the species of interest and click on Match.
3. From the dropdown menu, select analysis method (Search for TF binding sites). Use only high quality matrices: When enabled, this will exclude highly abundant matrices, which produce at minSUM more than 10 hits (false positives) per 1000 nucleotides.
4. Determine the range of nucleotide window upstream and downstream of the transcription start site.
5. Optional: Select the promoter of a certain tissue (e.g., immune cell-specific).
6. Select the cutoff criteria (e.g., minimize false positives, minimize false negatives, or minimize the sum of both error rates) (*see* **Notes 6–11**).
7. Submit the query (*see* **Note 12**).
8. Select the TFs of interest.
9. Copy DNA binding matrices to use for designing PCR primers.
10. Export the Results in csv or Excel format.

3.2 Cross-Linking

1. Gently add 100 μL of cross-linking solution to murine or human Th9 effector T cells cultured in 1 mL of culture medium in a 12-well plate (*see* introduction for discussion of Th9 culture conditions) (*see* **Notes 13** and **14**).
2. Leave the cells at room temperature for 15 min (*see* **Note 15**).
3. Quench the reaction by adding 107.5 μL of 1.4 M glycine.
4. Immediately transfer to microcentrifuge tubes and leave on ice for 2 min.
5. Spin down in microcentrifuge tube (300 × *g*, 4 °C) for 2 min.
6. While centrifuge is running, add 1 μL of protease inhibitor cocktail per 1 mL of PBS.
7. Discard supernatant and resuspend pellet in 500 μL of ice-cold PBS with protease inhibitors.
8. Repeat spin and resuspend in 500 μL of ice-cold PBS with protease inhibitors.
9. Repeat spin one more time.
10. Discard supernatant and resuspend in 500 μL of ice-cold Pierce IP Buffer with protease inhibitors.
11. Immediately freeze samples at −80 °C for long-term storage. Alternatively, proceed immediately to next heading.

3.3 Chromatin Shearing

1. Allow samples to thaw on ice.
2. Sonicate sample while on ice at a power setting of 4 with a 60% duty cycle for 30 cycles with 10 s pulse and 20 s rest (*see* **Note 16**).
3. To check DNA shearing, take 10 μL from each sample, reverse the formaldehyde-induced cross-linking of chromatin and DNA, and run on an agarose gel (*see* Subheading 3.4.2, **steps 26–28**).
4. Freeze the remaining 490 μL of sample at −80 °C.
5. Prepare a 1% agarose gel using TAE Buffer.
6. Run reverse cross-linked samples on gel alongside a 100 bp DNA ladder.
7. Visualize samples using UV transilluminator. Ideally, DNA should be fragmented between 800 and 200 bp.

3.4 Chromatin Immunoprecipitation

3.4.1 Sample Pre-clearing and Antibody Binding

1. Determine the number of samples and conditions to be tested. Each condition refers to a single immunoprecipitation (IP). For each sample, the minimum number of conditions are: Input, IgG, and the test condition(s). Technical replicates may be used if enough material is available, but biological replicates of sufficient quantity will also suffice.
2. Take 5 μL of Protein A magnetic beads per IP to be performed (for all samples) in a microcentrifuge tube and place on magnetic separator. Thus, to test six samples total (including positive and negative controls) 30 μL of beads is required. The beads may be washed without separating them for individual IPs (*see* **Note 17**).
3. Once beads have moved to the side of tube next to magnet, pipette out the supernatant and remove from magnet.
4. Resuspend the pellet in 900 μL of IP Dilution Buffer.
5. Repeat **steps 3** and **4** couple of times. On third wash, resuspend beads in same initial volume of IP Dilution Buffer. Thus in our earlier example of six IPs with 30 μL of beads initially, the beads would be resuspended in 30 μL of IP Dilution Buffer after the third wash.
6. Pipette volume of sample to be used into a microcentrifuge tube. For each condition, 50 μL of sample will usually be sufficient; however, for high cell concentrations or an abundant chromatin DNA interaction, less sample per condition may be used. If less than 50 μL is being used, bring volume up to 50 μL per condition with IP Dilution Buffer. The amount of sample required depends upon the abundance of the chromatin–DNA interaction being studied. A high strength abundant interaction (such as H3K4me) will require less sample to see signal. This must be determined empirically.

7. To each sample, add 5 μL of beads per condition. Thus, two samples with three conditions each would require 15 μL of beads.
8. Plate on a rotator at room temperature for 2 h.
9. After pre-clearing, place samples on magnetic separator. Save the supernatant (this is still your sample!) and discard the beads.
10. Samples may be frozen at −80 °C to continue the protocol later at this point.
11. Using 50 μL of sample per condition, transfer samples to a 96-well polystyrene plate with one well per condition. The input condition is not transferred to the 96-well plate and is instead saved at 4 °C until further use.
12. Dilute antibodies using IP Dilution Buffer to a total volume of 50 μL per IP. While antibody concentrations must be determined empirically, a basic starting point is 2 μg of antibody per IP.
13. Add 50 μL of antibody in IP Dilution Buffer to the appropriate well. This should include IgG as well as test conditions.
14. Place on microplate rotator overnight at 4 °C.

3.4.2 Chromatin IP and Washing

1. Take 5 μL of beads for every IP (not including input) in a microcentrifuge tube and place on magnetic separator.
2. Once beads have moved to the side of tube next to magnet, pipette out the supernatant and remove from magnet.
3. Resuspend in 900 μL of IP Dilution Buffer.
4. Repeat **steps 2** and **3** two times. On third wash, resuspend beads in same initial volume of IP Dilution Buffer. Thus in our earlier example of six IPs with 30 μL beads initially, the beads would be resuspended in 30 μL of IP Dilution Buffer after the third wash.
5. Remove plate from rotator.
6. To each well, add 5 μL of beads.
7. Place plate on rotator for 2 h at room temperature.
8. Place plate on magnetic separation rack and wait for 1 min.
9. Remove and discard supernatant.
10. Remove plate from magnetic separation rack and resuspend the beads in 190 μL of IP Dialysis Buffer per well.
11. Place on rotator for 15 min at room temperature.
12. Repeat **steps 8–11** once.
13. Place plate on magnetic separation rack and wait for 1 min.
14. Remove and discard supernatant and resuspend beads in 190 μL of Low-Salt Wash Buffer.
15. Place on rotator for 15 min at room temperature.

16. Place plate on magnetic separation rack and wait for 1 min.
17. Remove and discard supernatant and resuspend beads in 190 μL of High-Salt Wash Buffer.
18. Place on rotator for 15 min at room temperature.
19. Place plate on magnetic separation rack and wait for 1 min.
20. Remove and discard supernatant, resuspend beads in 100 μL of Elution Buffer (*see* **Note 18**).
21. Cover plate with microplate seal and incubate at 65 °C for 2 h with gentle shaking (150 rpm) in a shaking incubator.
22. Place plate on magnetic separation rack and wait for 1 min.
23. While plate is on separation rack, retrieve saved input conditions from earlier and transfer to 96-well polypropylene PCR plate. This is equivalent to 100% input for later calculation.
24. Add 50 μL of Elution Buffer to each input condition.
25. Transfer supernatant from magnetic separation rack to the PCR plate containing the input samples. The supernatant now contains eluted chromatin.
26. To each well, add 2 μL of 20 mg/mL of Proteinase K and 4 μL of 5 M NaCl.
27. Seal plate and incubate in thermocycler for 16 h at 62 °C, 15 min at 95 °C, and infinity at 4 °C.
28. Following thermocycler incubation, samples may be stored at −80 °C long term.

3.4.3 DNA Purification

1. Add 200 μL of DNA Binding Buffer to each well.
2. Add the entire chromatin–DNA Binding Buffer mixture (approximately 306 μL) to a Whatman UNIFILTER 96-well plate on top of a waste collection plate.
3. Remove liquid using vacuum manifold.
4. Return UNIFILTER plate to waste collection plate and add ethanol at appropriate ratio to Wash Buffer.
5. Add 500 μL of Wash Buffer to each well.
6. Remove liquid using vacuum. Allow to stay on vacuum an additional 2 min to completely dry membrane.
7. Plate UNIFILTER plate on top of a clean 96-well PCR plate (*see* **Note 19**).
8. Add 100 μL of TE Buffer and incubate at 37 °C for 15 min.
9. Spin UNIFILTER plate on top of 96-well PCR plate at 1100 × *g* for 10 min.
10. Purified DNA may be stored at −20 °C and is now ready for downstream applications.

4 Notes

1. A negative control such as IgG is to be included to account for background signal. When attempting to discover novel interactions, it is often difficult to include a separate antibody as a positive control and as a result, positive controls are often included downstream once purified DNA has been obtained.
2. The number of cells required to yield sufficient chromatin depends upon the antibodies being used and must be determined empirically. In our lab, we have found that we may reliably quantify the presence of chromatin marks such as H3K4me1 with as few as 200,000 cells per condition.
3. Fresh formaldehyde is preferable and Cross-Linking Buffer should be made new each use.
4. If using a vacuum, any plate is acceptable as a waste plate, but if using a centrifuge instead of vacuum the plate must accommodate up to 500 μL.
5. The Wash Buffer without ethanol may be stored long term and right before use, 40% Wash Buffer (without ethanol) should be added to 60% ethanol (v/v).

Select cutoff: The combobox has the following options:

6. Minimize false positives (minFP). The minFP cutoff can be used to reduce the number of false positives. The false positive rate is estimated by applying the MatchTM algorithm to upstream sequences. The minFP cutoff is defined as the score that gives one percent of hits in the used sequences relative to the number of hits received at the minFN cutoff.
7. Minimize false negatives (minFN). The minFN cutoff can be used to reduce the number of false negatives. The false negative rate is measured, as many as available, on known genomic binding sites for the transcription factors. In case not sufficient (less than 10) genomic binding sites are available, SELEX sites or sets of oligonucleotides based on the nucleotide distribution in the weight matrix are used for estimating the minFN cutoff. The minFN cutoff is defined as that score at which at least 90% of the positive test set are recognized, i.e., it equals a false negative rate of 10%.
8. Minimize the sum of both error rates (minSUM). The minSUM cutoff can be used to minimize the sum of both error rates. The sum of corresponding percentages for false positives and false negatives is computed for every cutoff ranging from minFN to minFP, whereby the false positive rate at minFN (10% false negative rate) is defined as 100%. The score at which this sum is minimal is used for the minSUM cutoff.

   Configure similarity score thresholds: It is possible to further limit the number of matches reported in the results by defining a threshold that is applied to all matches in the results. The similarity score measures the quality of a match (*see* 23 for details). The range is from 0.0 to 1.0 where a score of 1.0 is given to the exact match.

9. Define thresholds: This will enable similarity score thresholding.
10. Minimum matrix similarity score: The threshold to use for the similarity score computed for the whole matrix.
11. Minimum core similarity score: The threshold to use on the similarity score computed only on the core of the matrix (the five consecutive positions with highest similarity score).
12. The Workbench uses the Match program from BIOBASE to search for transcription factor binding sites. The binding sites can be reported as annotations that are either added to the input sequences or reported as a track of annotations (depending on the analysis).
13. The same protocol may be used for murine cells as well.
14. When adding cross-linking solution, it is normal for medium containing phenol red to become yellow.
15. The time for cross-linking depends upon the chromatin interaction being studied. It is possible that less time may be necessary as cross-linking can obscure epitopes recognized by antibodies.
16. Sonication conditions must be empirically determined for each cell type being studied. If necessary, add additional cycles to achieve a DNA spread between 200–800 bp.
17. The 96-well magnetic separation rack may easily be used with microcentrifuge tubes by holding the tube up to the side with hand. This makes it unnecessary to purchase multiple magnets.
18. It is important to remove as much of the supernatant as possible in this step.

Sometimes it is helpful to spin the plate down very quickly to remove any liquid from the plate seal and prevent loss as the seal is removed. Be sure to remove the seal very carefully.

19. Often, 200 μL PCR plates will be able to hold 300 μL. If the plate does not, the sample may be split into two plates and recombined into the Whatman UNIFILTER plate.

References

1. Dardalhon V, Awasthi A, Kwon H, Galileos G, Gao W, Sobel RA, Mitsdoerffer M, Strom TB, Elyaman W, Ho IC, Khoury S, Oukka M, Kuchroo VK (2008) IL-4 inhibits TGF-beta-induced Foxp3+ T cells and, together with TGF-beta, generates IL-9+ IL-10+ Foxp3(−) effector T cells. Nat Immunol 9(12):1347–1355. doi:10.1038/ni.1677
2. Beriou G, Bradshaw EM, Lozano E, Costantino CM, Hastings WD, Orban T, Elyaman W, Khoury SJ, Kuchroo VK, Baecher-Allan C, Hafler DA (2010) TGF-beta induces IL-9

production from human Th17 cells. J Immunol 185(1):46–54. doi:10.4049/jimmunol.1000356

3. Elyaman W, Bradshaw EM, Uyttenhove C, Dardalhon V, Awasthi A, Imitola J, Bettelli E, Oukka M, van Snick J, Renauld JC, Kuchroo VK, Khoury SJ (2009) IL-9 induces differentiation of TH17 cells and enhances function of FoxP3+ natural regulatory T cells. Proc Natl Acad Sci U S A 106(31):12885–12890. doi:10.1073/pnas.0812530106
4. Nowak EC, Weaver CT, Turner H, Begum-Haque S, Becher B, Schreiner B, Coyle AJ, Kasper LH, Noelle RJ (2009) IL-9 as a mediator of Th17-driven inflammatory disease. J Exp Med 206(8):1653–1660. doi:10.1084/jem.20090246
5. Kaplan MH, Hufford MM, Olson MR (2015) The development and in vivo function of T helper 9 cells. Nat Rev Immunol 15(5):295–307. doi:10.1038/nri3824
6. Kaplan MH (2013) Th9 cells: differentiation and disease. Immunol Rev 252(1):104–115. doi:10.1111/imr.12028
7. Staudt V, Bothur E, Klein M, Lingnau K, Reuter S, Grebe N, Gerlitzki B, Hoffmann M, Ulges A, Taube C, Dehzad N, Becker M, Stassen M, Steinborn A, Lohoff M, Schild H, Schmitt E, Bopp T (2010) Interferon-regulatory factor 4 is essential for the developmental program of T helper 9 cells. Immunity 33(2):192–202. doi:10.1016/j.immuni.2010.07.014
8. Chang HC, Sehra S, Goswami R, Yao W, Yu Q, Stritesky GL, Jabeen R, McKinley C, Ahyi AN, Han L, Nguyen ET, Robertson MJ, Perumal NB, Tepper RS, Nutt SL, Kaplan MH (2010) The transcription factor PU.1 is required for the development of IL-9-producing T cells and allergic inflammation. Nat Immunol 11(6):527–534. doi:10.1038/ni.1867
9. Goswami R, Jabeen R, Yagi R, Pham D, Zhu J, Goenka S, Kaplan MH (2012) STAT6-dependent regulation of Th9 development. J Immunol 188(3):968–975. doi:10.4049/jimmunol.1102840
10. Jabeen R, Goswami R, Awe O, Kulkarni A, Nguyen ET, Attenasio A, Walsh D, Olson MR, Kim MH, Tepper RS, Sun J, Kim CH, Taparowsky EJ, Zhou B, Kaplan MH (2013) Th9 cell development requires a BATF-regulated transcriptional network. J Clin Invest 123(11):4641–4653. doi:10.1172/JCI69489
11. Mullen AC, Orlando DA, Newman JJ, Loven J, Kumar RM, Bilodeau S, Reddy J, Guenther MG, DeKoter RP, Young RA (2011) Master transcription factors determine cell-type-specific responses to TGF-beta signaling. Cell 147(3):565–576. doi:10.1016/j.cell.2011.08.050
12. El Chartouni C, Schwarzfischer L, Rehli M (2010) Interleukin-4 induced interferon regulatory factor (Irf) 4 participates in the regulation of alternative macrophage priming. Immunobiology 215(9-10):821–825. doi:10.1016/j.imbio.2010.05.031
13. Elyaman W, Bassil R, Bradshaw EM, Orent W, Lahoud Y, Zhu B, Radtke F, Yagita H, Khoury SJ (2012) Notch receptors and Smad3 signaling cooperate in the induction of interleukin-9-producing T cells. Immunity 36(4):623–634. doi:10.1016/j.immuni.2012.01.020
14. Murugaiyan G, Beynon V, Pires Da Cunha A, Joller N, Weiner HL (2012) IFN-gamma limits Th9-mediated autoimmune inflammation through dendritic cell modulation of IL-27. J Immunol 189(11):5277–5283. doi:10.4049/jimmunol.1200808
15. Perumal NB, Kaplan MH (2011) Regulating Il9 transcription in T helper cells. Trends Immunol 32(4):146–150. doi:10.1016/j.it.2011.01.006
16. Bassil R, Orent W, Olah M, Kurdi AT, Frangieh M, Buttrick T, Khoury SJ, Elyaman W (2014) BCL6 controls Th9 cell development by repressing Il9 transcription. J Immunol 193(1):198–207. doi:10.4049/jimmunol.1303184
17. Olson MR, Verdan FF, Hufford MM, Dent AL, Kaplan MH (2016) STAT3 impairs STAT5 activation in the development of IL-9-secreting T cells. J Immunol 196(8):3297–3304. doi:10.4049/jimmunol.1501801
18. Tamiya T, Ichiyama K, Kotani H, Fukaya T, Sekiya T, Shichita T, Honma K, Yui K, Matsuyama T, Nakao T, Fukuyama S, Inoue H, Nomura M, Yoshimura A (2013) Smad2/3 and IRF4 play a cooperative role in IL-9-producing T cell induction. J Immunol 191(5):2360–2371. doi:10.4049/jimmunol.1301276
19. Purwar R, Schlapbach C, Xiao S, Kang HS, Elyaman W, Jiang X, Jetten AM, Khoury SJ, Fuhlbrigge RC, Kuchroo VK, Clark RA, Kupper TS (2012) Robust tumor immunity to melanoma mediated by interleukin-9-producing T cells. Nat Med 18(8):1248–1253. doi:10.1038/nm.2856
20. Wingender E (1988) Compilation of transcription regulating proteins. Nucleic Acids Res 16(5):1879–1902
21. Heinemeyer T, Chen X, Karas H, Kel AE, Kel OV, Liebich I, Meinhardt T, Reuter I, Schacherer F, Wingender E (1999) Expanding the TRANSFAC database towards an expert

system of regulatory molecular mechanisms. Nucleic Acids Res 27(1):318–322

22. Wingender E (1997) Classification of eukaryotic transcription factors. Mol Biol (Mosk) 31(4):584–600
23. Kel AE, Gossling E, Reuter I, Cheremushkin E, Kel-Margoulis OV, Wingender E (2003) MATCH: a tool for searching transcription factor binding sites in DNA sequences. Nucleic Acids Res 31(13):3576–3579

Chapter 10

Flow Cytometric Assessment of STAT Molecules in Th9 Cells

Lucien P. Garo, Vanessa Beynon, and Gopal Murugaiyan

Abstract

IL-9-producing Th9 cells are a novel subset of T helper cells that develop independently of other T helper subsets. Th9 cells have been implicated in the pathogenesis of allergic asthma and autoimmunity, while also serving as critical effector T cells in mediating antitumor immune responses. Concomitant presence of TGF-β and IL-4 lead to the differentiation of naïve $CD4^+$ T cells towards the Th9 phenotype. In addition, several cytokines, including IL-1β, IL-2, IL-25, and IL-33, further amplify Th9 responses. Negative regulators of Th9 cells include other cytokines such as IFN-γ, IL-23, and IL-27. Here, we describe a detailed protocol for the analysis of STAT molecules involved in the differentiation of Th9 cells and Th9 inhibition by IL-27.

Key words Th9, STAT3, STAT1, STAT6, IL-27, Flow cytometry, Asthma, Autoimmunity, Cancer

1 Introduction

Since their discovery as a new distinct T helper subset, IL-9-producing Th9 cells have been found to promote inflammation in a variety of contexts in both humans and mouse models [1–3]. Specifically, Th9 cells have been strongly associated with allergic diseases such as asthma and bronchial hyperreactivity [1–5]. They can be found at high frequencies in the peripheral blood of allergic and asthmatic patients [6]. Even before the Th9 subset was identified, IL-9 had been implicated in type 2 immunity-related processes. IL-9 enhances the IL-4-mediated production of IgE in human and murine B cells, promotes eosinophil maturation in synergy with IL-5, and stimulates mucin transcription in respiratory epithelial cells [7–9]. Studies using IL-9 transgenic mice or administration of recombinant IL-9 instilled into airways has shown that IL-9 is sufficient to produce a classical Th2-like response in vivo, and that IL-9 upregulates AHR, lung eosinophilia, and serum total IgE [10]. More recent studies further confirm the role of IL-9 in allergy and asthma [4, 5]. In an ovalbumin (OVA)-induced model of airway

Ritobrata Goswami (ed.), *Th9 Cells: Methods and Protocols*, Methods in Molecular Biology, vol. 1585,
DOI 10.1007/978-1-4939-6877-0_10,

inflammation, administration of IL-9-neutralizing antibodies ablated OVA-induced asthmatic responses [11]. Consistent with these findings, transfer of OVA-specific Th9 cells into T cell-deficient mice led to strong asthmatic symptoms following OVA challenge. These symptoms could again be ameliorated with anti-IL-9 antibodies. Th9 pathogenicity has also been implicated in other models of inflammatory disorders. Transfer of MOG-specific Th9 cells has been shown to induce EAE, the murine model of multiple sclerosis [12]. Accordingly, IL-9-deficiency or treatment with anti-IL-9 antibodies has been reported as protective against EAE [13–15]. Similarly, Th9 cells have been demonstrated to confer inflammation in a T cell transfer colitis model [16, 17]. Consistent with these results, another study has reported that neutralization of IL-9 has been shown to suppress inflammation in colitic mice [18].

Although Th9 cells appear to have pathogenic functions in a broad range of inflammatory and autoimmune diseases, recent studies suggest that these cells play a beneficial role in limiting tumor growth by promoting antitumor immune responses [19]. In fact, in a melanoma model, the antitumor effects of Th9 cells have been shown to be greater than the more widely acknowledged effects of Th1 and Th17 cells [20]. Interestingly, the authors also found lower numbers of Th9 cells in metastatic lesions of patients with progressive stage IV melanoma compared to healthy human skin and blood. It is thought that IL-9 may be directly involved in this process. Th9 cells have also been shown to induce protective antitumor immunity by eliciting a tumor-specific CTL response. Tumor-specific Th9 cells promote strong CTL activation via recruitment of DCs into tumor tissues and subsequent presentation of tumor antigens in tumor draining lymph nodes [21]. Very recently, it has been shown that Th9 cells differentiated in the presence of IL-1β produce high levels of IL-21 and exert IL-21-dependent antitumor effects [22]. Findings from human patients and murine disease models suggest that Th9 cells can be protective or pathogenic in different disease contexts, emphasizing the importance of defining the mechanisms that control the development and function of Th9 cells.

1.1 Positive Regulation of Th9 Cells

In the absence of IL-4, TGF-β promotes differentiation of Treg cells, and in the absence of TGF-β, IL-4 leads to the development of Th2 cells [23, 24]. However, the concomitant presence of TGF-β and IL-4 leads to the differentiation of naïve $CD4^+$ T cells towards the Th9 phenotype [16, 25]. Signal transducer and activator of transcription (STAT) molecules play critical roles in the differentiation and effector functions of almost all T helper subsets, including Th9 cells. STAT6 is a major signaling component of the IL-4 receptor (IL-4R) and is required for the in vitro differentiation of Th9 cells [16]. In fact, the majority of genes enriched in Th9 cells are STAT6 dependent. IL-4-induced STAT6 represses the expression of the Treg-specific transcription factor Foxp3 and the Th1-specific transcription factor T-bet, both of which repress IL-9 production [16, 26]. IRF4, a transcription factor downstream of the IL-4/STAT6 pathway, is required for the differentiation of Th9 cells [4]. Basic leucine zipper transcription factor (BATF) cooperates with IRF4 in binding to the *Il9* locus [27]. However, STAT6 also induces the Th2 transcription factors Gata3 and c-Maf in Th9 cells, and transduction of differentiating Th9 cells with GATA-3- and c-Maf-expressing retroviruses has been shown to decrease Th9 differentiation [26]. More recent work has identified a Th9-specific gene signature that distinguishes Th9 cells from their closely related Th2 cells. Specifically, the E26 transformation-specific (ETS) family transcription factor PU.1 promotes switching between the Th2 and Th9 phenotypes [11].

Although IL-4 has been shown to be a fundamental driver of Th9 differentiation, several other cytokines also work in synergy with IL-4/STAT6 signaling to amplify Th9 differentiation. For example, cell-intrinsic IL-2 has been shown to promote IL-9 production by Th9 cells [28]. IL-2 directly induces STAT5-dependent *Il9* transcription via binding sites in the *Il9* locus [29]. Consistent with this, inhibition of STAT5 and STAT6 by cytokine-induced SH-2 protein (CIS), a member of the suppressor of cytokine signaling (SOCS) family of proteins, has been shown to abrogate Th9 development [30]. In addition to IL-2, another STAT5-inducing cytokine, IL-25, has been shown to amplify IL-9 induction from Th9 cells; Th9 cells express higher levels of the IL-25R compared to other T helper cell subsets, making them particularly sensitive to this cytokine [31]. Furthermore, IL-1β and another IL-1 family member cytokine, IL-33, have been shown to promote Th9 differentiation, as well as the costimulatory molecule OX40 and the Notch1/Jagged2 pathway [17, 22, 32, 33].

1.2 Negative Regulation of Th9 Cells

Multiple cytokines have been identified as negative regulators of Th9 cells, including IFN-γ, IL-27, and IL-23. In fact, IFN-γ had been shown to downregulate IL-9 from T cells even before the Th9 subset was characterized [28]. Consistent with previous observations, our group has found that IFN-γ inhibits the differentiation of Th9 cells both in vitro and in vivo [34]. In addition to a direct inhibitory effect on Th9 differentiation, we found that IFN-γ suppresses Th9 cells by inducing IL-27 from dendritic cells [34]. IL-27 suppression of Th9 cells is partially dependent on STAT1 and T-bet. Although treatment of Th9 cells with IL-27 induces the Tr1 signature cytokines IFN-γ, IL-10, and IL-21; IL-9 inhibition by IL-27 is independent of these cytokines. Rather, we found this inhibition to be partially dependent on the transcription factors STAT1 and T-bet. However, the role of STAT1 and T-bet function downstream of IFN-γ and IL-27 in suppressing Th9 cells remains to be investigated. In addition to IFN-γ and IL-27, IL-23 also inhibits IL-9 production [20]. It has been shown that STAT3 attenuates the ability of STAT5 to promote the development of Th9 cells [35]. Although IL-23 has also been shown to signal via STAT3, the involvement of STAT3 in IL-23-mediated inhibition of Th9 cells has yet to be investigated. In addition to the cytokines described above, E3 ubiquitin ligase Cbl-b, an enzyme known to add ubiquitin to proteins negatively regulates Th9 differentiation by targeting STAT6 for ubiquitination and degradation [36].

Here, we describe a detailed protocol for the analysis of STAT molecules involved in the differentiation of Th9 cells, together with a protocol to determine the involvement of STAT molecules in Th9 inhibition by IL-27.

2 Materials

1. Six- to eight-week-old C57BL/6 mice.
2. Complete medium for murine Th9 generation: IMDM-GlutaMax supplemented with 5% fetal bovine serum (FBS) (v/v), 1% penicillin–streptomycin (v/v), 50 μM 2-mercaptoethanol.
3. ACK Lysing Buffer (Lonza).
4. Phosphate-buffered saline (PBS) without calcium and magnesium.
5. Fluorescence-activated cell-sorting (FACS) buffer: 2% FBS (v/v) in PBS.
6. CD4 (L3T4) microbeads.
7. LS Columns and magnet (Miltenyi Biotec, MidiMACS or QuadroMACS).
8. Purified anti-mouse CD3 and anti-mouse CD28 antibodies.
9. Cytokines: Recombinant mouse IL-4 and IL-27, recombinant human TGF-β.
10. Fluorescent-labeled antibodies: anti-CD4, anti-CD44, anti-CD62L, and anti-IL-9, anti-pSTAT3, anti-pSTAT1, and anti-pSTAT5, and anti-pSTAT6.
11. 7-amino actinomycin D (7-AAD).
12. Alternative to FACS antibodies for naïve T cell isolation: Naïve CD4+ T cell isolation kit, mouse (Miltenyi Biotec).
13. Cell Fixation/Permeabilization Kits for Intracellular Cytokine Analysis (BD Biosciences) containing Fixation/Permeabilization solution and 10× BD Perm/Wash™ Buffer.
14. Phorbol 12-myristate 13-acetate (PMA).
15. Ionomycin calcium Salt.
16. GolgiStop.
17. Methanol.

18. Distilled H_2O.
19. Petri dish.
20. 1 mL syringe.
21. 15 mL tubes.
22. Trypan blue.
23. Hemocytometer.
24. Flow cytometer.

3 Methods

3.1 Isolation of Naïve CD4+ T Cells

1. Euthanize mice and harvest spleens into 70 μm cell strainer in PBS-containing petri dish. Store it on ice.
2. Dissociate spleen with 1 mL syringe plunger to obtain single-cell suspension (*see* **Note 1**).
3. Collect cell suspension in 15 mL tube. Centrifuge at 420 × *g* for 5 min at 4 °C. Discard supernatant.
4. Enrich white blood cells by resuspending pellet in red blood cell (ACK) lysis buffer (approximately 3 mL/spleen). Incubate for 3 min at RT.
5. Wash by filling tube with PBS. Centrifuge at 420 × *g* for 5 min at 4 °C. Discard supernatant.
6. Resuspend cells in FACS sorting buffer and determine the number of viable cells using trypan blue and a hemocytometer or using automated cell counter (*see* **Note 2**).
7. Isolate total $CD4^+$ T cells using CD4 microbeads following manufacturer's instructions. Collect cell suspension in 15 mL tube. Centrifuge at 420 × *g* for 5 min at 4 °C. Discard supernatant.
8. Perform surface staining by resuspending pellet in 50 μL of FACS buffer cocktail containing 7AAD, CD4, CD62L, and CD44 antibodies. Incubate for 15–30 min at 4 °C in dark (*see* **Note 3**).
9. Wash cells by adding 10 mL of FACS buffer. Centrifuge at 420 × *g* for 5 min at 4 °C. Discard supernatant.
10. Resuspend pellet in 1–2 mL of FACS buffer (*see* **Note 4**). Prepare collection tubes for FACS machine by adding 1–2 mL of complete media or FACS buffer.
11. Sort naïve $CD4^+$ T cells by collecting the $7AAD^-CD4^+CD62L^+CD44^{lo}$ population using FACS machine. *See* Fig. 1 for gating strategy.
12. Wash collected cells with FACS buffer or complete media. Centrifuge at 420 × *g* for 5 min at 4 °C.
13. Alternatively, if FACS-sorting is unavailable or not preferable, naive $CD4^+$ T cells can be sorted using the Naïve $CD4^+$ T Cell

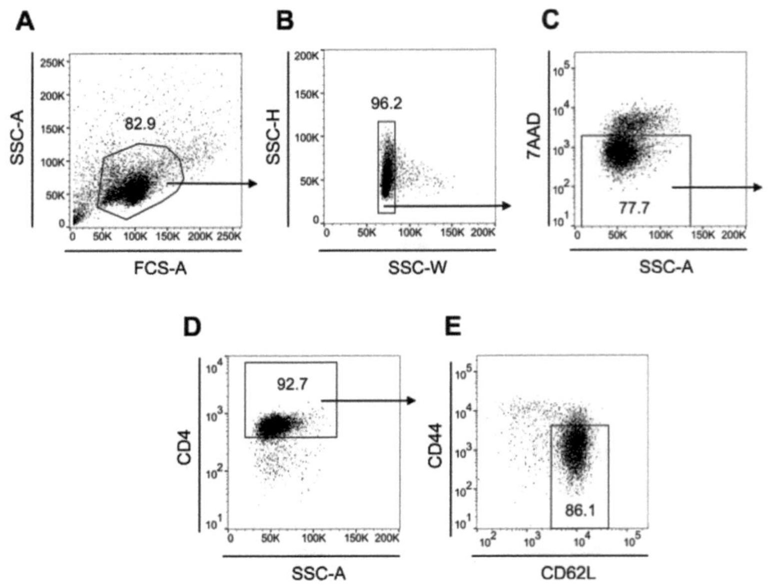

Fig. 1 Sorting strategy for the isolation of naïve CD4+ T cells by FACS sorting. (**a**) Lymphocytes are first gated by size and then (**b**) singlet cells are discriminated from doublets. (**c**) Gating on singlets reveals populations of mostly live (7AAD−) cells. (**d**) Gate on the CD4+ population and then sort (**e**) naïve (CD62L+CD44lo) from the effector/memory (CD62L−CD44hi) CD4+ T cells

Isolation Kit following the manufacturer's instructions. T cells are isolated through negative selection via two methods: (1) Depletion of non-CD4+ T cells using a biotin–antibody cocktail and anti-biotin microbeads, and (2) depletion of CD44hi cells using CD44 microbeads. Cells are separated using LS columns and a MidiMACS or QuadroMACS magnet. The negative fraction enriched in naïve CD4+ T cells (flow through) is collected.

14. Determine cell concentration from cell counts. Resuspend sorted naïve CD4+ T cells at 10^6 cells/mL in complete medium (*see* **Note 5**).

3.2 Th9 Differentiation/ Inhibition

1. Prepare culture plate by adding anti-CD3 and anti-CD28 mixture (2 μg/mL in 100 μL of sterile PBS) to wells in 96-well flat bottom tissue culture plate. Incubate at 37 °C for 2 h (*see* **Note 6**).
2. Carefully aspirate supernatant. Wash anti-CD3/CD28-coated wells by adding 200 μL of PBS and aspirate (*see* **Note 7**).
3. Plate cells in 100 μL of complete media at 2×10^4–1×10^5 cells/well.
4. To induce Th9 differentiation, add 100 μL of Th9 activation mix (20 ng/mL of IL-4 and 4 ng/mL of TGF-β in complete media).

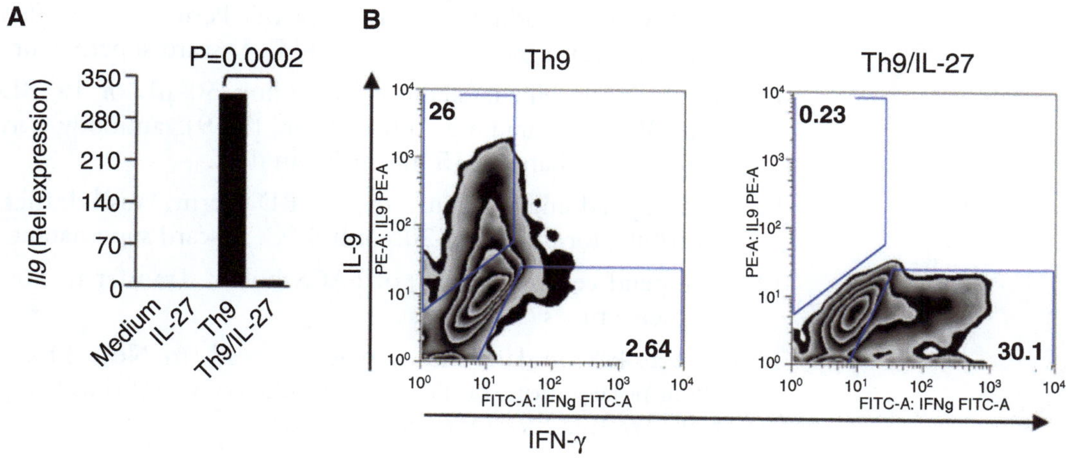

Fig. 2 Assessment of Th9 differentiation/inhibition. (**a**) Real-time PCR and (**b**) flow cytometric analysis of naïve $CD4^+$ T cells activated with anti-CD3 and anti-CD28 under Th9-polarizing conditions in the presence or absence of IL-27. IL-27 inhibits Th9 differentiation. Adapted from [34]

5. To inhibit Th9 differentiation, add 100 μL of Th9 inhibition mix (20 ng/mL of IL-4, 4 ng/mL of TGF-β, and 100 ng/mL of IL-27 in complete media).
6. Incubate cells at 37 °C for 4 days (*see* **Note 8**).
7. Assess Th9 differentiation/inhibition by measuring Il9 expression via intracellular staining and flow cytometric analysis, ELISA, or qPCR. *See* Fig. 2 for representative flow cytometry data.

3.3 Intracellular Staining for Cytokines (Th9)

1. Create master mix in PBS to add PMA (50 ng/mL), ionomycin (1 μM), and GolgiStop (1:1000 dilution) to each well (*see* **Note 9**).
2. Incubate at 37 °C for 4 h.
3. Transfer cells to 96-well round bottom plate to allow for effective pelleting. Centrifuge for 5 min at 420 × *g* at 4 °C. Discard supernatant.
4. Wash by adding 200 μL of FACS buffer. Centrifuge for 5 min at 420 × *g* at 4 °C. Discard supernatant.
5. Perform surface staining by resuspending cells in 50 μL of FACS buffer containing extracellular (CD4) antibody. Incubate for 15 min at RT or 30 min at 4 °C in dark (*see* **Note 10**). Centrifuge for 5 min at 420 × *g* at 4 °C. Discard supernatant.
6. Wash by adding 200 μL of FACS buffer. Centrifuge for 5 min at 420 × *g* at 4 °C. Discard supernatant.
7. Fix/permeabilize cells by resuspending in 100 μL of BD Cytofix/Cytoperm Buffer. Incubate for 20 min at RT in dark.
8. Wash by adding 100 μL of 1× BD Perm/Wash buffer (10× stock diluted in distilled water). Centrifuge for 5 min at 420 × *g* at 4 °C. Discard supernatant.

9. Wash again by adding 200 μL of 1× BD Perm/Wash buffer. Centrifuge for 5 min at 420 × *g* at 4 °C. Discard supernatant.
10. Perform intracellular staining by adding 50 μL of 1× BD Perm/Wash containing intracellular (IL-9) antibody (*see* **Note 3**). Incubate for 15 min at RT in dark.
11. Wash by adding 150 μL of 1× BD Perm/Wash buffer. Centrifuge for 5 min at 420 × *g* at 4 °C. Discard supernatant.
12. Resuspend cells in 200 μL of FACS buffer. Transfer to flow cytometry tubes.
13. Store at 4 °C in dark until samples are read (*see* **Note 11**). A strong polarization should result in >20% of $CD4^+$ cells being $IL\text{-}9^+$. *See* Fig. 2 for representative data.

3.4 Intracellular Staining for STAT Molecules

1. Transfer cells to 96-well round bottom plate to allow for effective pelleting (*see* **Note 8**).
2. Fix cells by resuspending in 100 μL of 2% PFA in PBS. Incubate at 37 °C for 10 min. Centrifuge for 5 min at 420 × *g* at 4 °C. Discard supernatant (*see* **Note 12**).
3. Resuspend cells in 100 μL of 90% ice-cold methanol. Incubate at −20 °C overnight.
4. Wash by adding 200 μL of FACS buffer. Centrifuge for 5 min at 420 × *g* at 4 °C. Discard supernatant.
5. Wash again by adding 200 μL of FACS buffer. Centrifuge for 5 min at 420 × *g* at 4 °C. Discard supernatant.
6. Perform intracellular staining by resuspending cells in 100 μL of FACS buffer containing 5 μL of each STAT antibody (*see* **Note 13**). Incubate for 60 min at RT.
7. Wash by adding 100 μL of FACS buffer. Centrifuge for 5 min at 420 × *g* at 4 °C. Discard supernatant.
8. Wash again by adding 200 μL of FACS buffer. Centrifuge for 5 min at 420 × *g* at 4 °C. Discard supernatant.
9. Resuspend cells in 200 μL of FACS Buffer. Transfer to flow cytometry tubes.
10. Store at 4 °C until samples are read (*see* **Note 11**). *See* Fig. 3 for representative data. *See* Fig. 4 for data assessing the STAT molecules involved in Th9 inhibition by IL-27.

4 Notes

1. Ensure there is no fur or other contaminating particulates. Wash syringe and strainer with PBS for maximum yield.

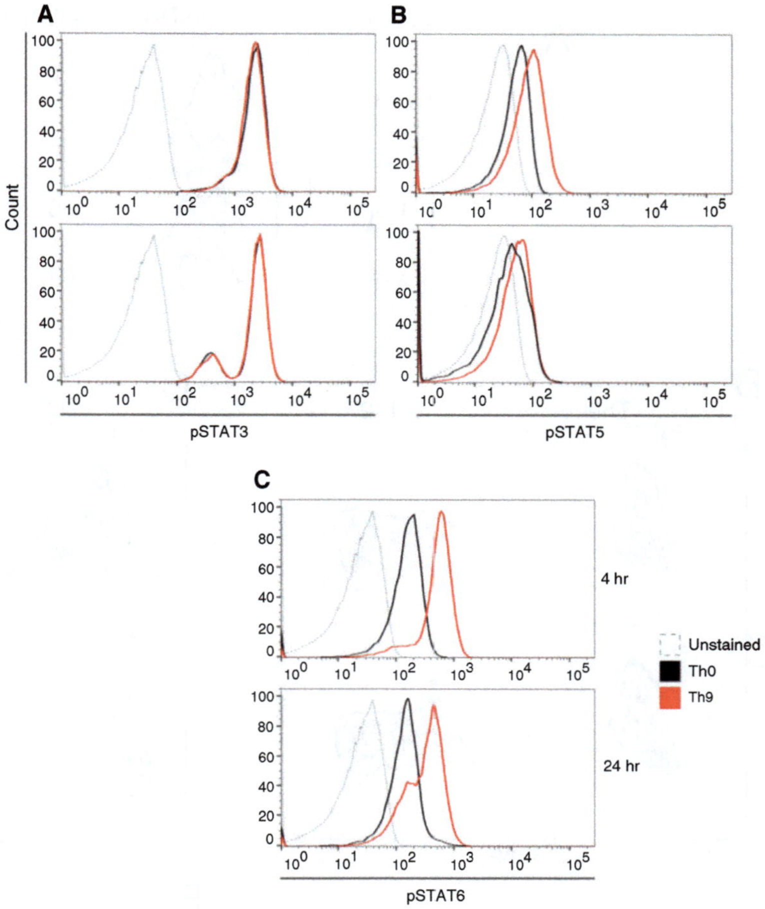

Fig. 3 Assessment of STAT molecules in Th9 differentiation. Flow cytometric analysis of (**a**) pSTAT3, (**b**) pSTAT5, and (**c**) pSTAT6 in T cells cultured under Th0 or Th9-polarizing conditions for 4 and 24 h. STAT5 and STAT6 phosphorylation is increased in Th9-polarized cells, while pSTAT3 is unaltered

2. Remove any visible cell clumps following red blood cell lysis before proceeding with bead isolation to prevent clogging the columns.
3. Select colors appropriate for your particular flow cytometer, maximizing brightness and minimizing overlap. All markers of interest in this experiment are highly expressed in the cell fraction and should be easily detectable. It is always ideal to experimentally optimize antibody dilutions for a particular FACS machine and experiment. Alternatively, one can follow the manufacturer's recommendations or use a common dilution of 1:100. Although cell death from freshly isolated cells is minimal, it is best to still exclude dead cells using 7AAD or another live–dead marker.

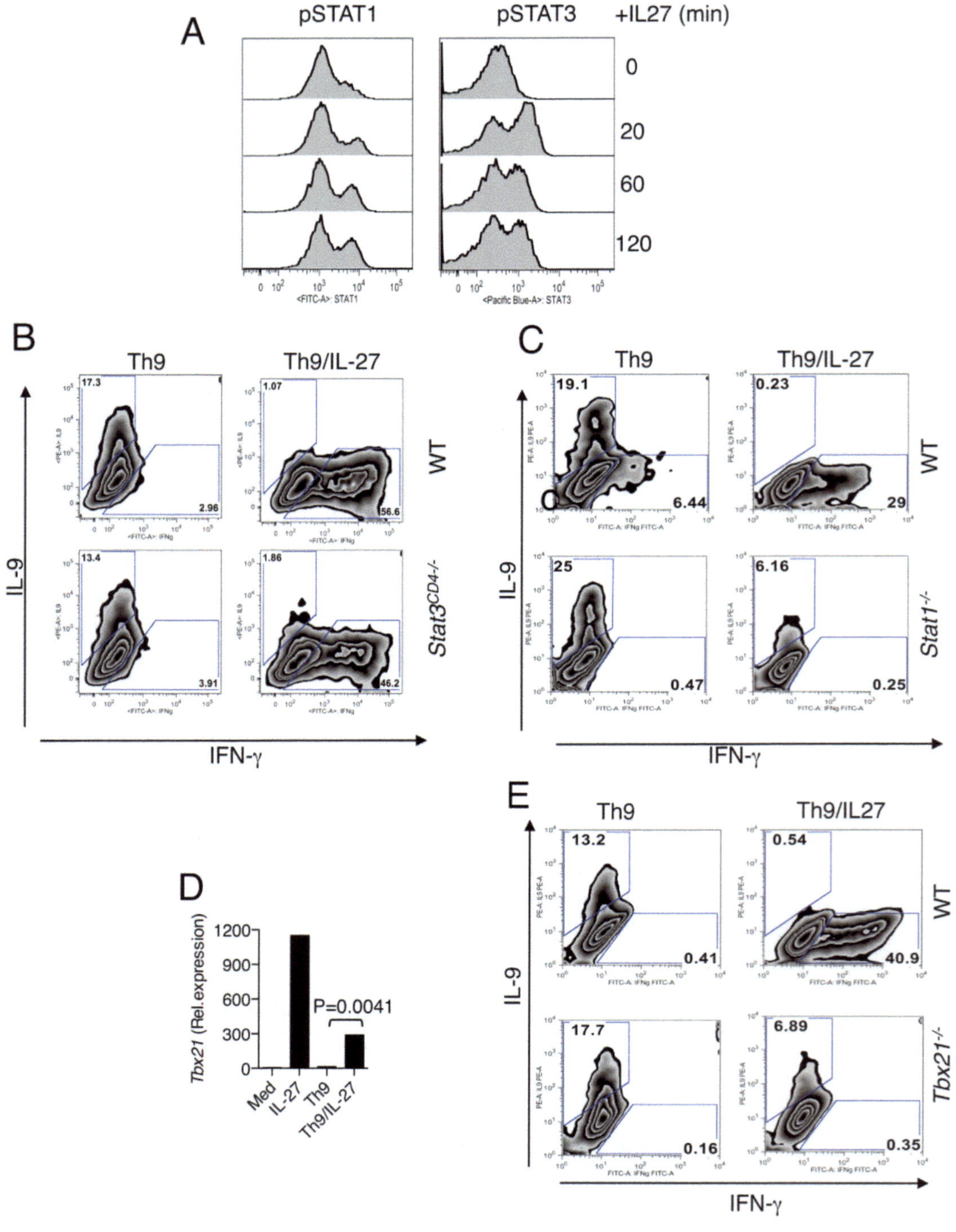

Fig. 4 Assessment of STAT molecules involved in Th9 inhibition by IL-27. IL-27 inhibition of Th9 cells is partially mediated by STAT1 and T-bet. (**a**) IL-27 induces STAT1 and STAT3 phosphorylation in Th9 cells. Naïve CD4+ T cells were stimulated with anti-CD3/28 for 24 h in the presence of TGF-β and IL-4 (Th9 conditions) to initiate the differentiation process. Thereafter, the cells were harvested, rested at 37 °C for 2 h in culture medium without cytokines and subsequently stimulated with IL-27 for 20, 60, and 120 min. Subsequently, the cells were analyzed by flow cytometry for STAT1 and STAT3 phosphorylation. (**b**) Flow cytometry of naïve

4. Keep cells on ice and protected from light until sorting begins.
5. Different groups have reported using both RPMI 1640 and IMDM complete medium for Th9 polarizations; however, we have found in our hands that IMDM works better. Subsequent protocols can also be followed using RPMI 1640 if preferred.
6. It is suggested to prepare plate-bound antibodies before beginning cell isolation to allow time for incubation and minimize time cells are on ice. Tap plate gently after applying antibody cocktail to ensure uniform coverage. If antibodies are left for more than 4 h, transfer plate to 4 °C until sorted cells are ready.
7. To avoid disturbing the bottom of the plate and antibody coating, tilt plate and aspirate slowly along well walls. *Do not* pipette up and down when adding PBS wash.
8. If the cells are plated and cultured as specified, it should not be necessary to change the media during this interval. Note that when assessing the Th9 differentiation via IL-9 expression by intracellular staining, one should wait the entire 4 days to allow for downstream changes and upregulation of IL-9 and other Th9-related proteins. However, assessment of STAT molecule involvement is time-dependent and the appropriate termination point may be earlier to detect changes in phosphorylation. For short incubation periods such as 4 h, a live–dead stain is unnecessary. For periods longer than 24 h, a fixable live–dead stain may be considered.
9. The PMA/ionomycin cocktail is sufficient to induce activation of T cells and cytokine production. Golgi stop is a protein transport inhibitor that will enhance intracellular retention of proteins and induced cytokines, increasing detectability in subsequent staining. This step is critical for successful intracellular T cell cytokine staining.
10. If naïve cells were FACS-sorted, residual florescence may remain. In this case, select the same fluorophores for markers used previously (i.e., CD4) for acquisition.
11. It is ideal to read cells immediately after they are stained, especially for less easily detectible markers like STAT molecules, but fixed cells can be read within 72 h if stored properly at 4 °C in the dark. If reading the same day as staining, start early to allow time for multiple incubation periods and washes.

Fig. 4 (continued) $CD4^+$ T cells from $STAT3^{flox/flox}$ (WT) and $STAT3^{flox/flox\ CD4\text{-}cre}$ (*$Stat3^{CD4-/-}$*) mice stimulated under Th9-polarzing conditions in the presence or absence of IL-27. (**c**) Flow cytometry of naïve $CD4^+$ T cells from WT and *$Stat1^{-/-}$* mice stimulated under Th9-polarizing conditions in the presence of absence of IL-27. (**d**) Real-time PCR and flow cytometry analysis of T-bet in Th9 cells in the presence or absence of IL-27. (**e**) Flow cytometry of naïve $CD4^+$ T cells from WT and *$Tbx21^{-/-}$* mice stimulated under Th9-polarizing conditions in the presence or absence of IL-27. Adapted from [34]

12. This permeabilization method is very harsh and makes surface staining difficult. Therefore, it is critical to start with a highly purified cell population.
13. The STAT molecules are more difficult to detect by flow cytometry and the fixation process dims any fluorophores used. It is ideal to reserve the brightest available fluorophores for these markers.

Acknowledgments

This work was supported by grants from the National Multiple Sclerosis Society (RG 4904A2/1 and RG 1507-05164).

References

1. Kaplan MH (2013) Th9 cells: differentiation and disease. Immunol Rev 252(1):104–115. doi:10.1111/imr.12028
2. Pan HF, Leng RX, Li XP, Zheng SG, Ye DQ (2013) Targeting T-helper 9 cells and interleukin-9 in autoimmune diseases. Cytokine Growth Factor Rev 24(6):515–522
3. Kaplan MH, Hufford MM, Olson MR (2015) The development and in vivo function of T helper 9 cells. Nat Rev Immunol 15(5):295–307. doi:10.1038/nri3824
4. Staudt V, Bothur E, Klein M, Lingnau K, Reuter S, Grebe N, Gerlitzki B, Hoffmann M, Ulges A, Taube C, Dehzad N, Becker M, Stassen M, Steinborn A, Lohoff M, Schild H, Schmitt E, Bopp T (2010) Interferon-regulatory factor 4 is essential for the developmental program of T helper 9 cells. Immunity 33(2):192–202. doi:10.1016/j.immuni.2010.07.014
5. Kerzerho J, Maazi H, Speak AO, Szely N, Lombardi V, Khoo B, Geryak S, Lam J, Soroosh P, Van Snick J, Akbari O (2013) Programmed cell death ligand 2 regulates TH9 differentiation and induction of chronic airway hyperreactivity. J Allergy Clin Immunol 131(4):1048–1057, 1057.e1041–1042. doi:10.1016/j.jaci.2012.09.027
6. Hoppenot D, Malakauskas K, Lavinskiene S, Bajoriuniene I, Kalinauskaite V, Sakalauskas R (2015) Peripheral blood Th9 cells and eosinophil apoptosis in asthma patients. Medicina (Kaunas) 51(1):10–17. doi:10.1016/j.medici.2015.01.001
7. Dugas B, Renauld JC, Pene J, Bonnefoy JY, Peti-Frere C, Braquet P, Bousquet J, Van Snick J, Mencia-Huerta JM (1993) Interleukin-9 potentiates the interleukin-4-induced immunoglobulin (IgG, IgM and IgE) production by normal human B lymphocytes. Eur J Immunol 23(7):1687–1692. doi:10.1002/eji.1830230743
8. Louahed J, Zhou Y, Maloy WL, Rani PU, Weiss C, Tomer Y, Vink A, Renauld J, Van Snick J, Nicolaides NC, Levitt RC, Haczku A (2001) Interleukin 9 promotes influx and local maturation of eosinophils. Blood 97(4):1035–1042
9. Longphre M, Li D, Gallup M, Drori E, Ordonez CL, Redman T, Wenzel S, Bice DE, Fahy JV, Basbaum C (1999) Allergen-induced IL-9 directly stimulates mucin transcription in respiratory epithelial cells. J Clin Invest 104(10):1375–1382. doi:10.1172/JCI6097
10. Temann UA, Geba GP, Rankin JA, Flavell RA (1998) Expression of interleukin 9 in the lungs of transgenic mice causes airway inflammation, mast cell hyperplasia, and bronchial hyperresponsiveness. J Exp Med 188(7):1307–1320
11. Chang HC, Sehra S, Goswami R, Yao W, Yu Q, Stritesky GL, Jabeen R, McKinley C, Ahyi AN, Han L, Nguyen ET, Robertson MJ, Perumal NB, Tepper RS, Nutt SL, Kaplan MH (2010) The transcription factor PU.1 is required for the development of IL-9-producing T cells and allergic inflammation. Nat Immunol 11(6):527–534. doi:10.1038/ni.1867
12. Jager A, Dardalhon V, Sobel RA, Bettelli E, Kuchroo VK (2009) Th1, Th17, and Th9 effector cells induce experimental autoimmune encephalomyelitis with different pathological phenotypes. J Immunol 183(11):7169–7177. doi:10.4049/jimmunol.0901906

13. Li H, Nourbakhsh B, Cullimore M, Zhang GX, Rostami A (2011) IL-9 is important for T-cell activation and differentiation in autoimmune inflammation of the central nervous system. Eur J Immunol 41(8):2197–2206. doi:10.1002/eji.201041125
14. Nowak EC, Weaver CT, Turner H, Begum-Haque S, Becher B, Schreiner B, Coyle AJ, Kasper LH, Noelle RJ (2009) IL-9 as a mediator of Th17-driven inflammatory disease. J Exp Med 206(8):1653–1660. doi:10.1084/jem.20090246
15. Li H, Nourbakhsh B, Ciric B, Zhang GX, Rostami A (2010) Neutralization of IL-9 ameliorates experimental autoimmune encephalomyelitis by decreasing the effector T cell population. J Immunol 185(7):4095–4100. doi:10.4049/jimmunol.1000986
16. Dardalhon V, Awasthi A, Kwon H, Galileos G, Gao W, Sobel RA, Mitsdoerffer M, Strom TB, Elyaman W, Ho IC, Khy S, Oukka M, Kuchroo VK (2008) IL-4 inhibits TGF-beta-induced Foxp3+ T cells and, together with TGF-beta, generates IL-9+ IL-10+ Foxp3(−) effector T cells. Nat Immunol 9(12):1347–1355. doi:10.1038/ni.1677
17. Gerlach K, Hwang Y, Nikolaev A, Atreya R, Dornhoff H, Steiner S, Lehr HA, Wirtz S, Vieth M, Waisman A, Rosenbauer F, McKenzie AN, Weigmann B, Neurath MF (2014) TH9 cells that express the transcription factor PU.1 drive T cell-mediated colitis via IL-9 receptor signaling in intestinal epithelial cells. Nat Immunol 15(7):676–686. doi:10.1038/ni.2920
18. Yuan A, Yang H, Qi H, Cui J, Hua W, Li C, Pang Z, Zheng W, Cui G (2015) IL-9 antibody injection suppresses the inflammation in colitis mice. Biochem Biophys Res Commun 468(4):921–926. doi:10.1016/j.bbrc.2015.11.057
19. Vegran F, Apetoh L, Ghiringhelli F (2015) Th9 cells: a novel CD4 T-cell subset in the immune war against cancer. Cancer Res 75(3):475–479. doi:10.1158/0008-5472.CAN-14-2748
20. Purwar R, Schlapbach C, Xiao S, Kang HS, Elyaman W, Jiang X, Jetten AM, Khy SJ, Fuhlbrigge RC, Kuchroo VK, Clark RA, Kupper TS (2012) Robust tumor immunity to melanoma mediated by interleukin-9-producing T cells. Nat Med 18(8):1248–1253. doi:10.1038/nm.2856
21. Lu Y, Hong S, Li H, Park J, Hong B, Wang L, Zheng Y, Liu Z, Xu J, He J, Yang J, Qian J, Yi Q (2012) Th9 cells promote antitumor immune responses in vivo. J Clin Invest 122(11):4160–4171. doi:10.1172/JCI65459
22. Vegran F, Berger H, Boidot R, Mignot G, Bruchard M, Dosset M, Chalmin F, Rebe C, Derangere V, Ryffel B, Kato M, Prevost-Blondel A, Ghiringhelli F, Apetoh L (2014) The transcription factor IRF1 dictates the IL-21-dependent anticancer functions of TH9 cells. Nat Immunol 15(8):758–766. doi:10.1038/ni.2925
23. Chatelain R, Varkila K, Coffman RL (1992) IL-4 induces a Th2 response in Leishmania major-infected mice. J Immunol 148(4):1182–1187
24. Chen W, Jin W, Hardegen N, Lei KJ, Li L, Marinos N, McGrady G, Wahl SM (2003) Conversion of peripheral CD4+CD25− naive T cells to CD4+CD25+ regulatory T cells by TGF-beta induction of transcription factor Foxp3. J Exp Med 198(12):1875–1886. doi:10.1084/jem.20030152
25. Veldhoen M, Uyttenhove C, van Snick J, Helmby H, Westendorf A, Buer J, Martin B, Wilhelm C, Stockinger B (2008) Transforming growth factor-beta 'reprograms' the differentiation of T helper 2 cells and promotes an interleukin 9-producing subset. Nat Immunol 9(12):1341–1346. doi:10.1038/ni.1659
26. Goswami R, Jabeen R, Yagi R, Pham D, Zhu J, Goenka S, Kaplan MH (2012) STAT6-dependent regulation of Th9 development. J Immunol 188(3):968–975. doi:10.4049/jimmunol.1102840
27. Jabeen R, Goswami R, Awe O, Kulkarni A, Nguyen ET, Attenasio A, Walsh D, Olson MR, Kim MH, Tepper RS, Sun J, Kim CH, Taparowsky EJ, Zhou B, Kaplan MH (2013) Th9 cell development requires a BATF-regulated transcriptional network. J Clin Invest 123(11):4641–4653. doi:10.1172/JCI69489
28. Schmitt E, Germann T, Goedert S, Hoehn P, Huels C, Koelsch S, Kuhn R, Muller W, Palm N, Rude E (1994) IL-9 production of naive CD4+ T cells depends on IL-2, is synergistically enhanced by a combination of TGF-beta and IL-4, and is inhibited by IFN-gamma. J Immunol 153(9):3989–3996
29. Liao W, Spolski R, Li P, Du N, West EE, Ren M, Mitra S, Leonard WJ (2014) Opposing actions of IL-2 and IL-21 on Th9 differentiation correlate with their differential regulation of BCL6 expression. Proc Natl Acad Sci U S A 111(9):3508–3513. doi:10.1073/pnas.1301138111
30. Yang XO, Zhang H, Kim BS, Niu X, Peng J, Chen Y, Kerketta R, Lee YH, Chang SH, Corry DB, Wang D, Watowich SS, Dong C (2013) The signaling suppressor CIS controls proallergic T cell development and allergic airway inflammation. Nat Immunol 14(7):732–740. doi:10.1038/ni.2633

31. Angkasekwinai P, Chang SH, Thapa M, Watarai H, Dong C (2010) Regulation of IL-9 expression by IL-25 signaling. Nat Immunol 11(3):250–256. doi:10.1038/ni.1846

32. Xiao X, Balasubramanian S, Liu W, Chu X, Wang H, Taparowsky EJ, Fu YX, Choi Y, Walsh MC, Li XC (2012) OX40 signaling favors the induction of T(H)9 cells and airway inflammation. Nat Immunol 13(10):981–990. doi:10.1038/ni.2390

33. Elyaman W, Bassil R, Bradshaw EM, Orent W, Lahoud Y, Zhu B, Radtke F, Yagita H, Khy SJ (2012) Notch receptors and Smad3 signaling cooperate in the induction of interleukin-9-producing T cells. Immunity 36(4):623–634. doi:10.1016/j.immuni.2012.01.020

34. Murugaiyan G, Beynon V, Pires Da Cunha A, Joller N, Weiner HL (2012) IFN-gamma limits Th9-mediated autoimmune inflammation through dendritic cell modulation of IL-27. J Immunol 189(11):5277–5283. doi:10.4049/jimmunol.1200808

35. Olson MR, Verdan FF, Hufford MM, Dent AL, Kaplan MH (2016) STAT3 impairs STAT5 activation in the development of IL-9-secreting T cells. J Immunol 196(8):3297–3304. doi:10.4049/jimmunol.1501801

36. Qiao G, Ying H, Zhao Y, Liang Y, Guo H, Shen H, Li Z, Solway J, Tao E, Chiang YJ, Lipkowitz S, Penninger JM, Langdon WY, Zhang J (2014) E3 ubiquitin ligase Cbl-b suppresses proallergic T cell development and allergic airway inflammation. Cell Rep 6(4):709–723. doi:10.1016/j.celrep.2014.01.012

Chapter 11

Transcription Factors Downstream of IL-4 and TGF-β Signals: Analysis by Quantitative PCR, Western Blot, and Flow Cytometry

Atsushi Sugimoto, Ryoji Kawakami, and Norihisa Mikami

Abstract

IL-9-producing Th9 cell is a novel Th cell subset involved in type II allergic inflammations such as asthma. Th9 cells can be induced from naïve Th cells in the presence of IL-4 and TGF-β. It is also well established that downstream signals of IL-4 and TGF-β, including STAT6, IRF4, Smad, and PU.1, directly mediate IL-9 production in Th9 cells. In this chapter we describe the methods of flow cytometry, qPCR and western blot analysis to determine the expression or activation of these transcription factors downstream of IL-4 and TGF-β.

Key words Th cell differentiation, Th9 cells, IL-9, Intracellular staining

1 Introduction

CD4⁺ T helper T (Th) cells have various roles in acquired immunity. Th cells classically can be divided into several subsets, including Th1, Th2, Th17, and regulatory T (Treg) cells. These subsets are polarized from naive T cells under specific cytokine conditions during the stimulation by antigen presentation cells (APC). Th cells are also involved in many allergic diseases, and thus its physiological regulation should be strict and complex [1, 2].

IL-9-producing Th9 cells have recently been described as a new Th cell subset [3, 4]. In vitro, Th9 cells differentiate from naive Th cells in the presence of both TGF-β and IL-4 or from Th2 cells in the presence of TGF-β [4]. The essential roles of IL-9 in the pathogenic progression of type I hypersensitivity have been demonstrated using IL-9 transgenic mice and IL-9-deficient mice. IL-9 transgenic mice exhibit accelerated hypersensitivity, and the condition is suppressed in IL-9-deficient mice [5, 6]. Although type I hypersensitivity has been considered a Th2 cytokine (IL-4, IL-5, and IL-13)-mediated reaction, recent findings showed that the blockade of IL-4 signaling had little effect on human asthma,

Ritobrata Goswami (ed.), *Th9 Cells: Methods and Protocols*, Methods in Molecular Biology, vol. 1585,
DOI 10.1007/978-1-4939-6877-0_11, © Springer Science+Business Media LLC 2017

and currently IL-9 is attracting attention as a new therapeutic target of type I hypersensitivity [7].

There are accumulating evidences of regulatory mechanisms of IL-9 production in Th9 cells. PU.1 is reported to be an important transcription regulator of Th9 cell differentiation [8]. In addition, other transcription factors, such as NF-κB, nuclear factor of activated T-cells (NFAT) c2, interferon regulatory factor 4 (IRF4), B-cell-activating transcription factor (BATF), and protein kinase A (PKA) also contribute to IL-9 production in Th9 cells [9–13]. In this chapter we describe in vitro protocols for the analysis of the transcription factors by quantitative polymerase chain reaction (qPCR), western blot, and flow cytometry after Th9 differentiation. These methods can assist in understanding how Th9 cell differentiation and functions are regulated.

2 Materials

2.1 Cell Preparation and Culture

1. BALB/c mice (*see* **Note 1**).
2. FACS buffer: calcium- and magnesium-free phosphate buffered saline (PBS), 2% fetal calf serum (FCS) (v/v), 1 mM EDTA.
3. 2R: Roswell Park Memorial Institute (RPMI) 1640 supplemented with 2% FCS.
4. Complete medium: RPMI 1640 supplemented with 10% FCS (v/v), 60 μg/mL penicillin G, 100 μg/mL streptomycin, 0.1 mM 2-mercaptoethanol.
5. Recombinant cytokines: human IL-2, murine IL-4, and human TGF-β.
6. Cell sorter: BD FACSAria II.
7. Antibodies: shown in Table 1.

2.2 RNA Extraction and qPCR

1. TRIzol reagent.
2. Chloroform.
3. Isopropyl alcohol.
4. 75% ethanol (in RNase-free water).
5. RNase-free water.
6. PrimeScript RT reagent kit: this kit includes 5× PrimeScript Buffer (for Real Time PCR), PrimeScript RT Enzyme Mix I, Oligo dT Primer, and Random 6 mers.
7. SYBR Green Real-Time PCR Master Mixes and a Lightcycler System.
8. Primers: shown in Table 2.

Table 1
The list of recommended antibodies to be used for cell stimulation, flow cytometry (FCM), and western blotting (WB)

Antibody	Clone	Application
Anti-mouse CD16/32	93	FCM
Anti-mouse CD4	RM4-5	FCM
Anti-mouse CD25	PC61	FCM
Anti-human/mouse CD44	IM7	FCM
Anti-mouse CD62L	MEL-14	FCM
Anti-mouse CD3ε	145-2C11	Stimulation
Anti-mouse CD28	37.51	Stimulation
Anti-mouse/human GATA3	TWAJ	FCM, WB
Anti-mouse/rat FoxP3	FJK-16s	FCM
Anti-mouse/human PU.1	9G7	WB
Anti-mouse IL-2	JES6-5H4	FCM
Anti-mouse IL-4	11B11	FCM
Anti-mouse IL-9	PM9A4	FCM
Anti-mouse p-Smad2 (S465/S467)/Smad3 (S423/S425)	O72-670	FCM, WB
Anti-mouse p-STAT6 (Y641)	J71-773.58.11	FCM, WB

2.3 Flow Cytometry Analysis

1. Phorbol 12-myristate 13-acetate (PMA).
2. Ionomycin.
3. Monensin solution: 0.26% monensin in ethanol (BD GolgiStop).
4. Fix/Perm Buffer: dilute the 4× Fix/Perm Buffer (BD Pharmingen Transcription Factor Buffer Set; containing 5% formaldehyde and 1.76% methanol) with the Diluent Buffer (provided in the buffer set) to the necessary volume of 1× Fix/Perm Buffer working solution (*see* **Note 2**).
5. Perm/Wash Buffer: dilute the 5× Perm/Wash Buffer (BD Pharmingen Transcription Factor Buffer Set) using H_2O to a 1× Perm/Wash Buffer working solution.
6. Lyse/Fix Buffer: dilute the 5× Lyse/Fix Buffer (BD Phosflow Lyse/Fix Buffer; containing 7.15% methanol, 20.35% formaldehyde, and 15.65% diethylene glycol) using H_2O to a 1× Lyse/Fix Buffer working solution.
7. Permeabilization Buffer: 87.68% methanol (BD Phosflow Perm Buffer III).

Table 2
The list of primers for qPCR analysis

Transcript	Forward	Reverse	Length (bp)	Annealing temp. (°C)
IL-9	CTGATGATTGTACCACACCGTGC	GCCTTTGCATCTCTGTCTTCTGG	237	60
T-bet	CTAACACACACGTCTTTACT	AACAGATGCGTACATGGA	146	57
GATA3	CGGAAACTCCGTCAGG	ACGTCTTGGAGAAGGG	170	59
NFATc1	GGAGCGGAGAAACTTTGCG	GTGACACTAGGGGACACATAACT	94	57
NFATc2	GCACATAAGGCCATCAGCTCA	TCGCCAGAGAGACTGGCAA	101	58
c-Maf	ACTTCGACGACCGCTTCTC	TTCTCGGAAGCCGTTGCTCAC	308	56
PU.1	AGAGCTATACCAACGTCCAATGC	TTCTCAAACTCGTTGTTGTGGAC	120	57
FoxP3	CCTCTTCTTCCTTGAACCC	GTGGACGCACTTGGAG	198	60
GAPDH	CATGCCATCACTGCCACCC	GGTAGGAACACGGAAGGCC	177	62

8. Flow cytometer: BD FACSCanto II.
9. Antibodies: shown in Table 1.

2.4 Western Blot

1. Sample buffer: 250 mM Tris–HCl (pH 6.8), 8% SDS, 20% 2-mercaptolethanol, 40% glycerol, 0.01% bromophenol blue (BPB) in H_2O (*see* **Note 3**).
2. Sonicator: Bioruptor UCD-200 TM.
3. Standard ladder: Precision Plus Protein Dual Color Standards.
4. 5–20% acrylamide gel: the concentration of acrylamide should be optimized according to the molecular weight of protein you want to detect (*see* **Note 4**).
5. Membrane: Millipore Immobilon-P PVDF membrane.
6. Methanol.
7. Running buffer: 25 mM Tris, 0.2 M glycine, and 0.1% SDS in H_2O.
8. Blotting buffer: 25 mM Tris, 0.2 M glycine, 0.01% SDS, and 20% methanol in H_2O.
9. Sponges, thick papers: fill a clearance between electrode and membrane and gel to stabilize electrophoresis.
10. Blocking solution: Blocking one-P (Nacalai), high molecular weight compounds, and bovine serum albumin (BSA) (*see* **Note 5**).
11. Solution I, Solution II according to manufacturer's instructions (TOYOBO) (*see* **Note 6**).
12. TBS-T: Tris Buffered Saline (TBS) supplemented with 0.5% Tween 20.
13. ECL Prime Western Blotting Detection Reagent: mix same volume of solution A and solution B to a working solution based on manufacturer's instructions.
14. CCD imager: FUJIFILM LAS-4000.
15. NP-40 lysis buffer: 20 mM Tris–HCl, pH 8.0, 150 mM NaCl, 2 mM $MgCl_2$, and 1% NP-40 in H_2O.
16. Protease inhibitor: PhosSTOP Phosphatase Inhibitor Cocktail (Sigma-Aldrich).
17. Phosphatase inhibitor: Complete, Mini in EASYPack Roche.
18. Antibodies: shown in Table 1.

3 Methods

3.1 Cell Sorting

1. Take lymph nodes and place in 60 mm dish with 3 mL of FACS buffer and prepare single cell suspension by gently grinding with frosted slides then filtering through a nylon mesh into a 15 mL tube.

2. Centrifuge for 5 min at 500 × *g* at 4 °C and remove the supernatants.
3. Suspend cells with anti-CD16/32, anti-CD4, anti-CD25, anti-CD44, and anti-CD62L antibodies in 200 μL of FACS buffer and incubate for 20 min at 4 °C in dark.
4. Add 1 mL of FACS buffer and centrifuge for 5 min at 500 × *g* at 4 °C. Remove the supernatants, wash again, and suspend cells with FACS buffer then filtering through a nylon mesh into 5 mL round tube.
5. Sort $CD4^{+}$ (FoxP3-GFP^{-}: optional) CD25–$CD44^{low}CD62L^{high}$ population by FACS sorter into 7 mL of 2R containing in 15 mL tube. Typical staining pattern is shown in Ref. [14].

3.2 Cell Stimulation

1. For Th cell stimulation, coat a 96-well flat plate with 5 μg/mL of anti-CD3ε antibody in 100 μL of PBS for 1 h at room temperature (*see* **Note 7**).
2. Remove antibody solution and then fill the well by 200 μL of 2R until just before plating cells.
3. Add naïve Th cells with 20 ng/mL of IL-4, 5 ng/mL of TGF-β, 100 U/mL of IL-2, and 1 μg/mL of anti-CD28 antibody in complete medium. The cell density should be adjusted to 1×10^6 cells/mL.
4. Remove 2R and plate 200 μL of cell suspensions into each well.
5. Incubate cells in 5% CO_2 incubator at 37 °C.
6. Then IL-4 and TGF-β-dependent IL-9 induction can be detected by the following methods. Stimulation period should be decided for each experiment (*see* **Note 8**).

3.3 RNA Extraction and Quantitative PCR

1. For RNA extraction, harvest cells by centrifugation (5 min at 500 × *g* at 4 °C) and remove media.
2. Add 500 μL of TRIzol Reagent.
3. Lyse cells by pipetting up and down several times and by vortex. Incubate the sample for 5 min at room temperature.
4. Add 100 μL of chloroform and mix by vortex for several minutes, and incubate for several minutes at room temperature. Centrifuge the sample for 10 min at 12,000 × *g* at 4 °C.
5. Remove 250 μL of aqueous phase and place into a new tube, and add 220 μL of 100% isopropanol to the aqueous phase. Incubate at room temperature for 10 min.
6. Centrifuge for 20 min at 15,000 × *g* at 4 °C and remove the supernatants from the tube by decantation (*see* **Note 9**).

7. Add 1 mL of 75% ethanol and vortex the sample briefly. Centrifuge the tube for 10 min at 15,000 × *g* at 4 °C and discard the wash.
8. Repeat **step** 7.
9. Vacuum-dry or air-dry the RNA pellet, and resuspend the RNA pellet in 20 μL of RNase-free water.
10. The RNA is reverse-transcribed into cDNA using a PrimeScript RT reagent kit by incubating the following mixture at 37 °C for 15 min.

    RT-PCR mixture: 2 μL of 5× PrimeScript Buffer (for Real Time), 0.5 μL of PrimeScript RT Enzyme Mix I, 0.5 μL of Oligo dT Primer (50 μM), 0.5 μL of Random 6 mers (100 μM), total RNA solution (up to 500 ng), and RNase-Free water. Total volume should be 10 μL.
11. Inactivate reverse transcriptase with heat treatment (85 °C, 5 s).
12. cDNA solution is prepared as tenfold dilution and used in real-time PCR.
13. Quantitative real-time RT-PCR can be performed using SYBR Green Real-Time PCR Master Mixes and a Lightcycler System. The primers used for PCR are indicated in Table 2.

    Real-time PCR mixture: 5 μL of 2× SYBR Green Master Mix, 0.2 μL of 10 μM forward primer, 0.2 μL of 10 μM reverse primer, 1 μL of cDNA solution, and RNase-free water. Total volume should be 10 μL.

3.4 Flow Cytometry Analysis for the Expression of Cytokines and Transcription Factors

1. For cytokine staining, suspend 1 × 10^6 cells in 1 mL of complete medium with 40 μg/mL PMA, 1 mM ionomycin, and 2.5 μM monensin (add 4 μL of monensin solution per 6 mL of complete medium) and put them into 24-well plate (*see* **Note 10**).
2. Incubate cells for 6 h in 5% CO_2 incubator at 37 °C.
3. After restimulation, collect cells into 1.5 mL tube and centrifuge for 5 min at 500 × *g* at 4 °C.
4. Remove the supernatants and add 1 mL of FACS buffer.
5. Centrifuge for 5 min at 500 × *g* at 4 °C, remove the supernatants, resuspend cells with 90 μL of Fix/Perm Buffer and incubate cells for 30 min on ice in dark (*see* **Note 11**).
6. Add 900 μL of Perm/Wash Buffer, centrifuge for 5 min at 500 × *g* at 4 °C, remove the supernatants, and add 1 mL of Perm/Wash Buffer.
7. Centrifuge for 5 min at 500 × *g* at 4 °C, remove the supernatants and suspend cells with antibodies against intracellular antigens, cytokines, and/or transcription factors in 50 μL of Perm/Wash Buffer.
8. Incubate cells for 30 min on ice in dark.

9. Add 900 µL of Perm/Wash Buffer, centrifuge for 5 min at 500 × *g* at 4 °C, and remove the supernatants.
10. Suspend cells with 1 mL of FACS buffer, centrifuge for 5 min at 500 × *g* at 4 °C, and remove the supernatants. Repeat this step once more.
11. Suspend cells with FACS buffer then filtering cells through 70 µm nylon mesh into 5 mL round tube and run on flow cytometer for analysis.
12. Typical IL-9 production can be observed as shown in Fig. 1c. Also the expression of transcription factors can be detected in similar specimen using this method (Fig. 2a and b).

3.5 Flow Cytometry Analysis for the Phosphorylation of Transcription Factors

1. Incubate cells in appropriate conditions (*see* **Notes 8** and **12**).
2. Warm Lyse/Fix Buffer at 37 °C.
3. After the stimulation period, fix cells immediately with 1 mL of pre-warmed Lyse/Fix Buffer in 1.5 mL tube.
4. Incubate cells for 10 min in 5% CO_2 incubator at 37 °C.

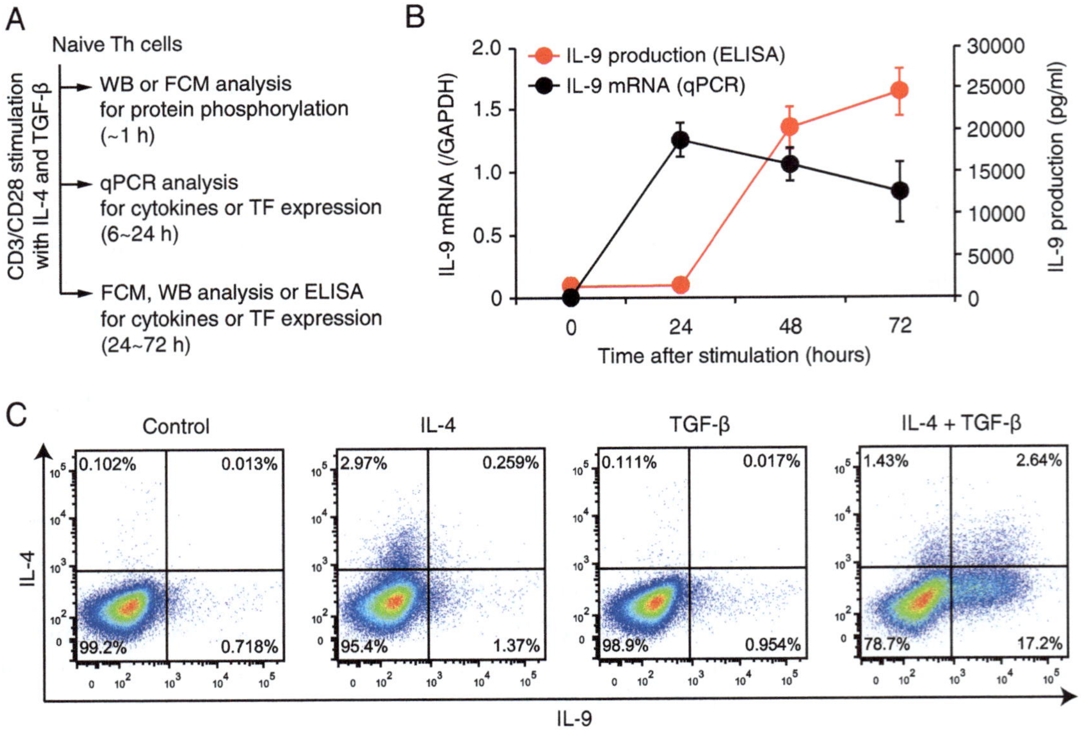

Fig. 1 Effects of IL-4 and TGF-*β* on Th9 differentiation from naïve Th cells. (**a**) Experimental timeline of Th9 differentiation and analysis of downstream signals after TCR, IL-4 and TGF-*β* stimulation. *TF* transcription factors. (**b**) Kinetic changes of *Il9* mRNA and IL-9 production of Th9 cells were confirmed by qPCR and ELISA respectively. (**c**) Flow cytometry analysis of IL-4 and TGF-*β*-dependent IL-9 production

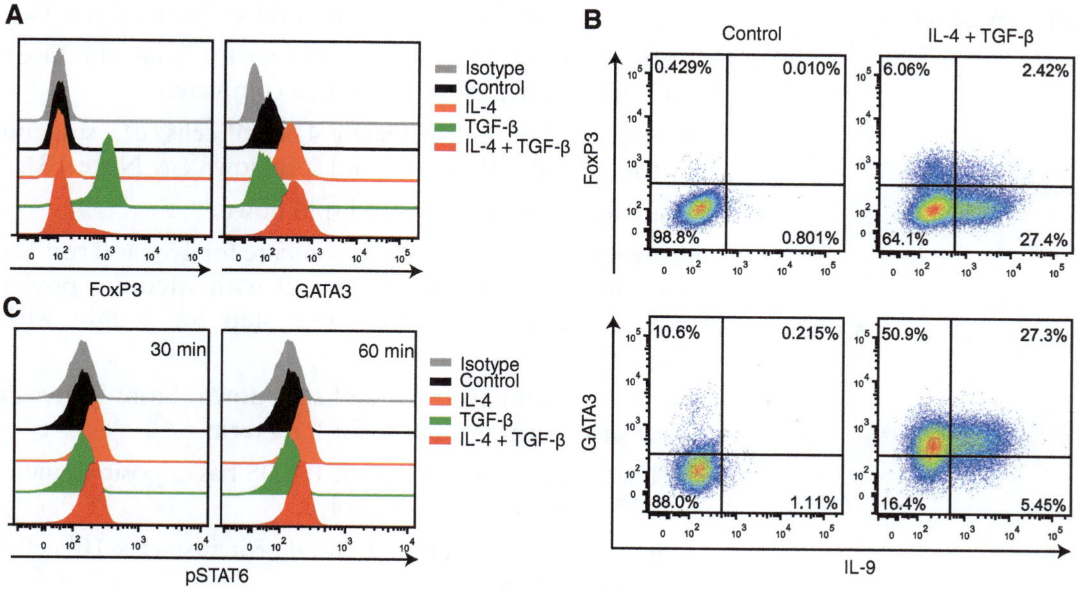

Fig. 2 Flow cytometry analysis of the expression and phosphorylation of transcription factors after IL-4 and TGF-*β* stimulation. (**a**) IL-4 and TGF-*β*-dependent expression of transcription factors was confirmed by flow cytometry analysis. (**b**) Co-staining of IL-9 and transcription factors after 72 h stimulation. (**c**) Phosphorylation of Stat6

5. Centrifuge for 6 min at 600 × *g* at 4 °C and remove the supernatants.
6. Suspend cells with 1 mL of FACS buffer, centrifuge for 6 min at 600 × *g* at 4 °C, and remove the supernatants.
7. Suspend cells with 500 μL of prechilled Permeabilization Buffer (Perm Buffer III) and incubate cells for 30 min on ice in dark.
8. Add 750 μL of FACS buffer, centrifuge for 6 min at 600 × *g* at 4 °C, and remove the supernatants.
9. Suspend cells with 1 mL of FACS buffer, centrifuge for 6 min at 600 × *g* at 4 °C, and remove the supernatants.
10. Suspend cells with antibodies against phosphorylated proteins in 100 μL of FACS buffer. Incubate for 1 h at room temperature in dark.
11. Add 1 mL of FACS buffer and centrifuge for 6 min at 600 × *g* at 4 °C and remove the supernatants.
12. Suspend cells with FACS buffer then filtering cells through 70 μm nylon mesh into 5 mL round tube and run on flow cytometer for analysis.
13. An example of analysis (phosphorylation of STAT6) is shown in Fig. 2c.

3.6 Western Blot

1. Incubate cells in appropriate conditions (*see* **Notes 8** and **12**).
2. Harvest cells into 1.5 mL tube, centrifuge for 5 min at 500 × *g* at 4 °C, and remove the supernatants completely.
3. Add sample buffer approximately 4 × 10^4 cells/μL, suspend pellet completely with vortex, and spin down (*see* **Note 13**).
4. Incubate for 5 min at 95 °C and spin down.
5. Sonicate samples. For example, sonicate for 15 s and rest for 15 s by using Bioruptor UCD-200 with Medium power setting (M, 200 W). Repeat this step for 5 min with cooling.
6. Incubate for 5 min at 95 °C and spin down. Store lysates at 4 °C for short term or −20 °C for long term.
7. Thaw lysates and standard ladder. If SDS has deposited, incubate at 37 °C to completely dissolve.
8. Load gel with samples (30 μg of total proteins: ~1 × 10^5 cells) and standard ladder.
9. Run gel for 60–90 min at 10–20 mA with running buffer.
10. Shake membrane gently with methanol more than 5 s and then blotting buffer for 5 min to activate membrane for preparing blotting.
11. For submarine blotting, prepare sponges and thick papers of appropriate size (larger than gel) and wet them with blotting buffer. Stack sponge, thick paper, gel, membrane, thick paper, and sponge from anode to cathode inside of the blotting apparatus and run at 90 V for 60 min with blotting buffer with cooling by ice.
12. Shake membrane gently with blocking solution for 1 h (*see* **Note 5**).
13. Incubate membrane with primary antibody for transcription factors (shown in Table 1) in Solution I overnight at 4 °C (*see* **Note 6**).
14. Shake membrane gently with TBS-T for more than 1 min. Discard TBS-T and repeat this step three times.
15. Incubate membrane with secondly HRP-conjugated antibody in Solution II for 1 h at room temperature (*see* **Note 6**).
16. Repeat **step 14**.
17. Incubate with ECL Prime Western Blotting Detection Reagent for 5 min at room temperature in dark.
18. Detect chemical luminescence of membrane with CCD imager and IL-4 and TGF-β-dependent signal activation can be observed.

4 Notes

1. In order to sort FoxP3⁻ cells strictly, FoxP3-IRES-GFP knock in reporter mice can be used.
2. The method and reagent of cell fixation should be selected as the proper one for the purpose of experiment. In order to see the phosphorylation of proteins, it is a good idea to use BD Phosflow lLyse/Fix Buffer. When the expression levels of transcription factors are analyzed, BD transcription factor buffer is suitable. Especially studying about Th9 cells, it is better to stain FoxP3 because TGF-β signal induce FoxP3 expression in naïve Th cells [15]. If it is not necessary to stain FoxP3 or other transcription factors, BD Cytofix/Cytoperm Kit can be used for only cytokine staining.
3. In this composition, sample buffer is 4× solution. Please adjust the concentration according to protein stability or localization.
4. Concentration of acrylamide in the gel influences running speed of loaded sample and capability of separation according to the molecular size of protein.

MW of target (kDa)	Recommended gel conc (%).
>100	5
50–130	7.5
30–60	10
15–40	12.5
<20	15

5. TBS-T supplemented with 2% BSA or 5% skim milk can be used as blocking buffer instead of Blocking one-P, but skim milk cannot be used for analysis of phosphorylated proteins.
6. TBS-T or TBS-T supplemented with 2% BSA can be used instead of Solution I and Solution II. It is recommended to use Solution I and Solution II because they are optimized to minimize nonspecific binding of antibodies and maximize signal–noise ratio for clear and consistent data.
7. The protocol of Th cell stimulation can be changed according to the purpose of each experiment. We used other options such as using anti-CD3/CD28 antibodies coated microbeads (Gibco Dynabeads Mouse T-Activator CD3/CD28; *see* Ref. 16). Antigen-specific stimulation can be also applied by using dendritic cells and antigen peptides (*see* Refs. 17, 18).

8. The period of cell stimulation should be proper for each experiment as shown in Fig. 1a and b. The pattern of expression or phosphorylation of each protein is variable, and thus it is recommended to perform kinetic analysis.
9. It is not recommended to use pipettes to remove the supernatants when the pellet cannot be seen.
10. If you are not staining cytokines, then PMA/ionomycin stimulation is not necessary.
11. It is possible to stain extracellular antigens at the same analysis. Before adding Fix/Perm Buffer, incubate cells with antibodies in 50 μL of FACS buffer for 20 min on ice in dark and wash them twice.
12. If the IL-4 or TGF-β-specific signals are weak, FCS-free medium (X-VIVO 15 Serum-free Hematopoietic Cell Medium—Chemically Defined) can be used.
13. Here, we don't use any enzyme inhibitors because the denaturing effect of sample buffer to proteins is so strong that enzyme activity gets lost immediately. If the protein extract is inefficient, extract protocol can be changed. Add 5×10^4 cells/μL NP-40 lysis buffer with protease and phosphatase inhibitor to cells to lose endogenic enzyme activity, incubate on ice for 5 min, sonicate for 5 min with cooling, centrifuge at $15{,}000 \times g$ at 4 °C and collect the supernatants. Protein concentration can be quantified if necessary (for example Lowry method or Bradford method). After these steps, add 1.6×10^5 cells/μL sample buffer. Then advance to **step 6**.

Acknowledgments

I would like to thank Wing JB (Osaka University) for his proofreading. This work is supported by Grant-in-Aid for Young Scientists (B) (15K21125) and Grant-in-Aid for JSPS Fellows (12J01417).

References

1. Zhu J, Paul WE (2008) CD4 T cells: fates, functions, and faults. Blood 112(5):1557–1569. doi:10.1182/blood-2008-05-078154
2. Mikami N, Fukada S, Yamamoto H, Tsujikawa K (2012) Neuronal derivative mediators that regulate cutaneous inflammations. Crit Rev Immunol 32(4):307–320
3. Veldhoen M, Uyttenhove C, van Snick J, Helmby H, Westendorf A, Buer J, Martin B, Wilhelm C, Stockinger B (2008) Transforming growth factor-beta 'reprograms' the differentiation of T helper 2 cells and promotes an interleukin 9-producing subset. Nat Immunol 9(12):1341–1346. doi:10.1038/ni.1659
4. Dardalhon V, Awasthi A, Kwon H, Galileos G, Gao W, Sobel RA, Mitsdoerffer M, Strom TB, Elyaman W, Ho IC, Khoury S, Oukka M, Kuchroo VK (2008) IL-4 inhibits TGF-beta-induced Foxp3+ T cells and, together with TGF-beta, generates IL-9+ IL-10+ Foxp3(−) effector T cells. Nat Immunol 9(12):1347–1355. doi:10.1038/ni.1677

5. Townsend JM, Fallon GP, Matthews JD, Smith P, Jolin EH, McKenzie NA (2000) IL-9-deficient mice establish fundamental roles for IL-9 in pulmonary mastocytosis and goblet cell hyperplasia but not T cell development. Immunity 13(4):573–583
6. Temann UA, Ray P, Flavell RA (2002) Pulmonary overexpression of IL-9 induces Th2 cytokine expression, leading to immune pathology. J Clin Invest 109(1):29–39. doi:10.1172/JCI13696
7. Corren J (2011) Cytokine inhibition in severe asthma: current knowledge and future directions. Curr Opin Pulm Med 17(1):29–33. doi:10.1097/MCP.0b013e3283413105
8. Chang HC, Sehra S, Goswami R, Yao W, Yu Q, Stritesky GL, Jabeen R, McKinley C, Ahyi AN, Han L, Nguyen ET, Robertson MJ, Perumal NB, Tepper RS, Nutt SL, Kaplan MH (2010) The transcription factor PU.1 is required for the development of IL-9-producing T cells and allergic inflammation. Nat Immunol 11(6):527–534. doi:10.1038/ni.1867
9. Angkasekwinai P, Chang SH, Thapa M, Watarai H, Dong C (2010) Regulation of IL-9 expression by IL-25 signaling. Nat Immunol 11(3):250–256. doi:10.1038/ni.1846
10. Jash A, Sahoo A, Kim GC, Chae CS, Hwang JS, Kim JE, Im SH (2012) Nuclear factor of activated T cells 1 (NFAT1)-induced permissive chromatin modification facilitates nuclear factor-kappaB (NF-kappaB)-mediated interleukin-9 (IL-9) transactivation. J Biol Chem 287(19):15445–15457. doi:10.1074/jbc.M112.340356
11. Staudt V, Bothur E, Klein M, Lingnau K, Reuter S, Grebe N, Gerlitzki B, Hoffmann M, Ulges A, Taube C, Dehzad N, Becker M, Stassen M, Steinborn A, Lohoff M, Schild H, Schmitt E, Bopp T (2010) Interferon-regulatory factor 4 is essential for the developmental program of T helper 9 cells. Immunity 33(2):192–202. doi:10.1016/j.immuni.2010.07.014
12. Jabeen R, Goswami R, Awe O, Kulkarni A, Nguyen ET, Attenasio A, Walsh D, Olson MR, Kim MH, Tepper RS, Sun J, Kim CH, Taparowsky EJ, Zhou B, Kaplan MH (2013) Th9 cell development requires a BATF-regulated transcriptional network. J Clin Invest 123(11):4641–4653. doi:10.1172/JCI69489
13. Mikami N, Miyagi Y, Sueda K, Takatsuji M, Fukada S, Yamamoto H, Tsujikawa K (2013) Calcitonin gene-related peptide and cyclic adenosine 5′-monophosphate/protein kinase A pathway promote IL-9 production in Th9 differentiation process. J Immunol 190(8):4046–4055. doi:10.4049/jimmunol.1203102
14. Mikami N, Watanabe K, Hashimoto N, Miyagi Y, Sueda K, Fukada S, Yamamoto H, Tsujikawa K (2012) Calcitonin gene-related peptide enhances experimental autoimmune encephalomyelitis by promoting Th17-cell functions. Int Immunol 24(11):681–691. doi:10.1093/intimm/dxs075
15. Chen W, Jin W, Hardegen N, Lei KJ, Li L, Marinos N, McGrady G, Wahl SM (2003) Conversion of peripheral CD4+CD25− naive T cells to CD4+CD25+ regulatory T cells by TGF-beta induction of transcription factor Foxp3. J Exp Med 198(12):1875–1886. doi:10.1084/jem.20030152
16. Yoshida Y, Mikami N, Matsushima Y et al (2016) Combination treatment with fingolimod and a pathogenic antigen prevents relapse of glucose-6-phosphate isomerase peptide-induced arthritis. Immun Inflamm Dis 4(3): 263–273
17. Mikami N, Sueda K, Ogitani Y, Otani I, Takatsuji M, Wada Y, Watanabe K, Yoshikawa R, Nishioka S, Hashimoto N, Miyagi Y, Fukada S, Yamamoto H, Tsujikawa K (2014) Calcitonin gene-related peptide regulates type IV hypersensitivity through dendritic cell functions. PLoS One 9(1):e86367. doi:10.1371/journal.pone.0086367
18. Yoshida Y, Mikami N, Tsuji T et al (2015) Mechanism of induction of immune tolerance in experimental autoimmune encephalomyelitis by combination treatment with fingolimod plus pathogenic autoantigen. Clin Exp Neuroimmunol 6:49–56. doi:10.1111/cen3.12140

Chapter 12

Retroviral Transduction and Reporter Assay: Transcription Factor Cooperation in Th9 Cell Development

Rukhsana Jabeen

Abstract

Naïve CD4+ T cells differentiate into different T helper subsets in response to specific cytokine environment and transcription factors. Th9 cells are induced in response to signals from cytokines, TGF-β and IL-4. Transcription factors that are downstream of these cytokines converge to drive the development of Th9 cells. Retroviral transduction allows the genetic modification in T cells thereby helping us to better understand the molecular mechanisms that control their development as well as function. In this chapter, an optimized protocol for retroviral transduction of murine Th9 cells as well as transient transfection of Th9 cells with luciferase reporter constructs is described.

Key words Retroviral gene transduction, Transcription factors, IRF4, BATF, Th9

1 Introduction

Differentiation of naïve CD4+ T cells into distinct T helper subsets allows the adaptive immune system to generate specific immune response to a variety of pathogens. These distinct T helper subsets are characterized by the expression of signature cytokines as well as the lineage-specific transcription factors [1]. Traditionally, differentiation of each T helper subset was believed to be dependent on a single transcription factor termed as "master regulator." However with subsequent studies as well as technological advances it became clear that interaction of multiple co-expressed lineage specific transcription factors, at both transcriptional and post translational levels drives the differentiation of CD4+ T helper cells towards a particular subset while suppressing alternative fates [2].

Among the T helper subsets recognized thus far, Th9 cells are one of the newer subsets. These cells are specialized for the secretion of cytokine IL-9 and are generated in the presence of balanced signals from cytokines IL-4 and TGF-β [3, 4]. Both IL-4 and TGF-β promote Th9 cell differentiation by inducing transcription factors that regulate IL-9 production, as well as expression of other

Ritobrata Goswami (ed.), *Th9 Cells: Methods and Protocols*, Methods in Molecular Biology, vol. 1585,
DOI 10.1007/978-1-4939-6877-0_12, © Springer Science+Business Media LLC 2017

multiple genes that are associated with the Th9 lineage. TGF-β induces PU.1, an ETS family transcription factor that plays a key role in the Th9 developmental program [5]. CD4+ T cells that are deficient in PU.1 have a diminished capacity to produce IL-9, while ectopic expression of PU.1 in either Th9 cells or Th2 cells enhances IL-9 production.

IL-4-activated transcription factors including STAT6 and IRF4 that are crucial for Th2 differentiation, also play essential role in the induction of Th9 cells [6]. STAT6 promotes Th9 development by suppressing the TGF-β-induced Foxp3 expression as well as the expression of T-bet, both of which would otherwise suppress IL-9 production [7]. STAT6 induces IRF4, another transcription factor that is required for the Th9 development [8]. IRF4 directly binds to and transactivates the *Il9* promoter and additionally contributes to Th9 cell development by blocking the expression of Th1-associated transcription factors. In addition to IRF4, Th9 cells also require BATF, an AP-1 family member that functions by dimerizing with Jun family members [9]. STAT6 is required for BATF expression and like IRF4; it binds to the *Il9* promoter and activates *Il9* transcription in reporter assays. Together these transcription factors that respond to stimuli from cytokines TGF-β and IL-4, form a specific transcriptional network that promotes and regulates Th9 development and function [10]. Investigating the regulation of different transcription factors during Th9 cell differentiation is critical for understanding of Th9 biology.

Genetic modification of T helper cells has become an important tool for both basic research as well as immunotherapy [11, 12]. Retroviral transduction is one of the simplest and efficient ways to introduce genetic elements into the murine T helper cells. It allows us to modify expression of key transcription factors in differentiating cells, thereby facilitating better understanding of multiple layers of gene regulation. For transducing murine Th9 cells in vitro efficiently and maintaining cell viability after the gene transfer, an optimized protocol is described herein.

Another excellent tool to study gene expression and cellular events coupled to gene expression is reporter assay. Briefly, a reporter gene is cloned with a DNA sequence of interest into an expression vector, which is then transfected into cells (cell lines or primary cells depending upon the experimental study). Following transfection, cells are then assayed for the presence of the reporter by directly measuring the reporter protein itself or the enzymatic activity of the reporter protein.

2 Materials

Prepare all the reagents under sterile conditions inside a biosafety cabinet.

2.1 Transfection of Plat-E Cells

1. Plat-E Retrovirus Packaging Cell Line.
2. Phosphate buffered saline (PBS): Dissolve 8 g NaCl, 0.2 g KCl, 1.44 g Na_2HPO_4 and 0.24 g KH_2PO_4 in 800 mL of distilled water. Adjust pH to 7.4 with HCl and adjust final volume 1 L with distilled water.
3. Dulbecco's Modified Eagle's Medium (DMEM): 20% fetal calf serum (FCS, v/v), 2 mM L-glutamine, and 100 U/mL penicillin–streptomycin.
4. 2 M $CaCl_2$ (filtered).
5. 0.1% gelatin solution.
6. 10 cm tissue culture dish.
7. 0.25% trypsin–EDTA.
8. 15 and 50 mL conical tubes.
9. Plasmid DNA.
10. Sterile, double distilled endotoxin-free water.
11. 0.45 μm syringe filters.
12. 2× HBS: 50 mM HEPES, 12 mM dextrose, 280 mM NaCl, 10 mM KCl, 1.5 mM Na_2HPO_4, adjust pH to 7.05, filter through a 0.2 μm filter, aliquot and store at −20 °C.

2.2 Transduction

1. 12- or 24-well flat-bottom tissue culture plates.
2. Polybrene (8 mg/mL).
3. Saran wrap.
4. Centrifuge.

2.3 Isolation of Naïve CD4+ T Cells

1. C57BL/6 or BALB/c mice.
2. Sterile scissors and forceps.
3. Naïve CD4+ T cell isolation kit.
4. MACS separators and MACS columns.
5. MACS buffer: To 976 mL of PBS, add 4 mL of 0.5 M EDTA and 20 mL of 2% FBS, store at 4 °C and keep sterile.

2.4 In Vitro Cell Culture

1. Complete RPMI 1640: RPMI 1640 (500 mL) with 10% v/v FBS (heat-inactivated and filtered), sodium pyruvate (10 mL), penicillin–streptomycin (5 mL), L-glutamine (2.5 mL), nonessential amino acids (2.5 mL), HEPES (2.5 mL), and 2-mercaptoethanol (2 μL). Mix well and filter through 0.45 μm membrane filter flasks.
2. Antibodies: Anti-mouse CD3, anti-mouse CD28, and anti-mouse IFN-γ.
3. Cytokines: Recombinant murine IL-4 and recombinant human IL-2, TGF-β.

4. PMA (phorbol 12-myristate 13-acetate) and Ionomycin.
5. 2 mM Monensin.
6. FACS buffer: 2% BSA and 0.05% sodium azide (NaN_3) in 1× DPBS.
7. Paraformaldehyde (4%): Add 4 g of paraformaldehyde to 50 mL of water in a conical flask, add 1 mL of NaOH, cover with Parafilm and transfer to a fume hood. Stir gently on a heating block at 60 °C until paraformaldehyde is dissolved. Add 10 mL of 10× PBS and allow the mixture to cool to room temperature. Adjust the pH to 7.4 with 1 M HCl and then adjust the final volume to 100 mL with water. Filter the solution through 0.45 μm membrane filter. Paraformaldehyde solution can be made fresh every time prior to use or aliquots could be stored at −20 °C for several months. Avoid repeated freezing–thawing.
8. Permeabilization buffer (Perm buffer): 1× DPBS, 0.1% saponin, 2% BSA and 0.01% sodium azide. Mix well, filter, and store at 4 °C.
9. Fluorochrome-labeled antibodies for cytokine staining.

2.5 Reporter Assay

1. Primary CD4+ T cells from BALB/c or C57BL/6 mice.
2. Plasmids of interest and luciferase reporter.
3. Amaxa Nucleofector device.
4. Amaxa T cell Nucleofector Kit (Lonza).
5. Dual-Luciferase Reporter Assay System.
6. Passive Lysis 5X buffer (Promega).
7. Luminometer.

3 Methods

3.1 Plat-E Cell Culture

Platinum-E or plat-E is a retrovirus packaging cell line that exhibits longer stability and produces higher yield of retrovirus than existing cell lines [13]. For efficient transduction of Th9 cells, utilizing this particular cell line is highly recommended.

1. Thaw frozen cells at room temperature and transfer to a 15 mL tube, add 10 mL of DMEM culture medium.
2. Centrifuge for 5 min at 500 × *g* at 4 °C.
3. Discard the supernatant, resuspend the pellet very gently by finger tapping and wash the cells one more time by adding 10 mL of DMEM.
4. Repeat **step 2** once more.

5. Discard the supernatant and resuspend the cells carefully in 10 mL of DMEM culture medium.
6. Transfer the cells to a 10 cm culture dish coated with 0.1% gelatin and incubate at 37 °C (*see* **Note 1**).
7. Split the cells once they are 80–90% confluent.

3.2 Splitting Plat-E Cells

1. Remove the old media from the plate.
2. Wash cells twice by adding 10 mL of 1× PBS.
3. Add 0.5 mL of 0.25% trypsin–EDTA solution to a 10 cm dish.
4. Incubate at 37 °C for 2–3 min.
5. Add 9.5 mL of DMEM medium and remove the cells from the dish surface by gently pipetting up and down few times.
6. Transfer the cells to a 50 mL tube and centrifuge for 5 min at 500 × *g* at 4 °C.
7. Discard the supernatant and dislodge the cell pellet by finger tapping.
8. Add 50 mL of DMEM medium and transfer the cells to 10 cm tissue culture dishes.
9. Keep the cells at 37 °C and when the cells are 70–80% confluent; they are ready for transfection.

3.3 Transfection of Plat-E Cells and Retrovirus Production

Described below is the calcium phosphate transfection method. It is a fairly simple method that does not require any specialized equipment and is very economical. However, this method results in relatively lower transfected cells (50–70% transfection efficiency). For experiments that require highly transfected cells (transfection efficiency 80–90%) and high levels of transgene expression, Lipofectamine may be used (*see* **Note 2**).

3.3.1 Calcium Phosphate Transfection

1. Prepare the Plat-E cells for transfection by removing the medium from the culture dish very carefully. Add 10 mL of fresh DMEM medium (10 mL medium + 10 μL of chloroquine), making sure not to disturb the adherent Plat-E cells (*see* **Note 3**).
2. Incubate at 37 °C for 2–3 h prior to transfection.
3. Prepare $CaCl_2$–DNA precipitates in 15 mL conical tubes. Transfection mix for 10 cm plate requires:
   (a) 36 μL of 2 M $CaCl_2$.
   (b) 15 μg of plasmid DNA (gene of interest).
   (c) Add sterile water to make final volume 300 μL.
4. Add 300 μL of 2× HBS buffer to the above mixture while swirling for 30 s (*see* **Note 4**).

5. Incubate above mentioned transfection mix at room temperature for 30 min.
6. Add 600 µL of transfection mixture dropwise over the cells.
7. Rock the plate back and forth gently and incubate the plate at 37 °C.
8. Replace the old media following day with 10 mL of pre-warmed DMEM media or complete RPMI (*see* **Note 5**).
9. Collect the retroviral supernatant 24 h later and filter the supernatant through 0.45 µM filters (*see* **Note 6**).
10. Slowly add 10 mL of fresh DMEM media to the transfected Plat-E cells.
11. 24 h later collect supernatant again and repeat **step 9**.
12. Transfection efficiency can be checked if desired by adding 5 mL of PBS to the plates and collecting Plat-E cells vigorously by pipetting PBS up and down and then looking at the cells by flow cytometry.

3.4 In Vitro Generation of Murine Th9 Cells

1. Isolate Naïve $CD4^+CD62L^+$ T cells from spleens and lymph nodes of mice by magnetic isolation following the manufacturer's instructions.
2. Coat a 12-well plate with 2 µg/mL of anti-CD3 diluted in PBS and incubate it for 2 h at 37 °C or overnight at 4 °C.
3. Remove anti-CD3 from the plate and add 500 µL of PBS to wash uncoated anti-CD3.
4. Seed 1×10^6 naïve $CD4^+$ T cells per mL complete RPMI to each well and culture under Th9 conditions:
   (a) Anti-CD28 (1 µg/mL).
   (b) IL-4 (20 ng/mL).
   (c) TGF-β (2 ng/mL).
   (d) Anti-IFN-γ (10 µg/mL).
   (e) Human IL-2 (50 U/mL).
5. Swirl the plate back and forth gently to mix well the cytokines. The day of seeding Th9 cells is designated as day 0.
6. Incubate the plate at 37 °C for 48 h.

3.4.1 Th9 Cell Transduction

Retroviral transduction is carried at 48 h after T cell stimulation, for a single gene transduction; add 2–3 mL of retroviral supernatant to the cells cultured under Th9 conditions. If double transduction is desired then add 2–3 mL of each retroviral supernatant.

1. On day 2, prior to the addition of retroviral supernatant, remove 500 µL of media from each well and keep at 4 °C.

2. Simultaneously take out retroviral supernatant (or thaw, if frozen) and pre warm in 37 °C water bath.
3. Add 8 µg of polybrene per 1 mL of pre-warmed retroviral supernatant. Next add the retroviral supernatant, treated with polybrene to cultured Th9 cells.
4. Wrap the plate with a saran wrap inside the hood and then spin the plate at 750 × *g* for 60–90 min with temperature set at 25 °C.
5. Aspirate the media containing retrovirus and add back 500 µL of media kept at 4 °C in **step 1**.
6. Add 50 U/mL of human IL-2 to the transduced cells and incubate the cells at 37 °C overnight.

3.4.2 Expansion of Transduced Th9 Cells

1. Following day, i.e., on day 3, expand the cells. Transfer the cells from 12-well plate to a 6-well plate by pipetting up and down. Add 3 mL of fresh complete RPMI to the cells. Also add cytokines, IL-4 (10 ng/mL), TGF-β (1 ng/mL), and IL-2 (50 U/mL) to the Th9 cultures at this point.
2. Incubate the cells at 37 °C for 2 more days.
3. On day 5, collect the cells in 15 or 50 mL conical tubes and spin down the cells at 500 × *g* for 5 min.
4. Count the cells and resuspend them at 1 × 10^6 cells/mL.
5. Depending upon the aim of the experiment, phenotype of transduced Th9 cells can be examined by flow cytometry at this point by restimulating the cells or transduced cells can be flow sorted for adoptive transfer based on the positive expression of reporter marker in the retroviral plasmid used, such as GFP or Thy1.1 or hCD4 (Fig. 1).

3.4.3 Intracellular Cytokine Staining

The following protocol describes intracellular staining of Th9 cells in 5 mL FACS tubes. However with adjustments in volume, the same protocol can be used for staining in a 96-well plate format.

1. Re-stimulate the transduced Th9 cells with PMA (50 ng/mL) and ionomycin (500 ng/mL) for 5 h and add monensin to cells for the final 2 h of stimulation (*see* **Note** 7).
2. Collect the cells, transfer to 5 mL FACS tubes and wash by adding 2 mL of FACS buffer, centrifuge at 500 × *g* for 5 min at 4 °C and then decant the FACS buffer from pelleted cells. Cells can be stained for surface markers and viability at this stage; however, if surface staining is not required then proceed to fixing the cells.
3. Resuspend cells in 100 µL of FACS buffer and fix the cells by adding 200 µL of 4 % paraformaldehyde. Vortex the cells and incubate for 10 min at room temperature in dark.

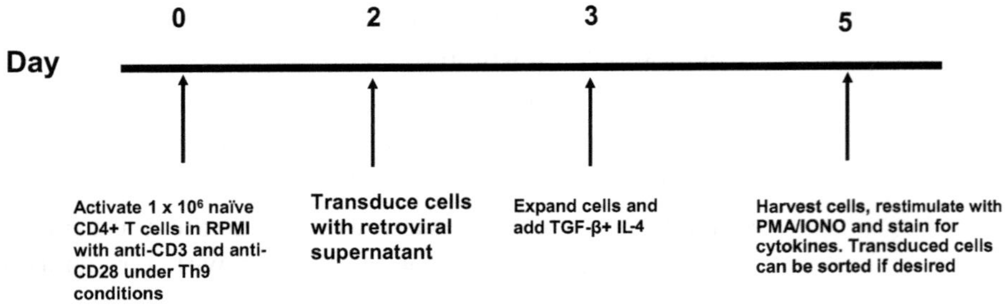

Fig. 1 Schematic representation of retroviral transduction of naïve CD4+ T cells primed under Th9 conditions

4. Wash the cells by adding 2 mL of FACS buffer, centrifuge at 500 × *g* for 5 min at 4 °C and decant the buffer. Cells can be stored at this stage in FACS buffer at 4 °C for 18–24 h if desired.
5. Following fixation, re-suspend the cells in 2 mL of Perm buffer. Mix well and incubate at room temperature for 10–15 min in dark (*see* **Note 8**).
6. Centrifuge the cells at 500 × *g* for 5 min at 4 °C and discard the supernatant.
7. Re-suspend the cells in 100 μL of Perm buffer. Cells can be stained for IL-9 or other secreted cytokines; this will depend on the specific goal of the experiment. For detection of IL-9, we recommend adding 0.5–1 μL of anti-IL9 antibody per million cells.
8. Incubate in dark at room temperature for 25–30 min and wash the cells by adding 2 mL of FACS buffer.
9. Centrifuge at 500 × *g* for 5 min at 4 °C and discard the supernatant.
10. Resuspend the stained cells in about 250–300 μL of FACS buffer. Cells can be acquired on flow cytometer immediately or in next 24–48 h (Fig. 2).

3.5 Reporter Assay

To identify the transcriptional factors or regulatory elements that upregulate *Il9* promoter, researchers have taken advantage of HEK293T cells as they can be transfected easily. On the other hand, primary T helper cells are difficult to transfect; however, Amaxa nucleofection technology has allowed efficient and rapid gene transfer into resting as well as activated murine and human T cells.

3.5.1 Th9 Cell Culture

1. Isolate and culture the naïve CD4+ T cells under Th9 culture conditions as described in Subheading 3.4.
2. On day 3, expand cells by splitting 1:4 in complete RPMI medium containing IL-4 (10 ng/mL), TGF-β (1 ng/mL), and IL-2 (50 U/mL) and transfer to 6-well plate. Incubate the cells at 37 °C for another 48 h.

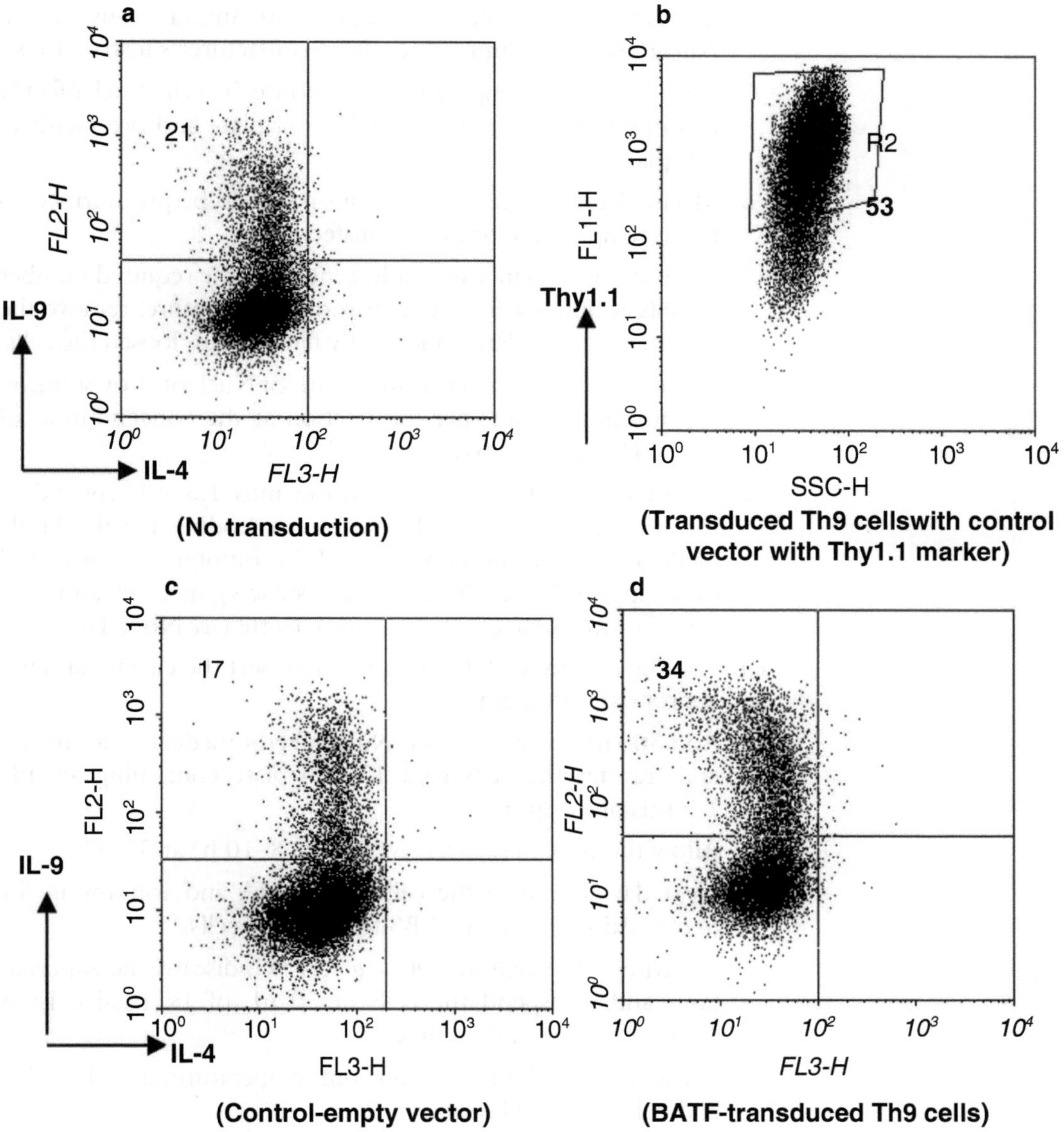

Fig. 2 (**a**, **b**) Representative FACS Analysis of Th9 cells stained with anti-mouse IL-9 and anti-mouse IL-4 antibodies. Differentiating Th9 cells were transduced on day 2 with either control empty vector expressing Thy1.1 or BATF expressing retrovirus. (**c**, **d**) Cells were restimulated on day 5 with PMA/ionomycin for 5 h before intracellular staining for cytokines IL-9 and IL-4. *Dot plots* are gated on cells that were positive for Thy1.1. Around 50% cells were positive for Thy1.1 (**b**)

3. Collect cells at day 5 by gently pipetting cells up and down and transfer the cells to 15 or 50 mL tubes. Wash the cells once in 10 mL of complete RPMI medium, count the cells using trypan blue and resuspend the cells in complete RPMI media at 1×10^6 cells per mL, without any cytokines.

4. Prepare the Nuclefection solution and Amaxa Mouse T cell Nucleofector medium as per the manufacturer's instructions.
5. Add 1.5 mL of supplemented Mouse T cell Nucleofector medium per transfection to a 12-well plate and pre-incubate this plate at 37 °C.
6. Mouse T cell Nucleofector solution should be pre-warmed to room temperature prior to transfection.
7. Once all the reagents are ready, centrifuge the required number of cells at 500 × *g* for 5 min at room temperature, remove the supernatant completely and gently flip pellet to loosen the cells.
8. Re-suspend the cells gently in 100 μL of Nucleofector Solution (room temperature) per transfection at the concentration of 2–4 × 10^6 cells (*see* **Note 9**).
9. Transfer 100 μL of cell suspension into 1.5 mL eppendorf tubes containing 20 μg DNA (Vector coding for the firefly luciferase under the control of *Il9* promoter, TK-pRL and other appropriate desired DNA constructs), mix well and then transfer into Amaxa Nucleofector cuvette (*see* **Note 10**).
10. Cap the cuvette with the sample and insert the cuvette and use appropriate program.
11. Add 500 μL of pre-warmed Nucleofector media to the cuvette and transfer the cells to a 12-well plate containing already 1.5 mL of medium.
12. Allow the cells to recover overnight (8–10 h) at 37 °C.
13. Next day, stimulate the cells with PMA and Ionomycin for 5–6 h. Add 5 mL of 1× PBS to wash the cells.
14. Centrifuge the cells at 500 × *g* at 4 °C, discard the supernatant, and resuspend the cells in 50 μL of 1× Passive Lysis Buffer in an eppendorf tube.
15. Shake the tube for 15 min at room temperature, spin the tube, and collect the cell lysate.
16. Perform the luciferase assay using the Dual-Luciferase kit from Promega following the manufacturer's instructions.

4 Notes

1. Coating plates with 0.1% gelatin enhances the adherence of Plate-E cell to the tissue culture plates.
2. Lipofectamine 2000 from Invitrogen can be used for transfecting Th9 cells if highly transient transfections are desired. The protocol is similar to calcium phosphate transfection protocol, and can be easily performed following manufacturer's instructions. The only difference between the two methods is that

Plate-E cells should be 90% confluent prior to transfection and only 1 mL of retroviral supernatant generated by this method is sufficient to transduce one million naïve CD4+ T cells primed under Th9 conditions. However, it can be expensive for large-scale applications.

3. The addition of chloroquine appears to increase retroviral titers, however duration of chloroquine treatment should not exceed more than 12 h as longer incubations with chloroquine can have toxic effects on cells resulting in cell death.
4. Do not reuse the buffer once thawed as it might cause changes in the pH of the buffer.
5. Retroviral supernatant in complete RPMI significantly increased the viability of transduced Th9 cells as opposed to retroviral supernatant generated in DMEM media.
6. Viral supernatant can be used immediately or frozen at −80 °C for future use, once thawed left over supernatant should be discarded.
7. Monensin is a protein transport inhibitor that leads to the accumulation of most cytokines at the Golgi complex/endoplasmic reticulum, thereby enhancing the intracellular cytokine staining.
8. Addition of permeabilization buffer is an essential step in intracellular staining as it creates pores in the cell membrane allowing antibodies to have access to the cytoplasm of the cell.
9. Do not store cells in this solution longer than 15 min as this might reduce cell viability as well as transfection efficiency.
10. Make sure that sample covers the bottom of the cuvette and avoid bubbles.

References

1. Zhu J, Yamane H, Paul WE (2010) Differentiation of effector CD4 T cell populations (*). Annu Rev Immunol 28:445–489. doi:10.1146/annurev-immunol-030409-101212
2. Zhou L, Chong MM, Littman DR (2009) Plasticity of CD4+ T cell lineage differentiation. Immunity 30(5):646–655. doi:10.1016/j.immuni.2009.05.001
3. Veldhoen M, Uyttenhove C, van Snick J, Helmby H, Westendorf A, Buer J, Martin B, Wilhelm C, Stockinger B (2008) Transforming growth factor-beta 'reprograms' the differentiation of T helper 2 cells and promotes an interleukin 9-producing subset. Nat Immunol 9(12):1341–1346. doi:10.1038/ni.1659
4. Dardalhon V, Awasthi A, Kwon H, Galileos G, Gao W, Sobel RA, Mitsdoerffer M, Strom TB, Elyaman W, Ho IC, Khoury S, Oukka M, Kuchroo VK (2008) IL-4 inhibits TGF-beta-induced Foxp3+ T cells and, together with TGF-beta, generates IL-9+ IL-10+ Foxp3(−) effector T cells. Nat Immunol 9(12):1347–1355. doi:10.1038/ni.1677
5. Chang HC, Sehra S, Goswami R, Yao W, Yu Q, Stritesky GL, Jabeen R, McKinley C, Ahyi AN, Han L, Nguyen ET, Robertson MJ, Perumal NB, Tepper RS, Nutt SL, Kaplan MH (2010) The transcription factor PU.1 is required for the development of IL-9-producing T cells and allergic inflammation. Nat Immunol 11(6):527–534. doi:10.1038/ni.1867

6. Jabeen R, Kaplan MH (2012) The symphony of the ninth: the development and function of Th9 cells. Curr Opin Immunol 24(3):303–307. doi:10.1016/j.coi.2012.02.001
7. Goswami R, Jabeen R, Yagi R, Pham D, Zhu J, Goenka S, Kaplan MH (2012) STAT6-dependent regulation of Th9 development. J Immunol 188(3):968–975. doi:10.4049/jimmunol.1102840
8. Staudt V, Bothur E, Klein M, Lingnau K, Reuter S, Grebe N, Gerlitzki B, Hoffmann M, Ulges A, Taube C, Dehzad N, Becker M, Stassen M, Steinborn A, Lohoff M, Schild H, Schmitt E, Bopp T (2010) Interferon-regulatory factor 4 is essential for the developmental program of T helper 9 cells. Immunity 33(2):192–202. doi:10.1016/j.immuni.2010.07.014
9. Jabeen R, Goswami R, Awe O, Kulkarni A, Nguyen ET, Attenasio A, Walsh D, Olson MR, Kim MH, Tepper RS, Sun J, Kim CH, Taparowsky EJ, Zhou B, Kaplan MH (2013) Th9 cell development requires a BATF-regulated transcriptional network. J Clin Invest 123(11):4641–4653. doi:10.1172/JCI69489
10. Kaplan MH, Hufford MM, Olson MR (2015) The development and in vivo function of T helper 9 cells. Nat Rev Immunol 15(5):295–307. doi:10.1038/nri3824
11. Lois C, Refaeli Y, Qin XF, Van Parijs L (2001) Retroviruses as tools to study the immune system. Curr Opin Immunol 13(4):496–504
12. Riviere I, Gallardo HF, Hagani AB, Sadelain M (2000) Retroviral-mediated gene transfer in primary murine and human T-lymphocytes. Mol Biotechnol 15(2):133–142. doi:10.1385/MB:15:2:133
13. Morita S, Kojima T, Kitamura T (2000) Plat-E: an efficient and stable system for transient packaging of retroviruses. Gene Ther 7(12):1063–1066. doi:10.1038/sj.gt.3301206

Chapter 13

Transcription Factor Binding Studies in CD4+ T Cells: siRNA Transfection, Chromatin Immunoprecipitation, and Liquid Luminescent DNA Precipitation Assay

Etienne Humblin, François Ghiringhelli, and Frédérique Végran

Abstract

CD4+ T cells are key components of the immune system that shape the anticancer immune response in animal models and in humans among other responses. The biology of CD4+ T cells is complex because naïve CD4+ T cells can differentiate into various subpopulations with various functions that depend on the milieu. Recently, a new population of IL-9-secreting T cells called Th9 cells has been described. These cells are characterized by their ability to produce IL-9. These cells were described to be involved in parasite infections and allergic inflammation. However, some reports described their presence in the tumor bed in mice and humans. Molecular events that account for Th9 effector properties and differentiation are still elusive. To study these mechanisms, downregulation of gene expression and protein–DNA interaction detection allow to identify and study putative transcription factors regulating gene expression during their differentiation. Here, we describe three methods allowing the study of transcription factors in Th9 cells: siRNA transfection for downregulation of gene expression, and chromatin immunoprecipitation and liquid luminescent DNA precipitation assay to detect protein–DNA interactions.

Key words Transcription factor, siRNA, ChIP, DNA interaction, CD4+ T cells

1 Introduction

CD4+ T cells are key components of the immune system that shape the anticancer immune response in animal models as well as in humans [1]. The biology of CD4+ T cells is complex because naive T cells can differentiate into various subpopulations with various functions. Th9 cells were described in 2008 [2, 3]. These cells are characterized by their capacity to produce copious amount of IL-9 and some amount of IL-10 and IL-21. They were first described in the context of parasite infections and allergic process; however, some reports described their presence in the tumor bed in mice and humans [4–8]. While the transcriptional program of Th9 cells is still incompletely described, a slew of transcription factors were shown to be essential for Th9 polarization including STAT6,

Ritobrata Goswami (ed.), *Th9 Cells: Methods and Protocols*, Methods in Molecular Biology, vol. 1585,
DOI 10.1007/978-1-4939-6877-0_13,

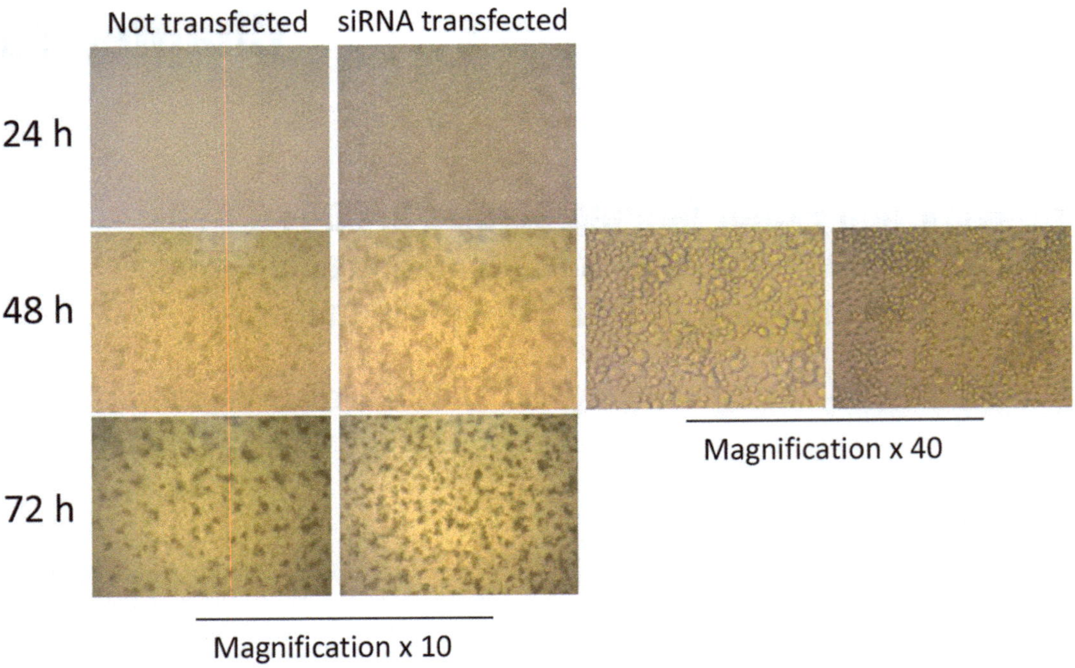

Fig. 1 siRNA transfection effect on CD4+ T cells. Naïve CD4+ T cells were isolated according to mouse $CD4^+CD62L^+$ T Cell Isolation Kit, mouse and transfected (or not) with siRNA targeting IRF1 using TransIT-TKO Transfection reagent (MiRus). Cells were pictured at 24, 48, and 72 h after transfection

GATA3, PU.1, and IRF4 [9]. In this chapter we describe three methods allowing exploring putative transcription factor functions during Th9 differentiation. In this chapter, we use the example of IRF1 that we have previously described to be induced when Th9 cells are differentiated in presence of IL-1β. In the first step we propose siRNA strategy to modulate its expression. After testing several transfection methods, the best option we have found is the use of *Trans*IT-TKO® Transfection Reagent from Mirus. This reagent allows good transfection efficiency without massive cell death (Fig. 1).

Furthermore, we propose two methods to analyze its interaction with DNA: chromatin immunoprecipitation (ChIP) and liquid luminescent DNA precipitation assay. These experiments are classically used to determine specific interactions of proteins and nucleic acids. For ChIP experiments, we have combined the very reproducible fragmentation method from Covaris TruChIP Chromatin Shearing Reagent Kit and the ChIP-IT Express Chromatin Immunoprecipitation Kit from Active Motif. Finally, instead of EMSA which is a lengthy process, rather heavy-handed because of radioisotopically labeled DNA and autoradiographic visualization that are required for the experiments, we used liquid luminescent DNA precipitation assay that provides rapid, reliable, and quantitative results to identify protein–DNA interactions.

This protein–DNA binding assay has been adapted from Iwasaki et al. protocol [10]. It is based on solution hybridization between digoxigenin-labeled (DIG) DNA and glutathione S-transferase (GST)-fused DNA binding protein bound to glutathione-Sepharose beads (Fig. 4), without electrophoresis [10]. *Digoxigenin* is a steroid found in plants. It is increasingly used as a label for nonradioactive detection of nucleic acids and proteins.

2 Materials

2.1 siRNA Transfection

1. RNAse-, DNAse-free water.
2. Transfection Medium: RPMI 1640, 10% FBS (fetal bovine serum) (v/v), MEM nonessential amino acids solution 10 mM (100 mL).
3. Opti-MEM I Reduced Serum Medium, liquid with GlutaMAX-I (Thermo Fisher Scientific).
4. TransIT-TKO Transfection reagent (MiRus).
5. Complete Medium: Transfection Medium, 5 mL antibiotics (penicillin–streptomycin 100,000 U/mL).
6. Anti-mouse CD3 (clone 145-2C11), anti-mouse CD28 (clone 37-51).

2.2 Chromatin Immunoprecipitation

1. PBS.
2. TruChIP Chromatin Shearing Reagent Kit (Covaris).
3. RNAse-, DNAse-free water.
4. Formaldehyde, methanol free.
5. ChIP-IT Express Chromatin Immunoprecipitation Kit (Active Motif; includes Proteinase K, Protease inhibitor, Shearing buffer, Elution buffer, Reverse cross-linking buffer, Lysis buffer).

2.3 Liquid Luminescent DNA Precipitation Assay

1. GST fusion protein.
2. Glutathione Sepharose 4B (GE Health Care Life Sciences).
3. Double-stranded oligonucleotides containing the putative binding sequence (Life Technologies).
4. DIG Gel Shift Kit (Roche Diagnostics).
5. Chemiluminescent substrate disodium 3-(4-methoxyspiro{1,2-dioxetane-3,2′-(5′-chloro)tricyclo[3.3.1.1^{3,7}] decan}[3]-4-yl) phenyl phosphate (CSPD) (*see* **Note 1**).
6. DNAse/RNAse-free water.
7. 1 M Tris–HCl.
8. Binding Buffer: 75 mM KCl, 50 mM NaCl, 1 mM EDTA, 1 mM DTT, 10% glycerol, 0.1% Triton X-100.

9. Washing Buffer: 0.1 M maleic acid, 0.15 M NaCl (pH 7.5), 0.3% Tween 20 (v/v).
10. Maleic Acid Buffer: 0.1 M maleic acid, 0.15 M NaCl, adjust with NaOH to pH 7.5.
11. 10× Blocking Solution: 10% (w/v) blocking reagent (DIG Gel Shift Kit) in Maleic Acid Buffer.
12. Detection Solution: 0.1 M Tris–HCl, 0.1 M NaCl (pH 9.5).

2.4 Th9 Cell Differentiation

1. Recombinant cytokines: TGF-β (human), IL-4 (mouse), IL-2 (human; can be used during expanding differentiating Th9 cells).
2. Blocking cytokines: Anti-IFN-γ, anti-IL-4.
3. T cell activation antibodies: Anti-CD3, anti-CD28.

3 Methods

For siRNA transfection and ChIP experiments, we isolate naive CD4+ T cells from C57BL/6 mice. We harvest spleen and lymph nodes and dissociate them through cell strainers (70 μm). We incubate cells with 1 mL red blood cell lysis buffer (commercially available) for 1 min to eliminate red blood cells. We add 10 mL medium to neutralize red blood cell lysis buffer and isolate naive CD4+ T cells using mouse CD4+CD62L+ T Cell Isolation Kit, mouse (Miltenyi Biotec) according to manufacturer guidelines.

3.1 siRNA Transfection (See Notes 2 and 3)

1. Prepare siRNA stock solution 10 μM in RNAse-, DNAse-free water.
2. Plate naïve CD4+ T cells at a density of 3×10^6 cells/mL in 12-well plates in “Transfection Medium.”
3. Place 100 μL (per well of transfected cells) of Opti-MEM I Reduced-Serum Medium in a 1.5 mL sterile tube.
4. Add 5 μL of *Trans*IT-TKO Reagent (*see* **Note 4**). Pipette gently to mix completely.
5. Add 2.8 μL of 10 μM siRNA stock solution (25 nM final concentration per well). Pipette gently to mix completely.
6. Incubate at room temperature for 15–30 min to allow sufficient time for complexes to form.
7. Add the *Trans*IT-TKO Reagent:siRNA complexes dropwise to different areas of the wells.
8. Gently rock the culture vessel back-and-forth and from side-to-side to evenly distribute the *Trans*IT-TKO Reagent–siRNA complexes.
9. Incubate for 24 h without replacing cell medium.

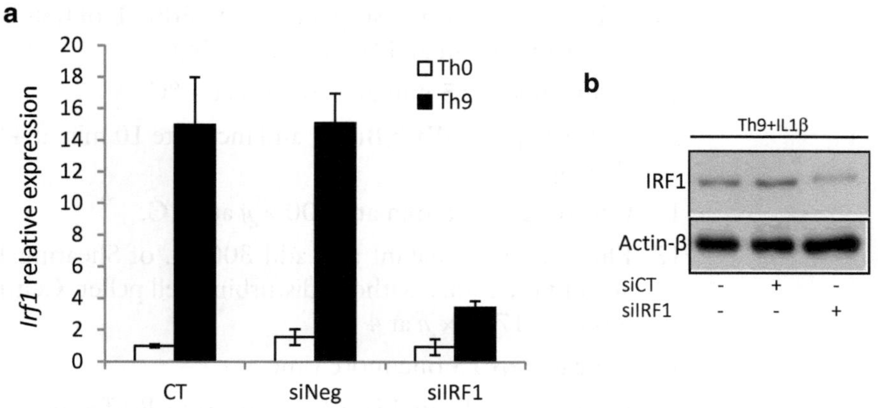

Fig. 2 Efficiency of siRNA transfection. Naïve CD4+ T cells were transfected with siRNA targeting IRF1. Twenty-four hours later, cells were harvested for PCR (**a**) and Western Blot (**b**) analysis. *CT* not transfected cells, *siNeg* cells transfected with Negative siRNA, *siIRF1* siRNA targeting IRF1

10. Harvest and count transfected cells.
11. Perform Th9 differentiation in complete medium on anti-CD3 and anti-CD28 (2 μg/mL each) coated culture plates (Fig. 1).

To ensure knockdown efficiencies, PCR and Western blot are required (Fig. 2a and b).

3.2 Chromatin Immunoprecipitation

40 × 10^6 naïve CD4+ T cells are isolated as previously described. Th9 differentiation was induced in anti-CD3 and anti-CD28 (2 μg/mL each) coated 12-well plates in the presence of IL-4 (20 ng/mL), TGF-β (2 ng/mL), and anti-IFN-γ (10 μg/mL). As a control, Th0 cells are induced in presence of anti-IFN-γ (10 μg/mL) and anti-IL-4 (10 μg/mL). According to putative transcription factor expression kinetics, ChIP experiment is performed 24 h later (*see* **Note 5**).

1. Harvest cells and centrifuge at 300 × *g* for 5 min.
2. Wash cells in 400 μL of PBS and transfer them in a 1.5 mL tube.
3. Heat Quenching Buffer from the kit to 55 °C.
4. Centrifuge cells for 5 min at 200 × *g* at room temperature.
5. Eliminate supernatant and add 400 μL of Fixing Buffer and 40 μL of 11.1% formaldehyde. Incubate for 5 min under shaking at room temperature.
6. Add 23 μL of preheated Quenching Buffer and agitate it for 5 min at room temperature.
7. Centrifuge for 5 min at 200 × *g*.
8. Wash cells twice with 300 μL of PBS.

9. After second wash, resuspend cells in 300 μL of lysis buffer and incubate 10 min at 4 °C under shaking.
10. Centrifuge for 5 min at 1700 × *g* at 4 °C.
11. Add 300 μL of Wash Buffer and incubate 10 min at 4 °C under shaking.
12. Centrifuge for 5 min at 1700 × *g* at 4 °C.
13. Eliminate supernatant and add 300 μL of Shearing Buffer all around tube walls without disturbing cell pellet. Centrifuge for 5 min at 1700 × *g* at 4 °C.
14. Repeat **step 13** one more time.
15. Resuspend cells in 130 μL of Shearing Buffer and transfer in Covaris Tube.
16. Samples are sonicated according to manufacturer recommendations and centrifuge for 5 min at 10,000 × *g* at 4 °C.
17. Measure DNA concentration with NanoDrop spectrophotometer.

For the subsequent steps, please use ChIP-IT Express Chromatin Immunoprecipitation kit.

18. Transfer 10 μL of DNA to a new tube. Store it at −20 °C. This sample will be used as "input DNA."
19. Set up ChIP reaction by adding in provided siliconized tubes 25 μL of Protein G Magnetic Beads (from the kit), 10 μL of ChIP Buffer 1 (from the kit), 7 μg of sheared chromatin, 1 μL of Protease Inhibitor Cocktail (PIC, from the kit), RNAse-, DNAse-free water (enough so that the final reaction volume will be 100 μL), and 3 μg of antibody or control Ig (*see* **Note 6**).
20. Cap the tubes and incubate using an end-to-end rotator overnight at 4 °C.
21. Spin the tubes briefly to collect liquid from the inside of the cap (*see* **Note 7**).
22. Place tubes on a magnetic stand to pellet beads on the tube side (*see* **Note 8**).
23. Carefully remove and discard supernatant.
24. Wash beads one time with 800 μL of ChIP Buffer 1 and two times with 800 μL of ChIP Buffer 2 (from the kit).
25. Resuspend the washed beads with 50 μL of Elution Buffer (from the kit).
26. Incubate for 15 min at room temperature using an end-to-end rotator.
27. Briefly spin tubes to collect liquid from the cap.
28. Add 50 μL of Reverse Cross-linking Buffer (from the kit) and mix by pipetting up and down.

29. Place tubes on a magnetic stand and allow beads to pellet to the sides of the tubes.
30. Transfer supernatant which contain the chromatin in a new tube.
31. Thaw "input DNA" on ice. Add 88 μL of ChIP Buffer 2 and 2 μL of 5 M NaCl (from the kit).
32. Incubate the ChIP and "Input DNA" samples at 95 °C for 15 min.
33. Return tubes to room temperature, spin the tubes briefly, and add 2 μL of Proteinase K (from the kit).
34. Mix well and incubate at 37 °C for 1 h. During this incubation, place the Proteinase K Stop Solution (from the kit) at room temperature.
35. Return the tubes to room temperature and add 2 μL of Proteinase K Stop Solution. The DNA can be used immediately for PCR (Fig. 3) or stored at −20 °C.

3.3 Liquid Luminescent DNA Precipitation Assay

GST fusion proteins are previously produced in BL21 bacteria (commercially available). GST alone is used as a control. Briefly, classical cloning method is used to insert GST-protein sequence in commercially available vector such as pGEX-4T-1 vector. BL21 are transformed by a heat shock according to supplier's instructions. Twenty-four hours later bacteria are lysed by 1% Triton solution (*see* **Note 9**).

Day 1

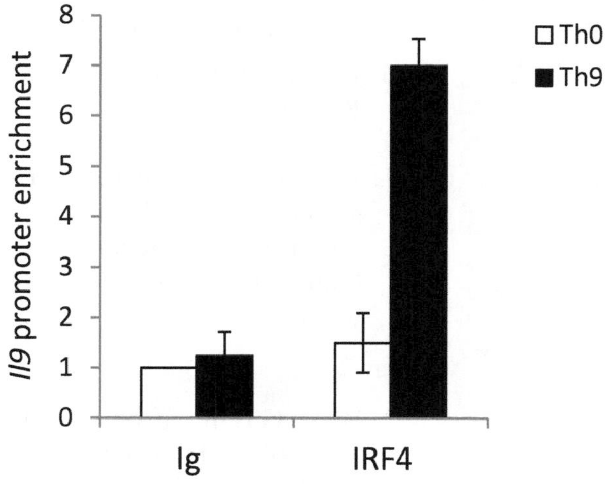

Fig. 3 Chromatin ImmunoPrecipitation of IRF4 to *Il9* promoter. Cells were differentiated for 24 h in Th0 (in presence of anti-IL-4 and anti-IFN-γ) or Th9 (TGF-β+ IL-4+ anti-IFN-γ) for 24 h. ChIP experiments were performed using either IRF4 specific antibody or control Ig (Ig Goat, Dako X0907). Data represent *Il9* promoter enrichment detected by qPCR

3.3.1 Binding of GST-Fusion Protein and Sepharose Beads

1. Wash Glutathione Sepharose 4B twice with 5 mL of Binding Buffer and centrifuge at 500 × *g* for 5 min at room temperature.
2. Purify GST fused proteins with Glutathione Sepharose 4B. Three milliliters of bacteria lysate obtained after BL21 sonication is added to the prepared Glutathione Sepharose 4B to a final volume of 30 mL of Binding Buffer, and incubate for 30 min at room temperature with gentle agitation.
3. Protein–Glutathione Sepharose 4B complexes are isolated by centrifugation at 500 × *g* for 5 min at room temperature and washed with 5 mL of Binding Buffer.
4. Centrifugation at 500 × *g* for 5 min allows collecting Protein–Glutathione Sepharose 4B complexes.

3.3.2 DNA Labeling Reaction

1. Label the Protein-binding sequence (PBS) with Digoxigenin by using DIG Gel Shift Kit (Liquid chemiluminescent DNA pull-down assays can also be performed using biotinylated DNA, alternatively).
2. Dilute double-stranded oligonucleotides at the concentration of 10 ng/μL with DNAse/RNAse-free water.
3. Put the tubes containing 100 ng of double-strand DNA (10 μL) on ice and add 4 μL of Labeling Buffer, 5 mM of $CoCl_2$ solution, 0.05 mM of DIG-ddUTP solution, and 200 U of Terminal transferase. Mix by pipetting up and down.
4. The tubes are centrifuged and incubated for 15 min at 37 °C in a dry block heater and put immediately on ice after this incubation.
5. Stop the reaction with 2 μL of EDTA.
6. Dilute the DNA to 4 ng/μL by adding 3 μL of water (DNAse/RNAse free).

3.3.3 Protein-DNA Binding Reaction

1. Incubate 2 μg of GST-proteins bound to Glutathione Sepharose 4B beads with 20 fmol DIG-labeled double-stranded DNA fragments in protein–DNA binding buffer from the DIG Gel Shift Kit in a final volume of 20 μL overnight at room temperature.

Day 2

1. Wash GST-fused proteins bound to DNA three times with 1 mL of Washing Buffer from the kit and centrifuge each time at 500 × *g* for 5 min at room temperature.
2. The beads are incubated in 500 μL of Blocking Solution beforehand. Dilute tenfold for 30 min at room temperature.
3. Incubate GST-fused proteins bound to DNA with anti-DIG Fab fragments conjugated with 75 mU/mL of alkaline

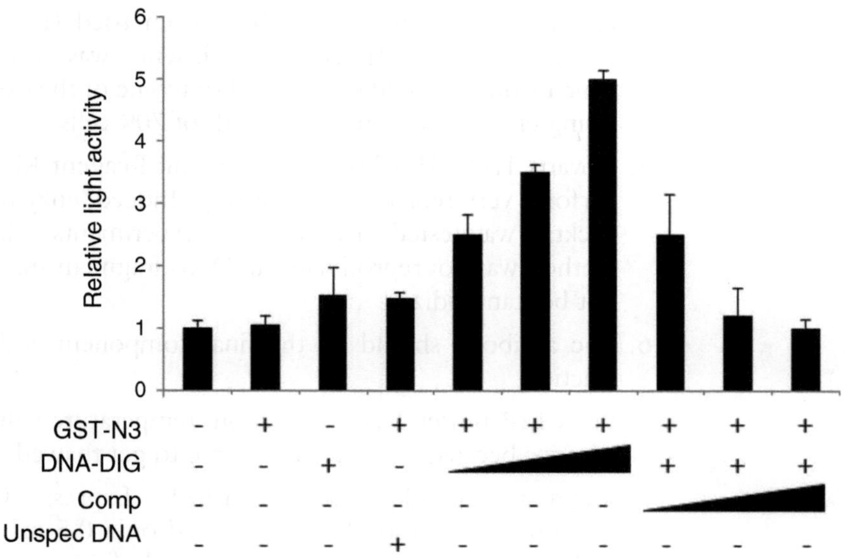

Fig. 4 Interaction analysis between DNA and protein. Liquid luminescent DNA precipitation Assay was performed to detect interaction between protein of interest coupled to GST (GST-N3) and its putative binding sequence coupled to DIG (DNA-DIG). An increased quantity of DNA-DIG was tested (15, 30, and 60 fmol). If not specified 30 fmol was used. Specific interaction between protein and DNA was determined by using an increased quantity of unlabeled-DNA (15, 30, 60, and 150 fmol) as a competitor (Comp). Unspecific DNA coupled to DIG (30 fmol) was used as a control

phosphatase (from DIG gel shit kit) for 30 min at room temperature in 1× Blocking Solution.

4. Wash GST-fused proteins bound to DNA twice with 5 mL of Washing Buffer.
5. After washes, transfer the complexes in detection solution to 96-well plate and incubate for 5 min at room temperature with 1 μg/mL chemiluminescent substrate disodium 3-(4-methoxyspiro{1,2-dioxetane-3,2′-(5′-chloro)tricyclo[3.3.1.1^{3,7}]decan}-4-yl) phenyl phosphate (CSPD), followed by incubation for 10 min at 37 °C.
6. Acquire light emission using a PerkinElmer Envision Reader during 1 s per well (Fig. 4).

4 Notes

1. It is included in DIG Gel Shift Kit. Alternatively, it can also be purchased from Sigma-Aldrich.
2. Before starting, all the reagents need to be thawed to room temperature and vortexed gently.
3. It is important to always use negative siRNA as a control of siRNA specificity.

4. Other transfection reagents have been tried (Lipofectamine 2000, Interferin). Transfection efficiency was about 10–20% which would depend on the final outcome of the experiments. Using electroporation led to death of 70% cells.
5. Covaris TruChIP Chromatin Shearing Reagent Kit is used to perform very reproducible shearing. Indeed, enzyme shearing cocktail was tested in preliminary experiments. This kind of method was not reproducible in DNA fragment size and could not be standardized.
6. The antibody should be the final component added to the reaction.
7. Put ChIP Buffer 1 and 2 at room temperature sufficiently in advance because they take long time to get thawed.
8. Magnetic spindle from ChIP-IT Express Chromatin Immunoprecipitation kit can be fixed on a 0.5 mL tube rack with adhesive paper and be used instead of magnetic stand.
9. All the buffers are not stable and have to be freshly prepared.

Acknowledgments

Liquid luminescent DNA precipitation assay protocol is adapted from Toshiharu et al. (2008) [10]. These projects were supported by the French National Research Agency ("Investissements d'Avenir" program; ANR-11-LABX-0021), the Ligue nationale contre le cancer (F.G. and F.V.), the Institut National du Cancer (F.G.), Fondation pour la Recherche Médicale (F.G.).

References

1. Mitchison NA (1995) T cell subsets in infectious and autoimmune diseases. Introduction. In: Ciba Foundation symposium 195, pp 1–5; discussion 5–6
2. Dardalhon V, Awasthi A, Kwon H, Galileos G, Gao W, Sobel RA, Mitsdoerffer M, Strom TB, Elyaman W, Ho IC, Khoury S, Oukka M, Kuchroo VK (2008) IL-4 inhibits TGF-beta-induced Foxp3+ T cells and, together with TGF-beta, generates IL-9+ IL-10+ Foxp3(−) effector T cells. Nat Immunol 9(12):1347–1355. doi:10.1038/ni.1677
3. Veldhoen M, Uyttenhove C, van Snick J, Helmby H, Westendorf A, Buer J, Martin B, Wilhelm C, Stockinger B (2008) Transforming growth factor-beta 'reprograms' the differentiation of T helper 2 cells and promotes an interleukin 9-producing subset. Nat Immunol 9(12):1341–1346. doi:10.1038/ni.1659
4. Forbes EE, Groschwitz K, Abonia JP, Brandt EB, Cohen E, Blanchard C, Ahrens R, Seidu L, McKenzie A, Strait R, Finkelman FD, Foster PS, Matthaei KI, Rothenberg ME, Hogan SP (2008) IL-9- and mast cell-mediated intestinal permeability predisposes to oral antigen hypersensitivity. J Exp Med 205(4):897–913. doi:10.1084/jem.20071046
5. Licona-Limon P, Henao-Mejia J, Temann AU, Gagliani N, Licona-Limon I, Ishigame H, Hao L, Herbert DR, Flavell RA (2013) Th9 cells drive host immunity against gastrointestinal worm infection. Immunity 39(4):744–757. doi:10.1016/j.immuni.2013.07.020
6. Purwar R, Schlapbach C, Xiao S, Kang HS, Elyaman W, Jiang X, Jetten AM, Khoury SJ, Fuhlbrigge RC, Kuchroo VK, Clark RA, Kupper TS (2012) Robust tumor immunity to melanoma mediated by

interleukin-9-producing T cells. Nat Med 18(8):1248–1253. doi:10.1038/nm.2856

7. Lu Y, Hong S, Li H, Park J, Hong B, Wang L, Zheng Y, Liu Z, Xu J, He J, Yang J, Qian J, Yi Q (2012) Th9 cells promote antitumor immune responses in vivo. J Clin Invest 122(11):4160–4171. doi:10.1172/JCI65459
8. Vegran F, Berger H, Boidot R, Mignot G, Bruchard M, Dosset M, Chalmin F, Rebe C, Derangere V, Ryffel B, Kato M, Prevost-Blondel A, Ghiringhelli F, Apetoh L (2014) The transcription factor IRF1 dictates the IL-21-dependent anticancer functions of TH9 cells. Nat Immunol 15(8):758–766. doi:10.1038/ni.2925
9. Vegran F, Apetoh L, Ghiringhelli F (2015) Th9 cells: a novel CD4 T-cell subset in the immune war against cancer. Cancer Res 75(3):475–479. doi:10.1158/0008-5472.CAN-14-2748
10. Iwasaki T, Miyazaki W, Rokutanda N, Koibuchi N (2008) Liquid chemiluminescent DNA pull-down assay to measure nuclear receptor-DNA binding in solution. Biotechniques 45(4):445–448. doi:10.2144/000112915

Chapter 14

Defining Epigenetic Regulation of the Interleukin-9 Gene by Chromatin Immunoprecipitation

Alla Skapenko and Hendrik Schulze-Koops

Abstract

Regulation of gene expression is essential for the differentiation of pluripotent precursor cells into specialized effector cells and, thus, for the evolution of multiorgan systems. The regulation of gene expression is controlled by a variety of "extra-genic" mechanisms, termed as epigenetic mechanisms. Obviously, alterations in such control mechanisms of gene expression may result in alterations of cellular effector functions, resulting in for example defects in cellular functions, but also, if immune cells are involved, leading to the development of immunologic disorders such as malignancies and autoimmune diseases. The analysis of epigenetic modifications is therefore pertinent not only for the understanding of the regular function of the immune system, but also for the understanding of the pathophysiology of such diseases. As interleukin-9 (IL-9) is the signature cytokine for Th9 cells, and since IL-9 plays important roles in the orchestration of a protective immune responses, the analysis of epigenetic mechanisms underlying the development of IL-9 producing effector T cells is of great interest. Here, we describe a protocol to analyze epigenetic regulation of the IL-9 gene by chromatin immunoprecipitation.

Key words Epigenetics, Chromatin immunoprecipitation (ChIP), Interleukin-9

1 Introduction

While the genetic information of any individual is encoded in its genome, the genome itself—with the notable exception of the haploid gametes—is identical in all nucleated cells of any given living being. Thus, the genome cannot account for differences in gene expression in different cells that are required to make these cells functionally specialized and to enable multicellular organisms to evolve. To foster such specific and characteristic gene expression (for example organ-specific cells), particular subsets of genes must be expressed in these cells, while most of the remaining genes would ideally be permanently silenced. The mechanisms controlling cell-specific gene expression are collectively termed as "epigenetic mechanisms" and they result in covalent modifications to the DNA that impacts gene expression without affecting the underlying genetic sequence. Epigenetic mechanisms comprise

Ritobrata Goswami (ed.), *Th9 Cells: Methods and Protocols*, Methods in Molecular Biology, vol. 1585,
DOI 10.1007/978-1-4939-6877-0_14, © Springer Science+Business Media LLC 2017

methylation of deoxyribonucleotides within the DNA, posttranscriptional chemical modifications of histone complexes, chromatin remodeling, and a variety of processed RNA including noncoding RNAs (ncRNA), such as micro RNAs, short interfering RNAs, piwi RNAs, and long non-coding RNAs [1].

Beside organ-specific gene expression, a particular aspect of cell specific gene expression that is controlled by epigenetic mechanisms is the differentiation of precursor cells into specialized effector cells within the three hematopoietic lineages [2]. In this regard, it has become apparent that epigenetic mechanisms control the differentiation of lymphocytes from naive precursor T or B cells into effector memory or plasma cells, respectively [3, 4]. Consequently, epigenetic abnormalities have been described in immunological disorders, both in malignancies such as leukemias as well as in autoimmune diseases [5, 6]. In order to gain insights into molecular mechanisms of these diseases as well as into mechanisms regulating the development of the normal immune system, it becomes fundamental to address epigenetic modifications in genes that define a particular status of a lymphocytes, such as an effector cell.

Effector CD4+ T (T helper) cells, in contrast to naïve T cells, are able to produce cytokines other than interleukin (IL)-2 upon T cell activation. Several effector T cell subsets have been identified, and these effector T cell subsets mediate a particular immune response by secreting characteristic signature cytokines. For example, Th1 cells express interferon (IFN)-γ, and they are involved in IFN-γ-dependent cellular immunity while Th2 cells are able to produce IL-4, thereby providing help for humoral immune responses [7]. The signature cytokine of Th17 cells is IL-17, which is instrumental for the host defense against certain pathogens at epithelial and mucosal barriers [7]. The function of a unique subset termed Th9 cells is still work in progress, but increasing evidence suggest that Th9 cells by virtue of their signature cytokine IL-9, play a role in T cell and mast cell proliferation, thereby contributing to parasite expulsion [8, 9]. However, if dysregulated, IL-9 might be involved in the development of tissue inflammation and autoimmune disorders, as shown in several animal models of asthma, allergy, colitis, diabetes, and experimental autoimmune encephalomyelitis [10–14]. In this regard, IL-9 has been suggested to enhance the pro-inflammatory Th17 cell-driven immune response by augmenting Th17 cell development or by recruiting mast cells to the sites of inflammation and subsequent triggering of their effector functions, such as release of cytokines, histamine and proteases [12–14]. Remarkably, however, IL-9 has also been implicated in preventing allograft rejection by modulating T cell tolerance via the very same effector mechanisms, which involve mast cell recruitment and activation [15].

It has been firmly established that the development of individual T helper cell subsets is triggered by key transcription factors, such as T-bet (for Th1 cells), GATA-3 (for Th2 cells), PU.1 (for Th9 cells), or RORC (for Th17 cells). As expression of these transcriptions factors are both, sufficient and necessary for the development of the respective T cell subsets, it must be tightly controlled and active only in T cells deemed to develop into the particular effector cells, while being silenced in the development of reciprocal T helper cells. Master regulators of effector T cell subsets are therefore the prime example of genes that could undergo epigenetic control. Analysis of permissive and repressive epigenetic alterations in these transcription factor genes enables a deep insight into the molecular dynamic of effector T cell development as well as into mechanisms of T cell-dependent diseases.

Various means to analyze epigenetic mechanisms have been developed. While ncRNAs can be best detected by one of the many sequencing techniques, DNA methylation is usually assessed by mass spectrometry, high-pressure liquid chromatography, enzyme linked immunosorbent assays (ELISA) with 5-methylcytosine antibodies, by bisulfite conversion, cloning, and sequencing, or by immunoprecipitation of methylated DNA. Histone modifications that comprise methylation/demethylation, phosphorylation, acetylation, ubiquitylation, and sumoylation, can be identified by extraction of nuclear proteins followed by appropriate immunoprecipitation techniques. The protocols for the analysis of ncRNAs, for DNA methylation and histone modifications are beyond the scope of this chapter and are described elsewhere. Here, we describe the protocols that enable the researcher to analyze target DNA sequence specific epigenetic modifications by use of chromatin immunoprecipitation (ChIP). In particular, we focus on ChIP assays to define epigenetic marks in the IL-9 gene.

2 Materials

1. Source of T cells (e.g., peripheral blood, umbilical blood, lymphoid tissue).
2. Naïve CD4+ T Cell Isolation Kit (Miltenyi Biotech), Memory CD4+ T Cell Isolation Kit (Miltenyi Biotech).
3. Fluorochrome-labeled anti-human CD4, CD45RA, CD45RO monoclonal antibodies.
4. EpiQuick Chromatin Immunoprecipitation (ChIP) Kit (for example from Epigentek; *see* **Note 1**).
5. Centrifuge.
6. Variable temperature water bath.
7. Vortex mixer.

8. Benchtop centrifuge.
9. Sonicator (Bioruptor Next Generation (Diagenode, Liége, Belgium)).
10. Orbital shaker.
11. Real-time PCR machine.
12. Pipettes and pipette tips.
13. 1.5 mL microcentrifugation tube.
14. 15 mL conical tube.
15. 50 mL conical tube.
16. 90% ethanol (v/v).
17. 70% ethanol (v/v).
18. Ficoll-Hypaque.
19. PBS.
20. 37% formaldehyde.
21. 1.25 M glycine solution.
22. 1× TE buffer (10 mM Tris–HCl, 1 mM EDTA, pH 8.0).
23. Normal rabbit IgG (*see* **Note 2**).
24. Parafilm M.
25. Antibodies to detect histone modification.

3 Methods

As IL-9 is produced by T cells and specifically by CD4+ T helper cells, the cells analyzed by the protocol described herein should comprise CD4+ T cells isolated by one of the various well-established approaches (e.g., magnetic beads, purification kits) to stringent purity (≥95%). As a source for CD4+ T cells, one may use—depending on the question to be addressed—peripheral blood or umbilical cord blood, but in animals spleen or other lymphatic tissues can be used. We routinely analyze human T cells and start from peripheral blood of healthy volunteers or patients with a defined immunological disease, or from umbilical cord blood. As a rule of thumb, 1–3 × 10^6 CD4+ T cells are sufficient for ChIP of the IL-9 gene and the analysis of epigenetic marks therein.

3.1 Chromatin Immunoprecipitation Assay (ChIP)

1. Always stain source cells for purity by flow cytometry using appropriate antibody mixes, for example CD4, CD45RA or CD4, CD3, CD45RO prior to starting the ChIP protocol.
2. The "EpiQuik Chromatin Immunoprecipitation (ChIP) Kit" (Epigentek) provides a simple, easy-to-use method for ChIP assays.

3.2 Preparation of Assay Plate for Immunoprecipitation

1. Wash plastic strips with washing buffer (strips and buffer provided in the kit).
2. Load wells with 200 μL of coating buffer and add negative and positive control (normal mouse IgG and anti-RNA polymerase, buffer, and both controls provided in the kit) (*see* **Note 2**).
3. Add antibodies against modified histones of interest (for example anti-H3K4me3, anti-H3ac, or anti-H3K27me3) into appropriate wells to a final concentration of 10–20 μg/mL coating buffer.
4. Cover strips/wells with Parafilm M and incubate at room temperature for 60–90 min.

3.3 Preparation of Cells and Cross-Link

1. Use freshly isolated CD4+ T cells or primed T cells (*see* **Note 3**). The differentiation protocol of both murine and human Th9 cells has been described in other chapters of this book.
2. Transfer 1–3 × 10^6 cells into a 15 mL canonical tube. Add PBS to a final volume of 10 mL.
3. Centrifuge the cells at 170 × *g* for 5 min at room temperature. Discard supernatant.
4. Wash once with 10 mL of PBS and centrifuge for 5 min at 170 × *g* at room temperature.
5. Discard supernatant.
6. Add 9 mL of fresh culture medium containing 1% formaldehyde (v/v).
7. Incubate at room temperature for 10 min in a rocking platform (50–100 rpm) (*see* **Note 4**).
8. Stop fixation by adding 1 mL of 1.25 M glycine, mix and centrifuge for 5 min at 100 × *g* at room temperature. Discard supernatant.
9. Wash once with 10 mL of PBS and centrifuge for 5 min at 170 × *g* at room temperature.
10. Discard supernatant.

3.4 Cell Lysis and DNA Shearing

1. Resuspend the cell pellet in pre-lysis buffer (provided in the kit; 100 μL/1 × 10^6 T cells).
2. Transfer suspension to a 1.5 mL microcentrifugation tube and incubate for 10 min on ice.
3. Prepare lysis buffer/protease inhibitor solution by adding 10 μL of proteinase inhibitor cocktail to 1 mL of lysis buffer (both provided in the kit).
4. Vortex the cells vigorously for 10 s and centrifuge for 5 min at 2650 × *g* at 4 °C.

5. Remove supernatant carefully. Add 50–100 µL of lysis buffer/proteinase inhibitor cocktail per 1 × 10^6 CD4+ T cells. Incubate for 10 min on ice. Vortex occasionally.
6. Shear DNA by sonication. We use 3–4 pulses of 10–12 s length at an energy level of 2 out of 4 (low energy) with the Bioruptor Next Generation, followed by 35 s of rest on ice in between each pulse (*see* **Note 5**).
7. Pellet cell debris by centrifugation for 10 min at 20,000 × *g* at 4 °C.

3.5 Immunoprecipitation of Protein–DNA Complexes

1. Transfer clear supernatant to a new 1.5 mL microcentrifuge tube (*see* **Note 6**). Add to required amount of supernatant (50 µL for each well from the assay plate) identical volume of ChIP dilution buffer (provided in the kit).
2. Transfer 5 µL of the diluted supernatant to a 0.5 mL vial. This is the “input DNA” control. Place on ice.
3. From the assay plate, remove the coating solutions with the antibodies, the positive and the negative control. Wash the wells three times with 150 µL of coating buffer (provided in the kit).
4. Transfer 100 µL of diluted supernatant to each well of the assay plate.
5. Cover the wells with Parafilm M and incubate at room temperature for 60–90 min on an orbital shaker (50–100 rpm).
6. Remove supernatant. Wash wells six times with 150 µL of washing buffer each (provided in the kit) while allowing 2 min on an orbital shaker (100 rpm) for each wash.
7. Wash wells once with 150 µL of 1× TE buffer.

3.6 Reversal of Cross Linked DNA and Purification of DNA

1. Prepare DNA release cocktail by adding 1 µL of Proteinase K into 40 µL of DNA release buffer (both provided in the kit).
2. Add 40 µL of DNA release cocktail containing Proteinase K to the wells (and to the “input DNA” vial). Cover the wells and incubate for 15 min in a water bath at 65 °C.
3. Add 40 µL of reverse buffer (provided in the kit) to all wells including the “input DNA” vial, mix, cover the wells and incubate for 90 min in a water bath at 65 °C.
4. Place a DNA-collecting spin column (provided in the kit) into a 2 mL collection vial. Add 150 µL of binding buffer (provided in the kit) to the samples, mix and transfer entire sample content to the column. Centrifuge for 20 s at 15,300 × *g* at room temperature.
5. Add 200 µL of 70% ethanol to the column. Centrifuge for 15 s at 15,300 × *g* at room temperature. Discard the flow-through.

Table 1
PCR primers for promoter regions of human T cell specific transcription factors

Transcription factor	Sense	Antisense
GATA-3	5′-CGCTCTCCCAAACACCCTG-3′	5′-AGCCAGCCTGCCCCATTC-3′
T-bet	5′-GGCAACCCGAAAGGTCACTTAG-3′	5′-TTCTCCTGGCACTCAGAGGCTC-3′
RORC	5′-TCTCCCCTATGCCTGTCACCTG-3′	5′-TGATTTTGCCCAAGGACTCACAC-3′
PU.1	5′-GAAGCAGGCATTTGTTGGGTTAG-3′	5′-TGGGACTGAGAGGGATGACTTTG-3′
PU.1 upstream enhancer	5′-TCTCTTTTCCTTGCTGTCCCTG-3′	5′-GTCACTCACCTCTTGCTTCTGGTC-3′
IRF-4	5′-CCCACCTCGCACTCTCAGTTTC-3′	5′-TATCAGCCTCACACCCCTCCTC-3′

6. Replace the column to the collection tube. Add 200 μL of 90% ethanol and centrifuge for 20 s at 15,300 × *g* at room temperature. Remove the column. Discard the flow-through.
7. Replace the column to the collection tube. Add 200 μL of 90% ethanol and centrifuge for 35 s at 15,300 × *g* at room temperature. Remove the column. Discard the flow-through.
8. Place the column in a new 1.5 mL tube. Add 20 μL of elution buffer (provided in the kit).
9. Centrifuge for 20 s at 15,300 × *g* at room temperature. The flow-through contains the eluted DNA (*see* **Note 7**).

3.7 Analysis of DNA

1. The eluted DNA can be analyzed by quantitative PCR.
2. The primers that we use for promoter sequences of transcription factors that are characteristic for defined CD4 effector T cell subsets in a SYBR Green PCR with the ABI PRISM 7000 Sequence Detection System are shown in Table 1.
3. Presentation of data should be done as a log to the base 2 of the ratio of the immunoprecipitated amounts of DNA of H3K4me3 to H3K27me3 and H3Ac to H3K27me3 pull-downs or as a ratio of immunoprecipitated DNA to that of input DNA.
4. Binding specificity is always verified via amplification of DNA precipitated with negative control Ig (for example mouse or rabbit IgG).

4 Notes

1. We have used with most success the products and the protocol provided with the kit mentioned in the text to ensure maximum reproducibility in particular for small sample sizes throughout

the entire period of a given project. The Epigentek Epiquick Chromatin Immunoprecipitation Kit contains buffers, such as the washing buffer for the plastic strips, antibody coating buffer, cell lysis buffers, dilution buffer for the ChIP, DNA release buffer, buffer for the reversal of DNA cross-link as well as binding and elution buffers for the final purification of the DNA fragments.

2. If the anti-histone antibodies that will be used for immunoprecipitation are of non-murine origin, we recommend the use of non-murine normal IgG for proper control.
3. Human T cells can be primed for example by incubation for 5 days with bead-coupled anti-CD3/anti-CD28 or by surface bound anti-CD3 (1 μg/mL anti-CD3 in the coating buffer) and anti-CD28 (1 μg/mL in the cell culture medium).
4. We have the best results in some settings when fixation is performed in ice rather than at room temperature.
5. The conditions of cross-linked DNA shearing should be optimized based on cells and sonicator equipment. Shearing efficacy can be verified by agarose gel analysis of 5 μL of sonicated cell lysate. The length of sheared DNA should be between 200 and 1000 bp.
6. At this point, supernatants can be stored at −80 °C for future analysis.
7. DNA can be stored at −20 °C for future analysis.

Acknowledgment

This work was supported by the Deutsche Forschungsgemeinschaft (Grants SK59/4-1, SK59/7-1, SCHU786/2-5, Schu786/8-1 and SCHU1683/9-1), by the DFG Sonderforschungsbereich Grant SFB571 (Autoimmunity), project D9, by the DFG Training Grant GK1202 (Oligonucleotides), project E2, and by the Verbundanträge "ArthroMark" (projects 1 and 7, OIEC100913 and 01EC1401B) and "Impam" (project 10, OIEC1008H) both by the Federal Ministry of Education and Research of Germany.

References

1. Holliday R (2006) Epigenetics: a historical overview. Epigenetics 1(2):76–80
2. Rodriguez RM, Lopez-Larrea C, Suarez-Alvarez B (2015) Epigenetic dynamics during CD4(+) T cells lineage commitment. Int J Biochem Cell Biol 67:75–85. doi:10.1016/j.biocel.2015.04.020
3. Ramming A, Druzd D, Leipe J, Schulze-Koops H, Skapenko A (2012) Maturation-related histone modifications in the PU.1 promoter regulate Th9-cell development. Blood 119(20):4665–4674. doi:10.1182/blood-2011-11-392589
4. Santos P, Arumemi F, Park KS, Borghesi L, Milcarek C (2011) Transcriptional and epigenetic regulation of B cell development. Immunol Res 50(2–3):105–112. doi:10.1007/s12026-011-8225-y
5. MacDonald IA, Hathaway NA (2015) Epigenetic roots of immunologic disease and new methods for examining chromatin regulatory pathways.

Immunol Cell Biol 93(3):261–270. doi: 10.1038/icb.2014.105

6. Zhou Q, Haupt S, Kreuzer JT, Hammitzsch A, Proft F, Neumann C, Leipe J, Witt M, Schulze-Koops H, Skapenko A (2015) Decreased expression of miR-146a and miR-155 contributes to an abnormal Treg phenotype in patients with rheumatoid arthritis. Ann Rheum Dis 74(6):1265–1274. doi:10.1136/annrheumdis-2013-204377
7. Zhu J, Paul WE (2008) CD4 T cells: fates, functions, and faults. Blood 112(5):1557–1569. doi:10.1182/blood-2008-05-078154
8. Van Snick J, Goethals A, Renauld JC, Van Roost E, Uyttenhove C, Rubira MR, Moritz RL, Simpson RJ (1989) Cloning and characterization of a cDNA for a new mouse T cell growth factor (P40). J Exp Med 169(1):363–368
9. Uyttenhove C, Simpson RJ, Van Snick J (1988) Functional and structural characterization of P40, a mouse glycoprotein with T-cell growth factor activity. Proc Natl Acad Sci U S A 85(18):6934–6938
10. Dardalhon V, Awasthi A, Kwon H, Galileos G, Gao W, Sobel RA, Mitsdoerffer M, Strom TB, Elyaman W, Ho IC, Khoury S, Oukka M, Kuchroo VK (2008) IL-4 inhibits TGF-beta-induced Foxp3+ T cells and, together with TGF-beta, generates IL-9+ IL-10+ Foxp3(−) effector T cells. Nat Immunol 9(12):1347–1355
11. Beriou G, Bradshaw EM, Lozano E, Costantino CM, Hastings WD, Orban T, Elyaman W, Khoury SJ, Kuchroo VK, Baecher-Allan C, Hafler DA (2010) TGF-beta induces IL-9 production from human Th17 cells. J Immunol 185(1):46–54. doi:10.4049/jimmunol.1000356
12. Elyaman W, Bradshaw EM, Uyttenhove C, Dardalhon V, Awasthi A, Imitola J, Bettelli E, Oukka M, van Snick J, Renauld JC, Kuchroo VK, Khoury SJ (2009) IL-9 induces differentiation of TH17 cells and enhances function of FoxP3+ natural regulatory T cells. Proc Natl Acad Sci U S A 106(31):12885–12890
13. Soussi-Gounni A, Kontolemos M, Hamid Q (2001) Role of IL-9 in the pathophysiology of allergic diseases. J Allergy Clin Immunol 107(4):575–582. doi:10.1067/mai.2001.114238
14. Nowak EC, Weaver CT, Turner H, Begum-Haque S, Becher B, Schreiner B, Coyle AJ, Kasper LH, Noelle RJ (2009) IL-9 as a mediator of Th17-driven inflammatory disease. J Exp Med 206(8):1653–1660. doi:10.1084/jem.20090246
15. Lu LF, Lind EF, Gondek DC, Bennett KA, Gleeson MW, Pino-Lagos K, Scott ZA, Coyle AJ, Reed JL, Van Snick J, Strom TB, Zheng XX, Noelle RJ (2006) Mast cells are essential intermediaries in regulatory T-cell tolerance. Nature 442(7106):997–1002. doi:10.1038/nature05010

Chapter 15

Allergic Inflammation and Atopic Disease: Role of Th9 Cells

Pornpimon Angkasekwinai

Abstract

IL-9-producing T helper cells (Th9) have recently emerged as an important T cell subset contributing to the pathogenicity of allergic diseases. Here, we describe the role of Th9 cells in allergic inflammation and provide detailed protocols to characterize IL-9-producing T cells and analyze OVA-specific IL-9 production in allergic lung diseases using a mouse model of OVA-induced chronic allergic lung inflammation.

Key words IL-9-producing T cells, Th9 cells, Th2 cells, Allergic inflammation, Flow cytometry, ELISA

1 Introduction

Allergic diseases, including atopic dermatitis, food allergy, allergic rhinitis, and asthma are recognized as a group of immune-mediated disorders. Since the discovery of T helper 1 (Th1) and T helper 2 (Th2) subsets in 1986, it has been thought that a shift in the balance toward Th2 cells is responsible for the pathogenesis of allergic diseases. Several lines of evidence indicated that Th2 cells accumulated in the airway of patients with asthma. Furthermore, the levels of IL-4, IL-5, and IL-13 were significantly elevated in bronchoalveolar lavage and sera of patients with allergic rhinitis and allergic asthma. In these patients, IL-9, a potent T cell and mast cell growth factor was also increased [1–3]. From human and animal studies, IL-9 expression is often associated with other Th2 cytokines, providing the initial support for categorizing IL-9 as a Th2 cytokine. Mice overexpressing IL-9 had markedly increased airway hyperresponsiveness to inhaled methacholine [4], supporting an important role for IL-9 in the pathogenesis of asthma. In addition, several studies indicate that IL-9 regulates the expression of mucus hyperproduction in asthmatic patients. Although some reports using IL-9-deficient mice demonstrated that IL-9 was not essential for the development of eosinophilia and AHR [5], IL-9 antibody

Ritobrata Goswami (ed.), *Th9 Cells: Methods and Protocols*, Methods in Molecular Biology, vol. 1585,
DOI 10.1007/978-1-4939-6877-0_15, © Springer Science+Business Media LLC 2017

treatment significantly reduced pulmonary eosinophilia, serum IgE levels, goblet cell hyperplasia, airway epithelial damage, and AHR [6]. Thus, these studies underline the importance of IL-9 in the regulation of allergic diseases.

Indeed, the regulation of IL-9 production differs from IL-4, IL-5, and IL-13 in Th2 cells. The production of IL-9 is significantly stimulated by TGF-β and further enhanced by supplementing with IL-4, which, by itself, has only a minimal influence [7]. Not until the discovery of a third subset of effector Th17 lineage, IL-9 producing T cells were reinvestigated and were then later recognized as a distinct T helper cell lineage, the Th9 subset. TGF-β and IL-4 were found to promote IL-9 and dampen IL-4 and IL-13 expression. In the presence of TGF-β, Th2 cells can further deviate to Th9 cells [8], supporting the close relationship between these two lineages. Several cytokines such as IL-1β, IL-6, IL-10, interferon (IFN)-α, IFN-β, and IL-21 can augment naïve T helper cell differentiation to Th9 cells [9]. Moreover, IL-25 has been shown to regulate Th9 cells in an allergen-induced allergic mouse model [10]. Further investigation demonstrated the requirement of the transcription factors PU.1 and IRF4 in the development of IL-9 producing T cells and the contribution in allergic inflammation [11, 12]. These reports strengthen the notion that the main sources of IL-9 during allergic inflammation are Th9 cells.

IL-9 derived from Th9 cells has multiple effects on immune cells, particularly lung epithelial cells, smooth muscle cells, mast cells, and lymphocytes [13], therefore playing crucial role in amplifying allergic pathology. Several approaches have been employed to investigate the role of IL-9-producing T cells in allergic lung diseases, including adoptive transfer of Th9 cells and neutralization of IL-9 during the induction of allergic inflammation [10–12, 14–17]. For adoptive transfer of Th9 cells, naïve T cells were often isolated from OVA peptide-specific TCR-transgenic (OTII) mice. Indeed, in vitro generated Th9 cell transfer could induce asthmatic phenotypes, including enhanced airway hyperresponsiveness, eosinophil infiltration, mucus hyperproduction, and histology scores [12, 14, 15]. These phenotypes could be attenuated by neutralizing IL-9 during the induction of allergic inflammation following Th9 cell transfer. Interestingly, in vitro generated Th9 cells and Th2 cells appeared to induce indistinguishable allergic asthma phenotype [12, 15]. However, recent studies have demonstrated that Th9 cells had greater ability than Th2 cells in activating and promoting mast cell accumulation during allergic inflammation [16]. These adoptive transfer and IL-9 blockade experiments provide important evidence that the induction of Th9 cells significantly augments the development of allergic asthma.

Since IL-9 appears to be produced not only by Th9 cells, but also by innate lymphoid cells type 2 (ILC2), natural killer T cells, mast cells, and other T helper cell subsets, including Th17 and regulatory T cells; investigation of Th9 cells by measuring the level of IL-9 may not completely reflect the function of Th9 cells in allergic diseases. With the availability of a commercial IL-9 staining antibody, the contribution of Th9 cells in allergic inflammation was demonstrated in mice and humans. In murine studies, models of ovalbumin induced chronic airway inflammation [10–12, 18, 19], *Aspergillus*-induced chronic airway hyperresponsiveness [20], and Th9-induced allergic inflammation [15] were used to address the function of Th9 cells in allergic pulmonary diseases. In humans, the increased Th9 cells were detected in peripheral blood of patients with peanut allergy [21], allergic asthma [17, 22], and patients with atopic dermatitis [23]. By activating human naïve T cells isolated from atopic infants under Th9 cell culture conditions in vitro, greater production of IL-9 was observed compared to those from non-atopic infants [11, 24], implicating the contribution of Th9 cells in individuals predisposed to atopy. In this chapter, we present methods used to characterize Th9 cells and their function during allergic diseases in murine model of allergic inflammation.

2 Materials

2.1 Induction of Allergic Lung Diseases for Th9 Cell Detection

1. Six- to eight-week-old female BALB/c or C57BL/6 mice.
2. Chicken Ovalbumin Grade V (Sigma) (1 mg/mL in phosphate buffered saline, pH 7.2, 0.45 μm filter-sterilized).
3. Alum (Imject Alum).
4. Phosphate buffered saline: 137 mM NaCl, 2.6 mM KCl, 10 mM Na_2HPO_4, 1.8 mM KH_2PO_4, pH 7.2.
5. Isoflurane (ready-to-use solution).

2.2 Bronchoalveolar Lavage Fluid Collection

1. 0.5 mM PBS–EDTA.
2. Needle catheter (BD Insyte, BD Biosciences).

2.3 Characterization of Th9 Cells in Allergic Lung Diseases in Murine Studies by Flow Cytometric Analysis

1. Complete RPMI: RPMI 1640, 10% fetal bovine serum (v/v), 2 mM L-glutamine, 1% penicillin–streptomycin in RPMI (v/v).
2. Red blood cell lysis buffer: 0.15 M NH_4Cl, 10 mM $KHCO_3$, 0.1 mM Na_2EDTA; sterilize by filtration.
3. Phorbol myristate acetate (PMA): 1 mg/mL in DMSO, store at −20 °C.
4. Ionomycin (Iono): 1 mg/mL in EtOH, store at −20 °C.
5. BD GolgiStop™ protein transport inhibitor solution (BD Biosciences).
6. Fixation and Permeabilization Solution (BD Biosciences).

7. 10× BD Perm/Wash™ Buffer (BD Biosciences).
8. 4% Paraformaldehyde: add 4 g paraformaldehyde in 100 mL of PBS at 60 °C, add 10 N NaOH dropwise with stirring until dissolved.
9. FACS staining buffer: 0.5% bovine serum albumin (w/v), 0.01% sodium azide (w/v), 3 mM EDTA in PBS, pH 7.2.
10. Antibodies
    - Purified anti-mouse CD16/32 (Fc blocker).
    - Fluorochrome-conjugated anti-CD3 antibody (clone 145-2C11).
    - Fluorochrome-conjugated anti-CD4 antibody (clone GK1.5).
    - Fluorochrome-conjugated anti-IL-4 antibody (clone 11B11).
    - Fluorochrome-conjugated anti-IL-9 antibody (clone RM9A4).
11. Cytospin.
12. Diff-Quick stain kit.

2.4 Analysis of Antigen-Specific IL-9 Production by Th9 Cells by ELISA

1. Recombinant murine IL-9 cytokine: 10 μg/mL in complete RPMI.
2. Antibodies
    - Purified anti-mouse IL-9 clone D8402E8 (1 mg/mL).
    - Biotin anti-mouse IL-9 clone D9302C12 (0.5 mg/mL).
3. Blocking buffer: 10% fetal bovine serum (v/v), sterile PBS, pH 7.2.
4. Wash buffer: 0.5% Tween 20 (v/v), PBS, pH 7.2.
5. Av-HRP (Horseradish Peroxidase Avidin D): 5 mg/mL.
6. SIGMAFAST™ OPD (o-phenylenediamine dihydrochloride) (Sigma-Aldrich): dissolve one OPD tablet and one urea hydrogen peroxide tablet (individually packaged in foil packets) in 20 mL of distilled water immediately before use.
7. Stop solution: 2 N H_2SO_4.

3 Methods

3.1 Induction of Allergic Lung Diseases for Th9 Cell Detection

To detect Th9 cells in allergic lung diseases, allergic lung inflammation in experimental models is often induced using ovalbumin with alum, including in our lab. In our experience, higher numbers of Th9 cells are detected in a chronic allergic lung model compared to an acute allergic lung inflammation model.

1. On day 1, prepare a 1:1 (v/v) mixture of ovalbumin (OVA, 1 mg/mL) and alum by dropwise adding alum to the OVA solution, vortex the mixture 1 h before use (to obtain a final

homogenous solution). Fill a 1 mL syringe with the mixture and use 200 μL of OVA–alum per mouse (100 μg OVA/mouse) for intraperitoneal injection.

2. On day 14, repeat the intraperitoneal injection with a freshly prepared OVA–alum solution.
3. On the same day (day 14), the mice are intranasally challenged with 50 μL of 1 mg/mL sterile ovalbumin per mouse (50 μg OVA/mouse) by pipetting the solution into the nose of each anesthetized mouse.
4. On days 25, 26, and 27, the anesthetized mice are intranasally injected with 50 μL of 1 mg/mL sterile ovalbumin (50 μg OVA/mouse). Within 24 h after the last challenge, the mice are sacrificed for analysis of Th9 cells. During allergic inflammation, the function of IL-9 can be investigated by intraperitoneal injection with 100 μg anti-IL-9 antibody (clone MAB4091) or a control rat antibody at the time of challenge.

3.2 Analysis of Th9 Cells in Allergic Lung Diseases in Murine Studies

Bronchoalveolar lavage (BAL) fluid, lung draining lymph nodes, and spleen are collected from the sacrificed mice for the analysis of IL-9 production and IL-9-producing T cells. BAL fluid cells can be analyzed for infiltrating cells by cell counting and flow cytometric analysis. BAL fluid supernatant can be used for measuring the level of IL-9 production and other Th2 cytokines, including IL-4, IL-5, and IL-13 during allergic inflammation. Lung draining lymph nodes and spleen can be used to analyze Th9 cells by intracellular cytokine staining or to assess antigen-specific IL-9 production by ELISA.

3.2.1 BAL Fluid Collection and Analysis

1. Insert the needle catheter into the trachea toward the lungs, remove the needle, and secure the catheter with ligature.
2. Inject 0.5 mL of PBS–EDTA using 1 mL syringe and slowly retrieve through the catheter to wash the lungs three times (*see* **Note 1**).
3. Centrifuge BAL fluid at 400 × *g* for 5 min at 4 °C.
4. Collect supernatant, aliquot and store at −70 °C until further analysis of IL-9 by ELISA.
5. For total cell number enumeration and analysis, resuspend cells in 100 μL of PBS. Count and adjust cell number to 1×10^5 cells in 200 μL of total volume. Centrifuge in cytospin, allow to air-dry and stain using Diff-Quick stain kit by dipping sequentially into each solution of fixative solution (methanol), solution I (eosinophilic) and solution II (basophilic) for 5–6 times (or left for 10–15 s in each solution), followed by a water rinse and drying.

Table 1
Cell stimulation by PMA and Ionomycin

Tissue culture plate	Cell suspension (µL)	PMA (50 µg/mL) (dilute 1:20 from 1 mg/mL stock PMA) (µL/well)	Ionomycin (1 mg/mL) (µL/well)	GolgiStop™ (µL/well)
96-well plate	200	0.2	0.2	0.14
24-well plate	1000	1	1	0.7

3.2.2 Characterization of Th9 Cells in Allergic Lung Diseases by Flow Cytometric Analysis

1. Harvest lung draining lymph nodes and spleen in 1 mL of complete RPMI and keep on ice for the preparation of single cell suspension.
2. Prepare a sterile tissue culture dish (60 × 15 mm) with 1 mL of complete RPMI. Place an individual lymph node or spleen on a nylon mesh in the tissue culture dish with complete RPMI and tear it apart into a single cell suspension using the plunger of a 3 or 5 mL syringe.
3. Collect the cell suspension in 5 mL of complete RPMI and pass through a cell strainer to remove debris. Centrifuge the cell suspension at 400 × *g*, at 4 °C for 5 min.
4. For draining lymph nodes, resuspend the cells in 5 mL of complete RPMI and adjust the cell concentration to 4×10^6 cells/mL. For spleen cells, red blood cell lysis step is required for preparing single cell suspension before resuspending cells in complete RPMI and adjust cell concentration by adding 5 mL of red blood cell lysis buffer, followed by 5 mL of complete RPMI and centrifuge at 400 × *g*, at 4 °C for 5 min.
5. For intracellular cytokine staining analysis, add 1 mL of the cell suspension in complete RPMI into each well of a 24-well tissue culture plate.
6. Stimulate the cells with 50 ng/mL of PMA and 1 µg/mL of ionomycin in the presence of GolgiStop™ by preparing a master mix of 1 mg/mL of PMA (1:20,000), 1 mg/mL of ionomycin (1:1000), and GolgiStop™ (0.7 µL/mL) and add into each well (Table 1). Incubate the cell suspension at 37 °C in a CO_2 incubator for 4–5 h (*see* **Note 2**).
7. After stimulation, collect the cell suspension, transfer into a FACS tube and centrifuge at 400 × *g*, at 4 °C for 5 min. Wash once with 1 mL of FACS staining buffer before staining (*see* **Note 3**).

8. Prepare a master mix of fluorochrome-conjugated anti-mouse CD3 antibody (1:200), anti-mouse CD4 antibody (1:200) and anti-mouse CD16/32 (1:100) in FACS staining buffer, vortex the master mix thoroughly, and add 50 μL of master mix into the cell pellet. Incubate on ice in the dark for 30 min (*see* **Notes 4** and **6**).
9. Add 1 mL of FACS staining buffer into each tube and centrifuge at 400 × *g*, at 4 °C for 5 min.
10. Discard the supernatant, vortex the cell pellet (to avoid cell aggregation) and add 250 μL of Fixation/Permeabilization solution per tube. Incubate the cells on ice, 30 min (or at 4 °C, 20 min).
11. Add 750 μL of 1× BD Perm/Wash™ Buffer (dilute 10× BD Perm/Wash™ buffer in sterile distilled water) into each tube and centrifuge at 400 × *g*, at 4 °C for 5 min.
12. Prepare a master mix of fluorochrome-conjugated anti-mouse IL-4 antibody (1:20) and fluorochrome-conjugated anti-mouse IL-9 antibody (1:20) in 1× BD Perm/Wash™ Buffer, vortex the master mix thoroughly, and add 50 μL of master mix into the cell pellet. Incubate on ice in the dark for 30 min (*see* **Notes 5** and **6**).
13. Add 1 mL of 1× BD Perm/Wash™ Buffer into each tube, centrifuge at 400 × *g*, at 4 °C for 5 min.
14. Discard the supernatant and resuspend the cells in 150 μL of ice cold FACS staining buffer. Keep the cells in the dark on ice or at 4 °C in refrigerator until FACS analysis. For extended storage (>16 h), add 50 μL of 4% paraformaldehyde to stabilize staining pattern and inactivate most biohazardous agents.
15. For flow cytometric analysis, determine IL-9 producing cells (Th9) compared to IL-4 producing cells (Th2) in gated $CD3^+CD4^+$ T helper cells.

3.2.3 Analysis of Antigen-Specific IL-9 Production by Th9 Cells by ELISA

Preparation of Culture Supernatant for OVA-Specific IL-9 Production Assay

1. Prepare cell suspensions from lymph nodes and spleens as described in Subheading 3.2.1, **steps 1–4** to obtain cell suspensions at concentrations of 4 × 10^6 cells/mL.
2. Add 1 mL of cell suspension in complete RPMI per well to a 24-well tissue culture plate.
3. Stimulate cells with 0, 10, 50, and 250 μg/mL of ovalbumin and incubate in a 5% CO_2 incubator at 37 °C for 72 h.
4. Collect supernatant for ELISA of OVA-specific IL-9 production and keep at −20 °C until further analysis to prevent cytokine degradation (avoid freeze–thaw cycles) (*see* **Note 7**).

ELISA of OVA-specific IL-9 production

Day 1:

5. Dilute the capture antibody (purified anti-mouse IL-9) to a concentration of 1 μg/mL in PBS and coat each well of a 96-well plate with 100 μL of the prepared capture antibody at 4 °C, overnight (*see* **Note 8**).

Day 2:

6. Aspirate wells and wash once with >250 μL/well wash buffer. Blot plate on absorbent paper to remove any residual buffer. Block wells with 200 μL/well of blocking buffer at 37 °C for 1 h.
7. During the blocking period, thaw the samples and prepare the standard. Dilute the IL-9 standard (10 μg/mL) 1:1000 in blocking buffer (10 ng/mL final concentration) and vortex briefly. Prepare eight 2-fold serial dilutions of IL-9 (10 ng/mL) in blocking buffer, which will be used for a standard curve.
8. Aspirate wells and wash once with >250 μL/well wash buffer. Blot plate on absorbent paper to remove any residual buffer. Add 100 μL/well of sample or standard and incubate either at 37 °C for 2 h or at 4 °C, overnight.
9. Prepare the detection antibody by diluting biotin-conjugated anti-mouse IL-9 (0.5 mg/mL) 1:500 in blocking buffer (1 μg/mL final concentration).
10. Aspirate wells and wash five times with >250 μL/well wash buffer. Blot plate on absorbent paper to remove any residual buffer. Add 100 μL/well of detection antibody and incubate at room temperature for 1 h.
11. Dilute Av-HRP 1:2000 in blocking buffer (2.5 μg/mL final concentration).
12. Aspirate wells and wash five times with >250 μL/well of wash buffer. Blot plate on absorbent paper to remove any residual buffer. Add 100 μL/well of Av-HRP and incubate at room temperature in the dark for 30 min.
13. Prepare OPD substrate by dissolving one OPD tablet and one urea hydrogen peroxide tablet to obtain 20 mL of ready-to-use substrate.
14. Aspirate wells and wash five times with >250 μL/well of wash buffer. Blot plate on absorbent paper to remove any residual buffer. Add 100 μL/well of ready-to-use substrate and leave plate in the dark until a staining gradient is observed in the standard rows.
15. Terminate the reaction by adding 50 μL of stop solution to each well.

    Measure absorbance on a plate reader at 490 nm within 20 min and analyze the data.

4 Notes

1. For BAL fluid collection, it is very important that sample is kept cold during sample processing. The measurement of cytokine should be performed at the same time to avoid repeated freeze–thaw cycles with sample. If only BAL cells will be collected, it is better to use 1 mL of PBS–EDTA for collection.
2. During PMA/ionomycin stimulation, it is recommended to not activate cells more than 6 h since longer incubation time can result in more cell death.
3. For flow cytometric analysis, polystyrene round-bottom 12 × 75 mm tubes or FACS cluster tubes are used in the cell staining steps. During staining, the samples should be incubated in the dark on ice or in the refrigerator to prevent antibody internalization and a loss of fluorescence intensity at all steps. If cell debris is observed, it is necessary to filter cells through a 70 μm cell strainer prior to analysis.
4. Besides characterization of Th9 cells in lung draining lymph nodes, Th9 cells could be observed in lung cell suspension. Including CD44 antibody at the surface staining step may improve detection and show a better population of IL-9 producing T cells.
5. The brightest set of fluorochrome for flow cytometric analysis, particularly for intracellular cytokine analysis, should be selected. For Th9 cell detection, allophycocyanin (APC)-conjugated anti-IL-9 is recommended.
6. Antibody dilution for surface and intracellular cytokine staining may be varied and optimized following the recommendations of the supplier.
7. An analysis of OVA-specific Th2 cytokines, including IL-4, IL-5, and IL-13, should be performed to investigate the function of both Th2 and Th9 cells.
8. For the analysis of antigen-specific IL-9 production by Th9 cells by ELISA, use only high affinity protein binding 96-well plates, e.g., Corning Costar 9018 or NUNC MaxiSorp™ 96-well plates to obtain optimal performance. Do not use tissue culture plates or low protein absorption plates.

Acknowledgment

We gratefully thank all members of the Angkasekwinai lab and the Chen Dong lab at the University of Texas MD Anderson Cancer Center for optimization of this protocol and for lab training.

References

1. Devos S, Cormont F, Vrtala S, Hooghe-Peters E, Pirson F, Snick J (2006) Allergen-induced interleukin-9 production in vitro: correlation with atopy in human adults and comparison with interleukin-5 and interleukin-13. Clin Exp Allergy 36(2):174–182. doi:10.1111/j.1365-2222.2006.02422.x
2. Erpenbeck VJ, Hohlfeld JM, Volkmann B, Hagenberg A, Geldmacher H, Braun A, Krug N (2003) Segmental allergen challenge in patients with atopic asthma leads to increased IL-9 expression in bronchoalveolar lavage fluid lymphocytes. J Allergy Clin Immunol 111(6): 1319–1327
3. Ying S, Meng Q, Kay AB, Robinson DS (2002) Elevated expression of interleukin-9 mRNA in the bronchial mucosa of atopic asthmatics and allergen-induced cutaneous late-phase reaction: relationships to eosinophils, mast cells and T lymphocytes. Clin Exp Allergy 32(6):866–871
4. Temann UA, Geba GP, Rankin JA, Flavell RA (1998) Expression of interleukin 9 in the lungs of transgenic mice causes airway inflammation, mast cell hyperplasia, and bronchial hyperresponsiveness. J Exp Med 188(7):1307–1320
5. McMillan SJ, Bishop B, Townsend MJ, McKenzie AN, Lloyd CM (2002) The absence of interleukin 9 does not affect the development of allergen-induced pulmonary inflammation nor airway hyperreactivity. J Exp Med 195(1):51–57
6. Kung TT, Luo B, Crawley Y, Garlisi CG, Devito K, Minnicozzi M, Egan RW, Kreutner W, Chapman RW (2001) Effect of anti-mIL-9 antibody on the development of pulmonary inflammation and airway hyperresponsiveness in allergic mice. Am J Respir Cell Mol Biol 25(5):600–605. doi:10.1165/ajrcmb.25.5.4533
7. Schmitt E, Germann T, Goedert S, Hoehn P, Huels C, Koelsch S, Kuhn R, Muller W, Palm N, Rude E (1994) IL-9 production of naive CD4+ T cells depends on IL-2, is synergistically enhanced by a combination of TGF-beta and IL-4, and is inhibited by IFN-gamma. J Immunol 153(9):3989–3996
8. Veldhoen M, Uyttenhove C, van Snick J, Helmby H, Westendorf A, Buer J, Martin B, Wilhelm C, Stockinger B (2008) Transforming growth factor-beta 'reprograms' the differentiation of T helper 2 cells and promotes an interleukin 9-producing subset. Nat Immunol 9(12):1341–1346. doi:10.1038/ni.1659
9. Wong MT, Ye JJ, Alonso MN, Landrigan A, Cheung RK, Engleman E, Utz PJ (2010) Regulation of human Th9 differentiation by type I interferons and IL-21. Immunol Cell Biol 88(6):624–631. doi:10.1038/icb.2010.53
10. Angkasekwinai P, Chang SH, Thapa M, Watarai H, Dong C (2010) Regulation of IL-9 expression by IL-25 signaling. Nat Immunol 11(3):250–256. doi:10.1038/ni.1846
11. Chang HC, Sehra S, Goswami R, Yao W, Yu Q, Stritesky GL, Jabeen R, McKinley C, Ahyi AN, Han L, Nguyen ET, Robertson MJ, Perumal NB, Tepper RS, Nutt SL, Kaplan MH (2010) The transcription factor PU.1 is required for the development of IL-9-producing T cells and allergic inflammation. Nat Immunol 11(6):527–534. doi:10.1038/ni.1867
12. Staudt V, Bothur E, Klein M, Lingnau K, Reuter S, Grebe N, Gerlitzki B, Hoffmann M, Ulges A, Taube C, Dehzad N, Becker M, Stassen M, Steinborn A, Lohoff M, Schild H, Schmitt E, Bopp T (2010) Interferon-regulatory factor 4 is essential for the developmental program of T helper 9 cells. Immunity 33(2):192–202. doi:10.1016/j.immuni.2010.07.014
13. Goswami R, Kaplan MH (2011) A brief history of IL-9. J Immunol 186(6):3283–3288. doi:10.4049/jimmunol.1003049
14. Niedbala W, Besnard AG, Nascimento DC, Donate PB, Sonego F, Yip E, Guabiraba R, Chang HD, Fukada SY, Salmond RJ, Schmitt E, Bopp T, Ryffel B, Liew FY (2014) Nitric oxide enhances Th9 cell differentiation and airway inflammation. Nat Commun 5:4575. doi:10.1038/ncomms5575
15. Yao W, Zhang Y, Jabeen R, Nguyen ET, Wilkes DS, Tepper RS, Kaplan MH, Zhou B (2013) Interleukin-9 is required for allergic airway inflammation mediated by the cytokine TSLP. Immunity 38(2):360–372. doi:10.1016/j.immuni.2013.01.007
16. Sehra S, Yao W, Nguyen ET, Glosson-Byers NL, Akhtar N, Zhou B, Kaplan MH (2015) TH9 cells are required for tissue mast cell accumulation during allergic inflammation. J Allergy Clin Immunol 136(2):433–440.e431. doi:10.1016/j.jaci.2015.01.021
17. Jones CP, Gregory LG, Causton B, Campbell GA, Lloyd CM (2012) Activin A and TGF-beta promote T(H)9 cell-mediated pulmonary allergic pathology. J Allergy Clin Immunol

129(4):1000–1010.e1003. doi:10.1016/j.jaci.2011.12.965

18. Li H, Edin ML, Bradbury JA, Graves JP, DeGraff LM, Gruzdev A, Cheng J, Dackor RT, Wang PM, Bortner CD, Garantziotis S, Jetten AM, Zeldin DC (2013) Cyclo-oxygenase-2 inhibits T helper cell type 9 differentiation during allergic lung inflammation via down-regulation of IL-17RB. Am J Respir Crit Care Med 187(8):812–822. doi:10.1164/rccm.201211-2073OC
19. Kara EE, Comerford I, Bastow CR, Fenix KA, Litchfield W, Handel TM, McColl SR (2013) Distinct chemokine receptor axes regulate Th9 cell trafficking to allergic and autoimmune inflammatory sites. J Immunol 191(3):1110–1117. doi:10.4049/jimmunol.1203089
20. Kerzerho J, Maazi H, Speak AO, Szely N, Lombardi V, Khoo B, Geryak S, Lam J, Soroosh P, Van Snick J, Akbari O (2013) Programmed cell death ligand 2 regulates TH9 differentiation and induction of chronic airway hyperreactivity. J Allergy Clin Immunol 131(4):1048–1057 . doi:10.1016/j.jaci.2012.09.0271057.e1041–1042
21. Brough HA, Cousins DJ, Munteanu A, Wong YF, Sudra A, Makinson K, Stephens AC, Arno M, Ciortuz L, Lack G, Turcanu V (2014) IL-9 is a key component of memory TH cell peanut-specific responses from children with peanut allergy. J Allergy Clin Immunol 134(6):1329–1338.e1310. doi:10.1016/j.jaci.2014.06.032
22. Hoppenot D, Malakauskas K, Lavinskiene S, Bajoriuniene I, Kalinauskaite V, Sakalauskas R (2015) Peripheral blood Th9 cells and eosinophil apoptosis in asthma patients. Medicina (Kaunas) 51(1):10–17. doi:10.1016/j.medici.2015.01.001
23. Ma L, Xue HB, Guan XH, Shu CM, Zhang JH, Yu J (2014) Possible pathogenic role of T helper type 9 cells and interleukin (IL)-9 in atopic dermatitis. Clin Exp Immunol 175(1):25–31. doi:10.1111/cei.12198
24. Yao W, Tepper RS, Kaplan MH (2011) Predisposition to the development of IL-9-secreting T cells in atopic infants. J Allergy Clin Immunol 128(6):1357–1360.e1355. doi:10.1016/j.jaci.2011.06.019

Chapter 16

Characterization of Th9 Cells in the Development of EAE and IBD

Sakshi Malik, Valerie Dardalhon, and Amit Awasthi

Abstract

Encephalitogenic and colitogenic effector T cells have been implicated in the induction of experimental autoimmune encephalomyelitis (EAE) and inflammatory bowel disease (IBD), respectively. Effector functions of Th1 and Th17 cells have been well characterized and described for the induction and development of EAE and IBD; however, the recently identified Th9 cells have also been suggested to play an important role in these autoimmune pathologies. Th9 cells, primarily characterized by their high level of production of IL-9, are not only essential in controlling extracellular pathogens but also contribute to the development of autoimmunity and allergic inflammation. Furthermore, it was also demonstrated that IL-9 promotes Th17 cell-mediated tissue pathology in EAE and it compromises the barrier functions of the gut in IBD. In vivo adoptive transfer of in vitro differentiated Th9 cells induces the development of autoimmune tissue inflammation in EAE and IBD. Here we describe methods for in vitro differentiation of naïve murine CD4+ T cells to generate IL-9-producing Th9 cells and follow their effector functions in EAE and IBD murine models.

Key words T cell differentiation, CD4, TGF-β, IL-4, Th9, IL-9, IL-2, EAE, IBD, Autoimmunity

1 Introduction

Upon TCR stimulation, naïve CD4+ T cells can differentiate into various subsets of helper T (Th) cells, depending on the nature of the antigen and cytokine milieu at the time of T cell priming. In this context, several Th subsets have initially been characterized based on the expression of specific transcription factors and their cytokine profile. Thus, Th1, Th2, and Th17 cells have been characterized phenotypically on the basis of their production of IFN-γ, IL-4/IL-5, and IL-17, respectively [1]. Furthermore, these Th subsets also induce distinct effector functions in various infections and autoimmune diseases [1]. In addition to their Th counterparts, Th9 cells were recently identified on the basis of their capacity to produce IL-9 and have been found to be associated with the pathogenesis of autoimmune inflammation such as EAE and IBD

Ritobrata Goswami (ed.), *Th9 Cells: Methods and Protocols*, Methods in Molecular Biology, vol. 1585,
DOI 10.1007/978-1-4939-6877-0_16, © Springer Science+Business Media LLC 2017

as well as allergic inflammation [2–6]. While Th9 cells have been shown to be essential for eliminating extracellular infections, they are also endowed with potent antitumor functions [7–9]. Before the identification of Th9 cells, IL-9-producing T cells were initially thought to be associated with Th2 cells, and therefore its effector functions have been characterized and validated in Th2-biased mouse models. In addition, CD4+ T cells were shown to produce IL-9 in response to IL-2, IL-4, and TGF-β [10]. While Th17 cells were shown to produce IL-9 in EAE, IL-9-producing Treg cells were also described in tolerant allograft and required for allograft survival [11]. Overall, IL-9 production appears to be associated with T cells and implicated in inflammation and immune-suppression. It is however important to note that although variable amounts of IL-9 can be produced by Th2 cells, Th17 and Treg cells in different disease conditions, specialized IL-9-producing cells have been identified [2, 3]. We and others have demonstrated that priming of naïve T cells in presence of TGF-β together with IL-4 induce their differentiation to a Th9 fate, which primarily produce IL-9 without inducing the production of signature cytokines of Th1, Th2 or Th17 cells [2, 3]. More precisely, TGF-β stimulation induces the expression of Foxp3 while IL-4 antagonizes the effect of TGF-β on Foxp3 induction and the resulting cells produce IL-9 [2]. Ever since the discovery of Th9 cells, both Th9 promoting and inhibiting cytokines were also identified in various disease conditions. While IL-1, IL-2, IL-25, TSLP, and IL-33 contribute to promote Th9 differentiation, IFN-γ and IL-27 limit their development [9, 10, 12–16].

It has been demonstrated that the transcription factors IRF4 and PU.1 are crucial for the development Th9 cells. IL-4-mediated induction of IRF4 is essential for the development of Th9 cells [17, 18]. IRF4 binds and transactivates *Il9* promoter in murine Th9 cells. Furthermore IRF4-deficient mice failed to develop IL-9-dependent pathologies in allergic inflammation [17]. Similarly, IRF4-deficient T cells failed to develop into Th9 cells under optimum polarization, suggesting IRF4 is critical for the development of Th9 cells [17]. In addition to IRF4, PU.1 (encoded by *Sfpi1* in mouse) has also been shown to be essential for the development of Th9 cells [19]. *Sfpi1*-deficient T cells failed to differentiate into Th9 cells, and demonstrated decreased airway inflammation [19]. More recently, the role of *Sfpi1* and IL-9 were found to be associated in ulcerative colitis [4, 5]. In addition, it has been shown recently that the transcription factors BATF and IRF1 also play an essential role in the development of Th9 cells [9, 20].

Current literature has suggested that the effector functions of Th9 cells are not only essential for the induction of allergic inflammation in asthma but also shown to be critical for skin inflammation and antitumor immunity [8–10, 21]. Although the functions of Th9 cells were described in EAE and IBD, the mechanism by which Th9 cells mediate disease pathology in these autoimmune diseases is not clearly understood. Especially, the potential interaction of Th9 cells with Th1 and Th17 cells, which are the major effector cells in these models, are not known. Nonetheless, Th9 cells might also be necessary to exacerbate immunopathology in EAE and IBD partly by enhancing the functions of other effector T cells. It is well established that myelin-reactive T cells can transfer EAE in various mouse models [22, 23]. Both myelin antigen specific Th1 and Th17 cells can transfer EAE with distinct immunopathology [6]. While transferring Th1 cells into mice induces monocyte/macrophage infiltration in the central nervous system (CNS), adoptive transfer of Th17 cells induces neutrophil infiltration [24]. In addition to Th1 and Th17 cells, Th9 cells are also essential for the development of EAE, as IL-9 neutralization and deficiency ameliorates EAE with abrogation of effector T cells functions, especially Th17 cells [6, 25]. TGF-β and IL-4-induced Th9 cells produce large amounts of IL-9 and IL-10 in mice [2]. The role of myelin-specific Th9 cells has been tested in the induction of EAE in an adoptive transfer model [6].

IBD such as Crohn's disease (CD) and ulcerative colitis (UC) are intestinal inflammation characterized by chronic destructive, inflammatory disorders of the gastrointestinal tract that lead to immunopathology-mediated weight loss, diarrhea, and abdominal pain. In addition to genetic and environmental triggers, infiltration of mucosal surfaces of intestine with colitogenic T cells leads to development of IBD. Classically, Th1/Th17 cells were associated with CD, while Th2 cells were found to be associated with UC [26–28]. On the other hand, Th17 cells were found to be associated with both forms of IBD. As observed in EAE, Th9 cells were also shown to be essential for the induction and development of IBD [5].

In this protocol, we outline the in vitro differentiation and characterization of Th9 cells and their use to induce the development of EAE and IBD in adoptive transfer models.

2 Materials

Mice: C57BL/6, 2D2 TcR transgenic (Generated in V. Kuchroo's laboratory) [22], Rag-1-deficient mice, Foxp3-GFP reporter mice (Generated in V. Kuchroo's laboratory) [29], and 2D2 × Foxp3-GFP reporter mice.

2.1 Cell Culture Media and Other Reagents

1. Dulbecco's phosphate buffered saline (DPBS; pH 7.4) (*see* **Note 1**), Hank's balanced salt solution (HBSS): NaCl 138 mM, KCl 5.33 mM, KH_2PO_4 0.44 mM, glucose 5.60 mM, phenol red 0.03 mM, NaH_2PO_4 0.3 mM.

2. T cell culturing clone medium: DMEM, 100 mM sodium pyruvate, 200 mM L-glutamine, 10,000 U/mL penicillin–streptomycin, nonessential amino acids (10 mM each of glycine, L-alanine, L-asparagine, L-aspartic acid, L-glutamic acid, L-proline, and L-serine), 10% fetal bovine serum (v/v), 1% Folic acid (w/v), 1% vitamin mix solution (v/v), 50 μM 2-mercaptoethanol, gentamicin (optional).
3. ACK lysing buffer.
4. Magnetic cell sorting (MACS) beads for CD4+ T cell purification (Miltenyi Biotec).
5. Phorbol myristate acetate (PMA), stock 0.1 mg/mL.
6. Ionomycin (stock 1 mg/mL).
7. Monensin.
8. MACS buffer: DPBS (pH 7.4), 2% BSA/FBS (v/v), 2.0 mM EDTA.
9. Seventy micron tissue culture grade cell strainer, 96-well flat bottom cell culture plate and 96-well V bottom plate.
10. Collagenase D.
11. DNase 1.
12. Percoll.

2.2 T Cell Stimulation/Polarization

1. Anti-mouse CD3 (clone: 145-2C11) and anti-mouse CD28 (clone: 37.51) antibodies for T cell stimulation.
2. Recombinant human TGF-β and recombinant murine IL-4.
3. Syngenic irradiated antigen presenting cells (APCs).
4. Fluorescent conjugated anti-mouse CD25 (clone: PC61), CD62L (clone: MEL-14), and CD4 (clone: RM4-5), IL-9 (RM9A4), IL-17 (TC11-18H10.1), IFN-γ (XMG1.2) antibodies for performing intracellular cytokine staining of differentiated Th9 cells.
5. Fixation/Permeabilization Solution Kit (BD Biosciences).

2.3 Realtime PCR Reagents

1. RNAeasy mini kit.
2. iScript cDNA synthesis kit (Bio-Rad).
3. TaqMan® Fast Universal PCR Master Mix (Life Technologies).
4. Taqman probe IL-9, GAPDH, IL-4, Foxp3, IRF4, Sfpi1 (Life Technologies).

3 Methods

3.1 Generation of Th9 Cells In Vitro

In this protocol, murine spleen and lymph node cells are isolated from either WT C57BL/6 or Foxp3-GFP.KI mice and enriched for total CD4+ T cells. Naïve CD4+ T cells were further sorted

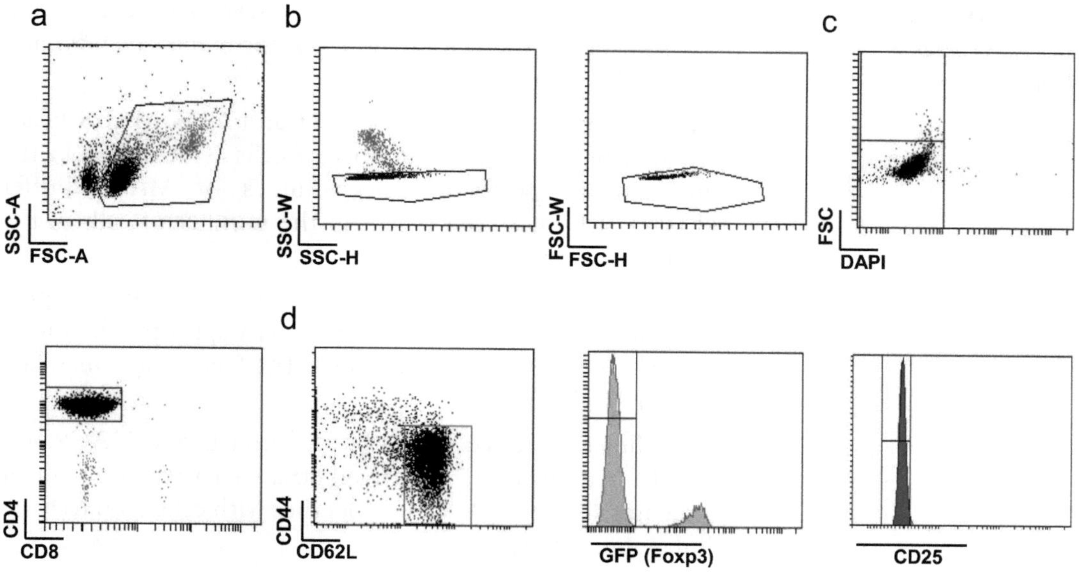

Fig. 1 Gating strategy for sorting mouse naïve CD4 T cells by FACS. (**a**) Forward scatter count (FSC) and side scatter count (SSC) of enriched CD4+ T cell population is shown. (**b**) Linear appearance of FSC/SSC-Area (A) and FSC/SSC-Height (H) to exclude doublet cells. (**c**) After doublet exclusion, live single cells are further gated. (**d**) CD4+ cells are then further gated on CD62L^{+}CD44^{-}Foxp3GFP^{-}(or CD25^{-})

using either CD44^{low}CD62L^{high}CD25^{-} or CD44^{low}CD62L^{high}Foxp3.GFP^{-} parameters (Fig. 1). After their purification, naïve CD4+ T cells are stimulated either on plate bound anti-CD3 and anti-CD28 antibodies or using syngenic irradiated APCs and soluble anti-CD3, in the presence of recombinant TGF-β and IL-4 for 3–4 days. Given below are the detailed steps of protocol from isolation of naïve CD4+ T cells to Th9 cell differentiation.

3.1.1 Preparation of Single Cell Suspension from Spleen and Lymph Nodes

1. Aseptically isolate spleen and lymph nodes (sub-iliac, axillary, mandibular, and parotid), make single cell suspension by mincing them.
2. Pass the single cell suspension through 70 μm cell strainer with cold (4 °C) DPBS/HBSS in a sterile manner. This step is essential to remove tissue debris and cellular aggregates.
3. Wash the cells with appropriate volumes of cold DPBS before proceeding to the next step.
4. Lyse red blood cell (RBC) using ACK lysing buffer. Mix the equivalent volume of ACK lysing buffer to the cell pellet, resuspend the cell pellet gently and swirl the tube for 30–60 s. Neutralize the ACK lysing buffer by filling the tube with serum free media and centrifuge it to collect the RBCs free cell pellet. Alternatively, RBC lysis is not essential, as RBCs will be eliminated in negative cellular fraction during magnetic enrichment of CD4+ T cells.

3.1.2 CD4+ T Cells Enrichment Using Magnetic Bead

1. Count the number of live cells (e.g., trypan blue exclusion dye), spin down the cells at 300 × *g* for 5 min at 4 °C and replace DPBS with MACS buffer.
2. Add appropriate volume/amount of anti-CD4 MACS beads calculated on the basis of number of cells as described in the manufacturer's protocol. The volumes of MACS buffer and beads are calculated based on the number of cells in the preparation.
3. Spleen and peripheral lymph nodes of one mouse usually provide about 1×10^8 mononuclear cells. Mix appropriate ratio of single cell suspension and anti-CD4 MACS beads and incubate them for 10–15 min at 4 °C.
4. Wash cellular suspension-CD4 beads with ice-cold MACS buffer and filter through 70 μm cell strainer to remove cellular aggregates, as anti-CD4 beads together with cells tend to form cellular aggregates.
5. Isolate CD4+ and CD4− cellular fractions using either manual MACS LS column or by an automated MACS separator according to manufacturer's protocol.
6. Wash and collect enriched CD4+ fraction of T cells before resuspending them into T culture medium and keep them on ice for further processing.
7. Collected CD4 negative fractions of cells in above protocol can be utilized as syngenic APCs (irradiated) in Th9 cell differentiation.

3.1.3 Purification of Naïve CD4+ Cells Using FACS Sorting

Following steps are followed to further isolate the naïve CD4+ T cells from total CD4+ T cells by high speed cell sorting:

1. Stain MACS-purified CD4+ T cells with anti-CD44, anti-CD62L, anti-CD25, and anti-CD4 mouse monoclonal antibodies in clone medium for 10–15 min at 4 °C.
2. Sort $CD4^+CD62L^{high}CD44^{low}CD25^-$/Foxp3GFP$^-$ naïve CD4+ T cells by flow cytometry using the detailed gating strategy (Fig. 1) and verify the purity after sorting (the purity needs to be over 95%) (*see* **Note 2**).
3. FACS-sorted naïve CD4+ T cells can be stored in T cell culture medium on ice for couple of hours until before adding cytokines for Th9 differentiation.

3.1.4 In Vitro Cell Culture for Th9 Differentiation

1. Coat cell culture-treated flat bottom plates with anti-CD3 (2 μg/mL) and anti-CD28 (2 μg/mL) mouse monoclonal antibodies diluted in sterile DPBS for 2–4 h at 37 °C or overnight at 4 °C (*see* **Notes 3** and **4**).
2. Remove the coating buffer by gently flipping the plate. To avoid drying the plate, add 100 μL of Th9 cell culture medium

to the plate with or without TGF-β at 4 ng/mL plus IL-4 at 40 ng/mL (Th9 and Th0 conditions, respectively) (*see* **Note 5**).

3. When performing the Th9 polarization of naïve CD4+ T cells into 96-well plate, add 100 μL of cell suspension at 1 × 10^6/mL of naïve CD4+ T cells per well, which will bring the final concentration of cytokines to the following (TGF-β: 2 ng/mL and IL-4: 20 ng/mL). Keep the cultured cells in CO_2 incubator (10% CO_2) at 37 °C for up to 4 days.
4. To obtain higher amount of Th9 cells, the polarization can also be performed in 24- or 6-well tissue culture plates as well as in 90 mm petri dish.
5. Alternatively, depending on the requirements of the downstream experiments, Th9 cell polarization from naïve CD4+ T cells can also be performed using either soluble anti-CD3 (1.0 μg/mL; in case of WT CD4+ T cells) or MOG35-55 peptide (20 μg/mL; in case of 2D2 CD4+ T cells) in the presence of irradiated syngenic splenocytes/APCs (3400 rads; five times the amount of naïve CD4+ T cells being used) (*see* **Note 6**).

3.2 Characterization of Th9 Cells

The efficiency of polarization of naïve CD4+ T cells towards Th9 fate can be evaluated by assessing the level of IL-9. IL-9 production can be measured by multiple and complementary approaches such as intracellular cytokine staining, ELISA (or bead-based cytokine array) and qPCR (Figs. 2a, b and 3a). However, Th9 differentiation efficiency can also be estimated by monitoring the expression level of several transcription factors (PU1, IRF4, BATF) (Fig. 3b and c), which are specifically associated with Th9 cell differentiation (*see* **Note 7**).

1. Quantitative PCR analysis, flow cytometric analysis, cytokine bead array, and enzyme-linked immunosorbent assay (ELISA) are straightforward methods to assess expression and quantify the production of IL-9 protein.
2. The level of IL-9 can be assessed at per cell basis by intracellular cytokine staining following T cell restimulation using phorbol 12-myristate 13-acetate (PMA) and ionomycin (*see* **Note 8**).
3. Complementary, IL-9 production can be detected in the supernatant of polarized T cells collected at 48 h following stimulation by sandwich ELISA (or bead-array based cytokine assays), which is a quantitative method using anti-mouse IL-9 antibodies.

3.2.1 Expression Analysis by qPCR

1. Briefly, harvest the cells at 48–72 h to isolate RNA using RNAeasy kit.
2. Reverse-transcribe mRNA into cDNA with cDNA synthesis kit.

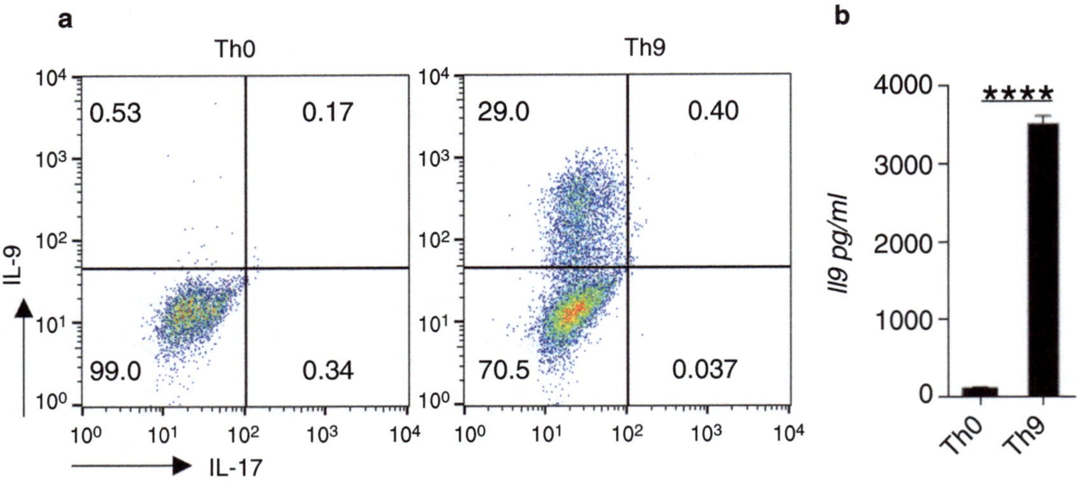

Fig. 2 Determination of IL-9 by FACS and ELISA. (**a**) A flow cytometric strategy using forward scatter count (FSC) and side scatter count (SSC) to identify the cell population is shown. Exclude dead cells by appropriate marker (Live/Dead marker in this case). After dead cell exclusion, live cells are further analyzed for the expression of IL-17 and IL-9. In vitro differentiated Th9 cells will appear as IL-9 single positive and will be negative for IL-17. (**b**) Collect supernatant after 48 h of stimulation of naïve CD4+ T cells into Th0 and Th9 culture condition, measure the quantity of IL-9 by IL-9 ELISA as mentioned

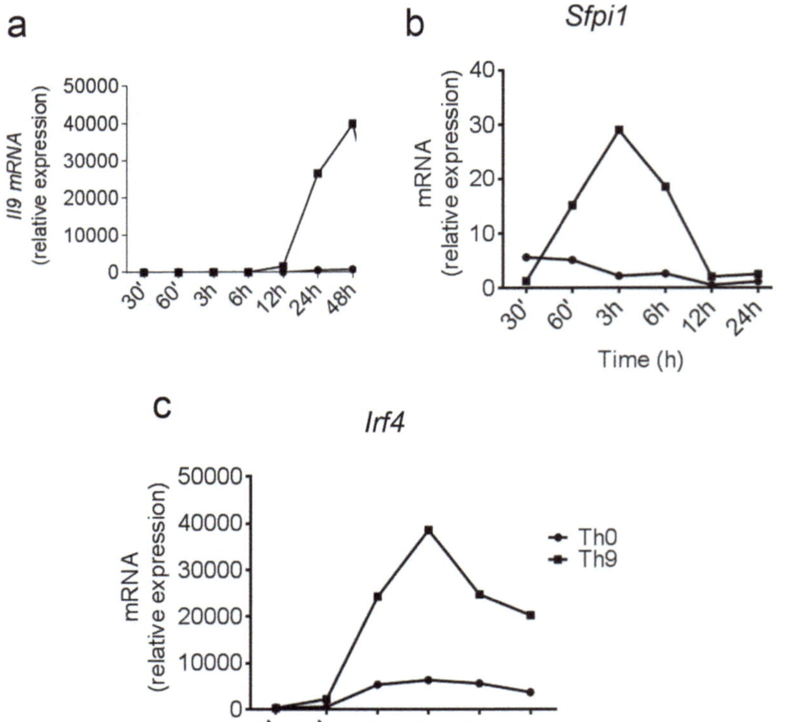

Fig. 3 Expression profiles of in vitro differentiated Th9 cell. Sorted naïve CD4+ T cells were sorted and cultured into Th0 and Th9 culture condition, 48 h later total RNA was extracted and converted into cDNA, expression of IL-9, PU.1 and IRF4 was determined as mentioned in the protocol

3. Determine the relative level of expression of *Il9* mRNA (normalized to housekeeping gene expression level) using IL-9 Taqman probe and appropriate buffer mix (Life Technologies) (*see* **Note 9**).

3.2.2 Intracellular Cytokine Staining Protocol

1. Restimulate differentiated Th9 cells by PMA (50 ng/mL), ionomycin (1.0 μg/mL), and GolgiStop (1.0 μL/mL) for 4–5 h at 37 °C.
2. Wash the cells with ice cold DPBS and transfer them into V-bottom 96-well plate or FACS tubes and perform the following steps for intracellular cytokine staining.
3. Stain the cells with anti-CD4 (1:200 final dilution as mentioned above). Use 50–100 μL of antibody mix for the staining in plate or FACS tube, respectively, and incubate for 15–20 min at 4 °C.
4. Wash the cells in flow buffer by centrifuging them at 300 × *g* for 5 min at 4 °C.
5. Discard the wash buffer and add 100 μL of 2% PFA or Cytofix (BD Biosciences) on top of cells and mix it gently, fix the cells by incubating them at RT for 10 min or at 4 °C for 20–30 min. Remove and discard the fixing solution by centrifuging the cells at 300 × *g* for 5 min.
6. Resuspend the cells in flow buffer and keep them at 4 °C overnight or process directly with the intracellular staining as per the requirement.
7. Wash the fixed cells once in Wash/Perm buffer.
8. Centrifuge cells for 5 min at 300 × *g* and resuspend the cell pellet in 100–200 μL of Perm/Wash buffer, incubate them for 15 min at RT.
9. Centrifuge cells for 5 min at 300 × *g* and resuspend the cells in Perm/Wash buffer containing:

   Anti-mouse IL-9 (clone RM9A4): 1:200 dilution.

   Anti-mouse IFN-γ (clone XMG1.2): 1:200 dilution.

   Anti-mouse IL-4: (clone 11B11): 1:200 dilution.

   Anti-mouse IL-2: (clone JES6-1A12): 1:200 dilution.

   Anti-mouse IL-10: (clone JES5-16E3): 1:200 dilution).

   All antibody dilution must be prepared in the Perm/Wash buffer.
10. Stain the cells by incubating the antibodies cocktail for 30 min at 4 °C *(see* **Note 10**). Wash twice in Perm/Wash buffer and resuspend cells in flow buffer and run the samples on a flow cytometer and record 20,000–30,000 events/sample.

11. *Il9* mRNA and IL-9 protein are usually easily detectable within the cell and in the culture supernatant, respectively. However, the detection of intracellular IL-9 cytokine by intracellular cytokine staining is challenging (*see* **Notes 11** and **12**).

3.3 Adoptive Transfer of In Vitro Differentiated Th9 Cells and EAE, IBD Induction

The role Th9 cells has been suggested in EAE [25], as Th9 cells can adoptively transfer EAE [6]. Here, we briefly describe the essential steps of the protocol of Th9 cells mediated induction of EAE in adoptive transfer model.

3.3.1 Th9 and EAE Induction

1. For EAE induction, isolate naïve CD4+ T cells from 2D2 TcR transgenic mice as described earlier (*see* **Note 13**).
2. Stimulate and differentiate cells into Th9 cells as described above, harvest and wash the cells on day 4 and rest them on non-coated tissue culture plate.
3. After 48 h of resting, restimulate the cells with plate bound anti-CD3 (1.0 μg/mL) and anti-CD28 (1.0 μg/mL) for 48 h.
4. Harvest the cells and extensively wash them with ice cold PBS to remove the traces of media and estimate the amount of live cells number using trypan blue dye to exclude the dead population.
5. Transfuse 3–5 × 10^6 differentiated Th9 cells intravenously into C57BL/6 host. On day 0 and 2 after Th9 cell transfer, inject 150–200 ng of pertussis toxin per mouse, intravenously.
6. Follow the mouse every day and score them for the sign of EAE development according to the following criteria: 0, no disease; 1 decrease tail tone; 2, hind limb weakness and partial paralysis; 3, complete hind limb paralysis; 4, forelimb and hind limb paralysis; 5, moribund state (*see* **Note 14**) (Fig. 4).
7. To further understand the in vivo phenotype of transferred Th9 cells to induce EAE, cells can be recovered from the CNS to test their cytokine profile and CNS tissue can be harvested and monitored for counting the CNS lesions.
8. Sacrifice the recipient mice at the peak of disease (or at other time points as per the requirement of the experiment), perfuse them through the left cardiac ventricle with DPBS. This step is very critical to flush out all the blood from the CNS.
9. After perfusion, CNS tissue can be stored into formalin buffer for further histopathological analysis (*see* **Note 15**).
10. Remove the brain and spinal cord carefully and chop them into small pieces and digest them for 30 min and 37 °C with continuous rotation with collagenase D and DNase 1 at concentrations of 2.5 mg/mL and 1.0 mg/mL, respectively.
11. After the digestion, the tissue is disintegrated and can easily mesh by rubbing them with the shaft of the syringe on 70 μm

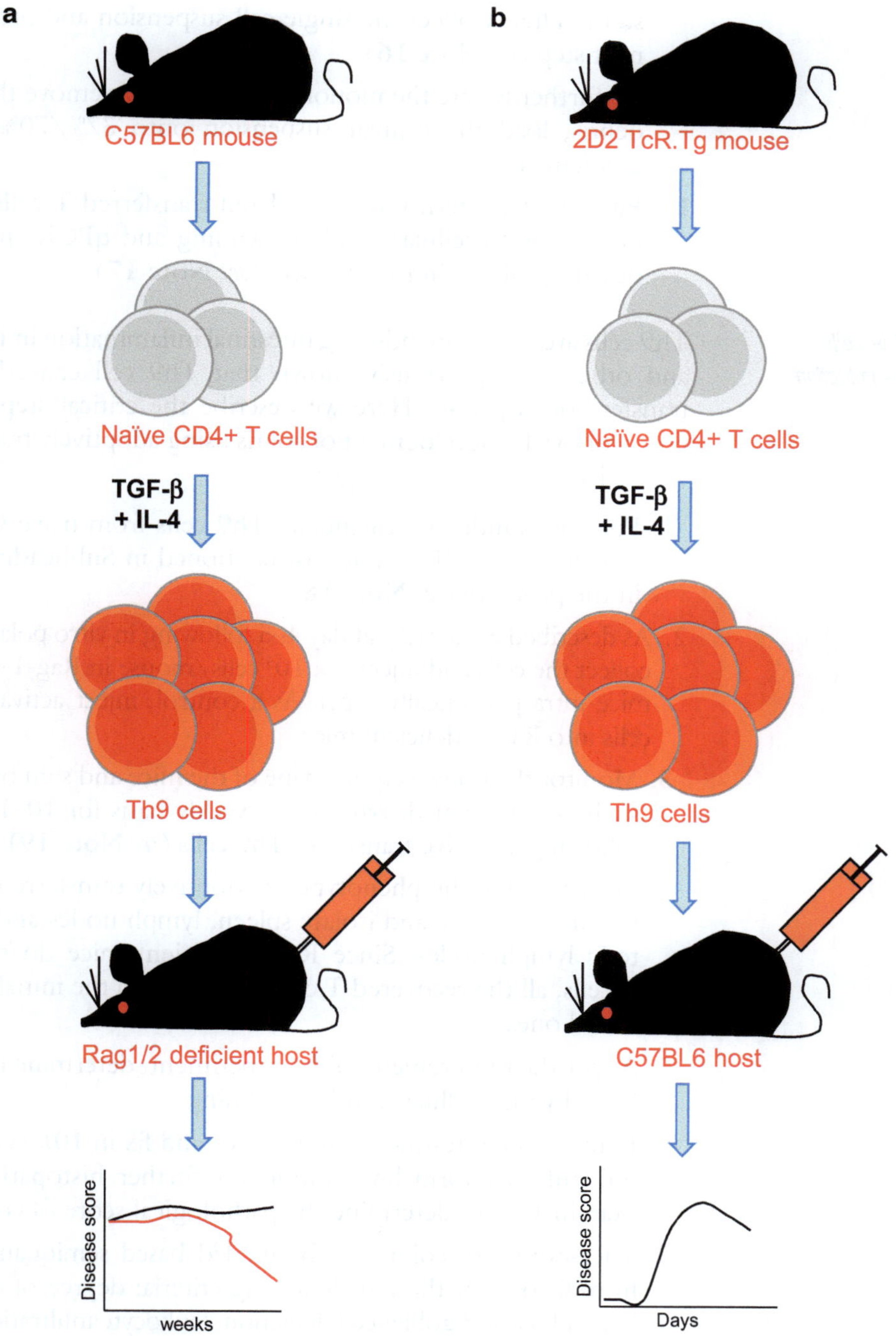

Fig. 4 Schematic and diagrammatic presentation of Th9 differentiation and adoptive transfer to induce EAE and colitis. Naïve CD4+ T cells sorted either from C57BL/6 (WT) mouse (for colitis induction) or 2D2 transgenic mouse (for EAE induction). Sorted naïve T cells are differentiated into Th9 cells with TGF-β/IL-4. Differentiated Th9 cells transferred into either Rag-1-deficient mice (for colitis induction) or C57BL/6 WT mice (for EAE induction). In case of EAE induction, PT was injected (200 ng/mouse) on day 0 and 2. Mice were followed for the signs of EAE and colitis

sterile filter. Collect the single cell suspension and proceed to next step (*see* **Note 16**).

12. To further isolate the mononuclear cells and remove the tissue debris, load the cellular suspension onto 37%/70% percoll gradient [6].
13. Further characterization of isolated transferred T cells can be done by intracellular cytokine staining and qPCR analysis as described above in the protocol (*see* **Note 17**).

3.3.2 Th9 Cells and Colitis Induction

Th9 cells are critical in inducing intestinal inflammation in IBD. We and other have previously shown that Th9 cell can adoptively transfer colitis [2, 4]. Here we describe the critical steps of the protocol for the development of colitis using adoptively transferred Th9 cells.

1. For colitis induction, generate Th9 cells from naïve CD4+ T cells from C57BL/6 mice as mentioned in Subheading 3.1.4 in the protocol (*see* **Note 18**).
2. As described earlier [2], at day 4–5 following in vitro polarization, collect the cells and inject 5×10^5 cells/mouse in Rag-1-deficient mice intra peritoneally (i.p.). As a control, inject activated Th0 cells into Rag-1-deficient mice.
3. Monitor the body weight, shape of the mice and sign of distress such as persistent diarrhea on a weekly basis for 10–12 weeks following adoptive transfer of Th9 cells (*see* **Note 19**).
4. To determine the phenotype of adoptively transferred T cells, sacrifice the mice and isolate spleen, lymph nodes and mesenteric lymph nodes. Since Rag-1-deficient mice do not have T cells, all the recovered T cells will be from the initially transferred ones.
5. As per the requirement of the experiment, determine the cytokines by intracellular cytokine staining.
6. Remove the intestine from the mice and fix in 10% (v/v) neutral buffered formalin solution for further histopathological examination to determine the pathological score of colitis.
7. The severity of colitis can be graded based semiquantitatively from 0 to 3 for the four following criteria: degree of epithelial hyperplasia and goblet cell depletion, leukocyte infiltration in the lamina propria, area of tissue affected, and the presence of markers of severe inflammation such as crypt abscesses, submucosal inflammation, and ulcers (*see* **Note 20**). Although the comparative analysis of Th1, Th9, and Th17-meidated colitis and disease pathology is not clearly understood, it is interesting to note that Th1, Th9, and Th17 cells can transfer EAE independently with distinct disease pathology. Th17 cells were shown to transfer very severe EAE as compared to Th1 and Th9 cells.

4 Notes

1. It is recommended to use DPBS, which contains glucose, while isolating and sorting naïve CD4+ T cells. Having glucose in the DPBS maintains the viability of T cells during the process of their isolation.
2. It is essential to have highly purified population of naïve CD4+ T cells for the optimal differentiation of Th9 cells, as contamination of memory CD4+ T cells affects the differentiation of Th9 cells.
3. It is recommended to coat anti-CD3 and anti-CD28 antibodies overnight at 4 °C for activating naïve CD4+ T cells in Th9 culture condition (*see* Ref. [16]).
4. It is also important to note that different protocols of activation/polarization of Th9 cells have been described in the literature. Therefore, the concentration of anti-CD3 used for polarization of Th9 cells can be titrated for the best result. Soluble anti-CD28, instead of plate bound, at lower concentration (1.5 μg/mL) can also be advisable for getting expected outcome in IL-9 intracellular staining (*see* Ref. [16]).
5. Th9 differentiating cytokines, TGF-β and IL-4, should be stored at −80 °C in small aliquots to avoid their freeze–thaw. Once thawed and used in differentiation, leftover cytokines can be reused within a week and stored at 4 °C instead of freezing back at −80 °C.
6. In absence of irradiated APCs, Mitomycin C-treated APCs can be used in Th9 cell differentiation. It is essential to remove traces of Mitomycin C from APCs by extensively wash them with DPBS, as small amounts of Mitomycin C can affect the differentiation of Th9 cells.
7. In vitro differentiation of Th9 cells can be completed in 4 days for various in vitro analysis and adoptive T cell transfers in vivo (*see* Ref. [16]).
8. Polarized Th9 cells can be analyzed after 3 days of their culture (*see* Ref. [16]). The PMA/ionomycin stimulation polarized Th9 cells for intracellular cytokines staining usually takes about 4–5 h (*see* Ref. [16]).
9. Gene profiling of Th9 cells usually takes 2–3 h, which include RNA extraction, cDNA preparation and qPCR. However, timings for expression profiling of Th9 cells can vary greatly depending on the sample size and number of genes of interest to be tested.
10. It is advisable to perform appropriate controls such as IL-9 staining on non-PMA/iono restimulated T cells (but in presence of monensin), or isotype control staining on restimulated polarized Th9 cells.

11. If intracellular IL-9 detection proves to be problematic using flow cytometry, one can try to decrease the PMA concentration (down to 20 ng/mL) and increase the time of stimulation prior to intracellular staining. IL-9 reporter mice can be used to bypass the hurdle of IL-9 intracellular staining (*see* Ref. [16]).
12. Determining the concentration of IL-9 in culture supernatant by ELISA typically takes 3 days (*see* Ref. [16]). However, the 3 day time period of ELISA protocol can be shortened by advance preparation of the first two steps (coating plates with capture antibodies and blocking) before polarization is complete.
13. For EAE adoptive transfer model, it is recommended to differentiate Th9 cells from naive CD4+ T cells of 2D2 × Foxp3-GFP reporter mice. Resort Foxp3- Th9 cells after differentiation before transferring in vivo.
14. For EAE experiments; the mean onset of disease is usually reached between day 10–15 (*see* Ref. [16]). Depending on the requirements of the experiment and the setting used, it is important to follow the clinical score of the mice for 30–40 days (*see* Ref. [1]). During EAE development, mice have to be carefully followed on a daily basis and gel food needs to be provided to support their daily food and water requirements.
15. Histopathological analysis can be performed on brain and spinal cords harvested and fixed in 10% (v/v) neutral buffered formalin solution (*see* Ref. [1]).
16. It is recommended to remove the traces of collagenase and DNAse by washing the single cell suspension with ice cold DPBS extensively (*see* Ref. [16]).
17. In vivo cytokines profile of adoptively transferred Th9 cells can be identified by staining anti-Vα3.2 TcR, as 2D2 mice express Vα3.2 TcR (*see* Ref. [16]).
18. TGF-β induces Foxp3 expression and converts naïve T cells into iTreg cells (induced Foxp3+ cells). Under Th9 polarization condition, although IL-4 suppresses TGF-β-induced Foxp3 expression, a small fraction of Foxp3+ T cells can be detected in differentiated Th9 cells. It is therefore advisable to FACS sort the differentiated Th9 cells to remove small fraction of Foxp3+ cells in order to obtain a homogenous population of Th9 cells prior to adoptive transfer into immune-deficient host to induce colitis (*see* Ref. [16]).
19. For colitis experiments; the first signs of disease (weight loss) usually appear at around 5–6 weeks following adoptive transfer. It is important to note that mice losing more than 20% of the original weight should be sacrificed. Clinical signs of the disease (weight loss, overall shape of the mice) are followed for 9–10 weeks (*see* Ref. [16]).

20. It is important to note that overall incidence and first signs of disease for IBD can be modulated by the environment of the animal facility where experiments are performed, as changing in microbiota composition may affect overall impact of intestinal inflammation and disease pathogenesis in IBD.

Acknowledgment

This work was supported by grant from Department of Biotechnology (DBT), Govt. of India and DBT-Wellcome Trust intermediate fellowship (IA/I/12/1/500524). S.M. is supported by a senior research fellowship from Council of Scientific and Industrial Research (CSIR), Govt of India. V.D. is supported by CNRS.

References

1. Bettelli E, Korn T, Kuchroo VK (2007) Th17: the third member of the effector T cell trilogy. Curr Opin Immunol 19(6):652–657. doi:10.1016/j.coi.2007.07.020
2. Dardalhon V, Awasthi A, Kwon H, Galileos G, Gao W, Sobel RA, Mitsdoerffer M, Strom TB, Elyaman W, Ho IC, Khoury S, Oukka M, Kuchroo VK (2008) IL-4 inhibits TGF-beta-induced Foxp3+ T cells and, together with TGF-beta, generates IL-9+ IL-10+ Foxp3(−) effector T cells. Nat Immunol 9(12):1347–1355. doi:10.1038/ni.1677
3. Veldhoen M, Uyttenhove C, van Snick J, Helmby H, Westendorf A, Buer J, Martin B, Wilhelm C, Stockinger B (2008) Transforming growth factor-beta 'reprograms' the differentiation of T helper 2 cells and promotes an interleukin 9-producing subset. Nat Immunol 9(12):1341–1346. doi:10.1038/ni.1659
4. Gerlach K, Hwang Y, Nikolaev A, Atreya R, Dornhoff H, Steiner S, Lehr HA, Wirtz S, Vieth M, Waisman A, Rosenbauer F, McKenzie AN, Weigmann B, Neurath MF (2014) TH9 cells that express the transcription factor PU.1 drive T cell-mediated colitis via IL-9 receptor signaling in intestinal epithelial cells. Nat Immunol 15(7):676–686. doi:10.1038/ni.2920
5. Nalleweg N, Chiriac MT, Podstawa E, Lehmann C, Rau TT, Atreya R, Krauss E, Hundorfean G, Fichtner-Feigl S, Hartmann A, Becker C, Mudter J (2015) IL-9 and its receptor are predominantly involved in the pathogenesis of UC. Gut 64(5):743–755. doi:10.1136/gutjnl-2013-305947
6. Jager A, Dardalhon V, Sobel RA, Bettelli E, Kuchroo VK (2009) Th1, Th17, and Th9 effector cells induce experimental autoimmune encephalomyelitis with different pathological phenotypes. J Immunol 183(11):7169–7177. doi:10.4049/jimmunol.0901906
7. Licona-Limon P, Henao-Mejia J, Temann AU, Gagliani N, Licona-Limon I, Ishigame H, Hao L, Herbert DR, Flavell RA (2013) Th9 cells drive host immunity against gastrointestinal worm infection. Immunity 39(4):744–757. doi:10.1016/j.immuni.2013.07.020
8. Purwar R, Schlapbach C, Xiao S, Kang HS, Elyaman W, Jiang X, Jetten AM, Khoury SJ, Fuhlbrigge RC, Kuchroo VK, Clark RA, Kupper TS (2012) Robust tumor immunity to melanoma mediated by interleukin-9-producing T cells. Nat Med 18(8):1248–1253. doi:10.1038/nm.2856
9. Vegran F, Berger H, Boidot R, Mignot G, Bruchard M, Dosset M, Chalmin F, Rebe C, Derangere V, Ryffel B, Kato M, Prevost-Blondel A, Ghiringhelli F, Apetoh L (2014) The transcription factor IRF1 dictates the IL-21-dependent anticancer functions of TH9 cells. Nat Immunol 15(8):758–766. doi:10.1038/ni.2925
10. Schmitt E, Klein M, Bopp T (2014) Th9 cells, new players in adaptive immunity. Trends Immunol 35(2):61–68. doi:10.1016/j.it.2013.10.004
11. Lu LF, Lind EF, Gondek DC, Bennett KA, Gleeson MW, Pino-Lagos K, Scott ZA, Coyle AJ, Reed JL, Van Snick J, Strom TB, Zheng XX, Noelle RJ (2006) Mast cells are essential

intermediaries in regulatory T-cell tolerance. Nature 442(7106):997–1002. doi:10.1038/nature05010

12. Schmitt E, Beuscher HU, Huels C, Monteyne P, van Brandwijk R, van Snick J, Ruede E (1991) IL-1 serves as a secondary signal for IL-9 expression. J Immunol 147(11): 3848–3854
13. Schmitt E, Germann T, Goedert S, Hoehn P, Huels C, Koelsch S, Kuhn R, Muller W, Palm N, Rude E (1994) IL-9 production of naive CD4+ T cells depends on IL-2, is synergistically enhanced by a combination of TGF-beta and IL-4, and is inhibited by IFN-gamma. J Immunol 153(9):3989–3996
14. Murugaiyan G, Beynon V, Pires Da Cunha A, Joller N, Weiner HL (2012) IFN-gamma limits Th9-mediated autoimmune inflammation through dendritic cell modulation of IL-27. J Immunol 189(11):5277–5283. doi:10.4049/jimmunol.1200808
15. Yao W, Zhang Y, Jabeen R, Nguyen ET, Wilkes DS, Tepper RS, Kaplan MH, Zhou B (2013) Interleukin-9 is required for allergic airway inflammation mediated by the cytokine TSLP. Immunity 38(2):360–372. doi:10.1016/j.immuni.2013.01.007
16. Angkasekwinai P, Chang SH, Thapa M, Watarai H, Dong C (2010) Regulation of IL-9 expression by IL-25 signaling. Nat Immunol 11(3):250–256. doi:10.1038/ni.1846
17. Staudt V, Bothur E, Klein M, Lingnau K, Reuter S, Grebe N, Gerlitzki B, Hoffmann M, Ulges A, Taube C, Dehzad N, Becker M, Stassen M, Steinborn A, Lohoff M, Schild H, Schmitt E, Bopp T (2010) Interferon-regulatory factor 4 is essential for the developmental program of T helper 9 cells. Immunity 33(2):192–202. doi:10.1016/j.immuni.2010.07.014
18. Goswami R, Jabeen R, Yagi R, Pham D, Zhu J, Goenka S, Kaplan MH (2012) STAT6-dependent regulation of Th9 development. J Immunol 188(3):968–975. doi:10.4049/jimmunol.1102840
19. Chang HC, Sehra S, Goswami R, Yao W, Yu Q, Stritesky GL, Jabeen R, McKinley C, Ahyi AN, Han L, Nguyen ET, Robertson MJ, Perumal NB, Tepper RS, Nutt SL, Kaplan MH (2010) The transcription factor PU.1 is required for the development of IL-9-producing T cells and allergic inflammation. Nat Immunol 11(6):527–534. doi:10.1038/ni.1867
20. Jabeen R, Goswami R, Awe O, Kulkarni A, Nguyen ET, Attenasio A, Walsh D, Olson MR, Kim MH, Tepper RS, Sun J, Kim CH, Taparowsky EJ, Zhou B, Kaplan MH (2013) Th9 cell development requires a BATF-regulated transcriptional network. J Clin Invest 123(11):4641–4653. doi:10.1172/JCI69489
21. Schlapbach C, Gehad A, Yang C, Watanabe R, Guenova E, Teague JE, Campbell L, Yawalkar N, Kupper TS, Clark RA (2014) Human TH9 cells are skin-tropic and have autocrine and paracrine proinflammatory capacity. Sci Transl Med 6(219):219ra218. doi:10.1126/scitranslmed.3007828
22. Bettelli E, Pagany M, Weiner HL, Linington C, Sobel RA, Kuchroo VK (2003) Myelin oligodendrocyte glycoprotein-specific T cell receptor transgenic mice develop spontaneous autoimmune optic neuritis. J Exp Med 197(9):1073–1081. doi:10.1084/jem.20021603
23. Kuchroo VK, Sobel RA, Laning JC, Martin CA, Greenfield E, Dorf ME, Lees MB (1992) Experimental allergic encephalomyelitis mediated by cloned T cells specific for a synthetic peptide of myelin proteolipid protein. Fine specificity and T cell receptor V beta usage. J Immunol 148(12):3776–3782
24. Kroenke MA, Carlson TJ, Andjelkovic AV, Segal BM (2008) IL-12- and IL-23-modulated T cells induce distinct types of EAE based on histology, CNS chemokine profile, and response to cytokine inhibition. J Exp Med 205(7):1535–1541. doi:10.1084/jem.20080159
25. Li H, Nourbakhsh B, Ciric B, Zhang GX, Rostami A (2010) Neutralization of IL-9 ameliorates experimental autoimmune encephalomyelitis by decreasing the effector T cell population. J Immunol 185(7):4095–4100. doi:10.4049/jimmunol.1000986
26. Neurath MF, Weigmann B, Finotto S, Glickman J, Nieuwenhuis E, Iijima H, Mizoguchi A, Mizoguchi E, Mudter J, Galle PR, Bhan A, Autschbach F, Sullivan BM, Szabo SJ, Glimcher LH, Blumberg RS (2002) The transcription factor T-bet regulates mucosal T cell activation in experimental colitis and Crohn's disease. J Exp Med 195(9): 1129–1143
27. Boirivant M, Fuss IJ, Chu A, Strober W (1998) Oxazolone colitis: a murine model of T helper cell type 2 colitis treatable with antibodies to interleukin 4. J Exp Med 188(10): 1929–1939
28. Harbour SN, Maynard CL, Zindl CL, Schoeb TR, Weaver CT (2015) Th17 cells give rise to Th1 cells that are required for the pathogenesis of colitis. Proc Natl Acad Sci U S A 112(22):7061–7066. doi:10.1073/pnas.1415675112
29. Bettelli E, Carrier Y, Gao W, Korn T, Strom TB, Oukka M, Weiner HL, Kuchroo VK (2006) Reciprocal developmental pathways for the generation of pathogenic effector TH17 and regulatory T cells. Nature 441(7090):235–238. doi:10.1038/nature04753

Chapter 17

B16 Lung Melanoma Model to Study the Role of Th9 Cells in Cancer

Alka Dwivedi, Sushant Kumar, and Rahul Purwar

Abstract

T cell therapy has shown remarkable promise in multiple malignancies including melanoma and Acute Lymphoblastic Leukemia (ALL). Recent data demonstrated the differential efficacy of various subsets of T-helper cells in tumor regression. Th9 cells, the new member of T helper cell family, possess superior anti-tumor activity compared to Th1 and Th2 cells in murine model of melanoma. Therefore, it is important to examine the anti-tumor activity of specific subsets of Th cells in tumor models. Here, we describe the methodology of examining the anti-tumor activity of Th9 cells in murine model of melanoma.

Key words T cell therapy, Th9 cells, Murine model of melanoma, B16-OVA

1 Introduction

Th9 cells are recently discovered subsets of CD4+ T cells, which predominantly secrete IL-9 [1]. While T cell therapy of hematological malignancies has been very successful in recent clinical trials, it gained limited success in solid tumor [2, 3]. Specific subsets of CD4+ T cells like Tregs are described to inhibit anti-tumor activity of effector T cells [4]. Similarly, other subsets of Th cells, including Th17 and Th9 cells, have demonstrated increased anti-tumor activity compared to Th1 and Th2 cells [1, 5]. In this chapter, we demonstrate the method of examining the anti-tumor activity of Th9 cells in murine model of melanoma (Fig. 1) [6]. Ovalbumin expressing B16F10 is a murine tumor cell line that expresses ovalbumin. Tumor-specific Th9 cells are generated from naive CD4+ T cells (CD4+CD62L+) from OT-II mice [7, 8]. OT-II mice are TCR transgenic mice whose all CD4+ T cells are specific to ovalbumin [1]. Here, we describe the methodology of examining the anti-tumor properties of Th9 cells in B16-murine model of melanoma.

Ritobrata Goswami (ed.), *Th9 Cells: Methods and Protocols*, Methods in Molecular Biology, vol. 1585,
DOI 10.1007/978-1-4939-6877-0_17, © Springer Science+Business Media LLC 2017

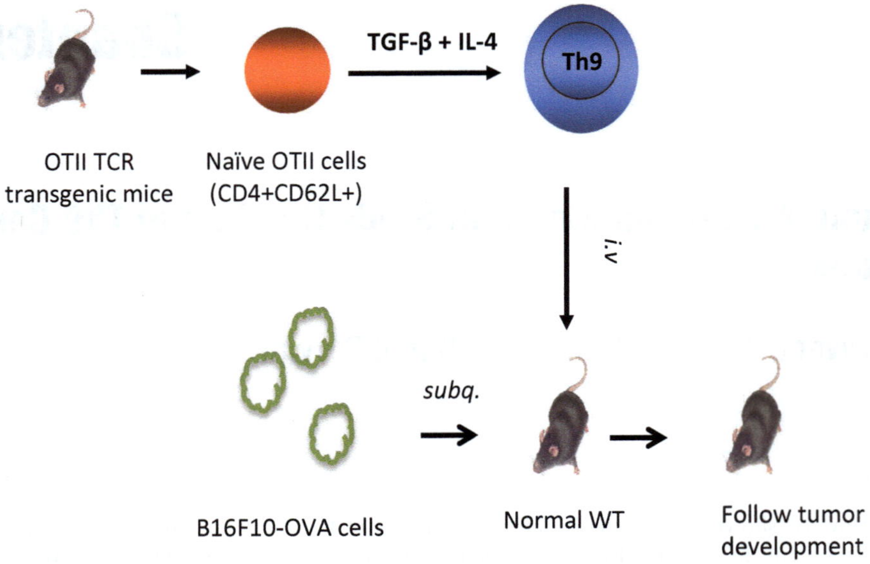

Fig. 1 Overview of Methodology of inducing tumor in mice: Naive CD4+CD62L+ cells are isolated and differentiated into Th9 cells. Next, Th9 cells are injected (i.v.) into B16F10-OVA bearing mice. Tumor development is followed overtime

2 Materials

2.1 Generation of Th9 Cells

1. T cell culture media (complete DMEM):
   DMEM high glucose, 1 mM sodium pyruvate, Nonessential amino acids, 1× L-Arginine plus + L-Asparagine (to be prepared from 10×: add 1.16 g of L-Arginine and 0.36 g of L-Asparagine into ddH$_2$O and make and make up to 100 mL), 4 mM L-glutamine, 0.6 mg/mL Folic Acid, MEM Eagle vitamin mixture (1 mg/L D-Ca panothenate, 1 mg/L choline chloride, 1 mg/L folic acid, 2 mg/L i-Inositol, 1 mg/L nicotinamide, 1 mg/L pyridoxine·HCl, 0.1 mg/L riboflavin, 1 mg/L Thiamine·HCl) (Lonza: 13-607C), 1% Penicllin-Streptomycin (v/v), β-mercaptoethanol (2.5 μL per 500 mL), 10% FBS (v/v).
2. Reagents and chemicals:
   (a) CD4+CD62L+ naive T cells isolation kit.
   (b) RBC lysis buffer (8.3 g/L ammonium chloride in 0.01 M Tris–HCl buffer, pH 7.5).
   (c) Antibodies: anti-mouse CD4, anti-mouse CD62L, anti-mouse CD3, anti-mouse CD28, anti-mouse IFN-γ (clone XMG1.2).
   (d) PBS (137 mM NaCl, 2.7 mM KCl, 8 mM Na_2HPO_4, and 2 mM KH_2PO_4).

(e) Cytokines: rTGF-β (human) (1 ng/mL), rIL-4 (murine) (10 ng/mL), rIL-2 (human) (10 ng/mL).

(f) 24-well plate.

(g) 70 μm nylon cell strainer.

2.2 B16F10-OVA Cells as a Mouse Model

1. B16F10 OVA cell line culture medium: 10% FBS (v/v) in RPMI 1640 medium supplemented with 1% (v/v) pen/strep.
2. RPMI 1640 medium.
3. 0.25% Trypsin/EDTA.
4. HBSS (400 mg/mL potassium chloride, 60 mg/mL potassium phosphate monobasic anhydrous, 350 mg/mL sodium bicarbonate, 8000 mg/mL sodium chloride, 90 mg/mL sodium phosphate dibasic heptahydrate, 1000 mg/mL dextrose anhydrous).
5. Trypan blue.
6. Ethanol.
7. T-75 flask.
8. 50 mL falcon tube.
9. 45 μm cell strainer.
10. 1 mL syringe.
11. 27^{1}/2-G needle.

3 Methods

3.1 Generation of Th9 Cells

3.1.1 Isolation of CD4+ T Cells from OT-II Mice

1. Collect spleen and lymph nodes of a 6–8 weeks old OT-II B6 mouse, mash both the organs and pass through 70 μm nylon mesh (cell strainer) to make single cell suspension.
2. Lyse the red blood cells with RBCs lysis buffer. Count the cells, centrifuge the cell suspension at 300 × *g* for 10 min, and discard the supernatant.
3. Isolate the naive CD4+ T cells either using naive CD4+ T cell isolation kit using manufacturer's instruction or by staining naive CD4+ T cells with anti-CD4 and anti-CD62L and sort CD4+CD62L+ naive T cells with a FACS sorter (*see* **Note 1**).
4. Resuspend the sorted cells as 2 × 10^6 cells/mL in complete T cell media.

3.1.2 Differentiation of Th9 Cells

1. Coat a 24-well plate with anti-mouse CD3 mAb (2 μg/mL in PBS) and anti-mouse CD28 mAb (1 μg/mL in PBS) overnight at 4 °C (*see* **Note 2**) [1]. After incubation time, wash the coated wells twice with 2 mL of PBS.

2. Transfer the 0.5 × 10^6 cells/mL sorted naive T cells into coated wells.
3. Irradiate the splenocytes (non-naive T cells) by a gamma-irradiator (200 Gy) and resuspend in complete T cell media (2.5 × 10^6 cells/mL) (*see* **Note 3**).
4. Transfer 1 mL of 2.5 × 10^6 cells/mL irradiated splenocytes (1:5 ratio) into each well.
5. Add rTGF-β (1 ng/mL), rIL-4 (10 ng/mL), and anti-IFN-γ (10 μg/mL) in the culture wells and keep for 48 h [9].
6. After 48 h, feed the cells with replacing approximately half of the media with fresh media containing rIL-2 (10 ng/mL) and split the cells into two parts and add fresh 1.5 mL of T cell media containing rIL-2 into each well.
7. On day 5, harvest the Th9 cells (*see* **Note 4**). Stain the differentiated cells for IL-9 and IL-10 to examine the frequency of Th9 cells using flow cytometry.

3.2 B16F10-OVA Cells as Mouse Model for Melanoma

3.2.1 Preparation of B16F10-OVA Cells

1. Grow B16F10-OVA cells in T-75 flask in 15 mL of B16-OVA culture medium (*see* **Note 5**).
2. Once cells are 50–60% confluent, discard used media from the flask, trypsinize the cells with 4 mL of 0.25% trypsin/EDTA, keep at 37 °C for 3 min to detach the cells properly.
3. Neutralize the trypsin/EDTA with 20 mL of ice-cold complete RPMI 1640 medium, make single cell suspension, transfer it into 50 mL tube, and pellet down at 400 × *g* (Swinging bucket rotor) for 5 min at 4 °C.
4. Discard the supernatant and resuspend properly in ice-cold HBSS to keep final concentration 1–5 × 10^6 cells/mL, pass single cell suspension through 45 μm cell strainer to remove cell clumps.
5. Count the cells with trypan blue in hemocytometer by preparing solution of 5:4:1 ratio of culture, HBSS, and trypan blue respectively. Viability should be >90%.

3.2.2 Subcutaneous Injection of B16F10-OVA Cells into Mice

1. Maintain the cells on ice and inject into the mouse (100 μL of 1 × 10^6 cells/mL in each mouse) as soon as possible because viability will go down as time passes.
2. Take 6–12 week old wild-type C57BL/6 mice, wet abdominal area, and fur with 70% ethanol and rub it, make skin clearly visible by moving the fur sideways with needle.
3. Mix the cell suspension by inverting the tube and by tapping, take 1 mL sterile syringe with $27^1/2$-G needle, and fill the cells in it very carefully.
4. Inject needle in the skin very superficially so that you can visualize the needle, insert needle 5–8 mm subcutaneously and

dispense 100 μL cell suspension into the skin, observe for bleb formation, bleb indicates appropriate inoculation (*see* **Note 6**).

5. If no clear appearance of bleb, sacrifice the inoculated mice and take new mice for another inoculation (*see* **Note** 7). This work needs proper practice because needle should be always superficial.
6. Withdraw the needle from the skin very gently.
7. Every day observe the mice and tumor should appear within 5–10 days. Once tumor size exceeds greater than 3000 mm^3 or there is necrosis in tumor, euthanize the mice (*see* **Notes 8** and **9**).

4 Notes

1. To achieve highly differentiated Th9 cells, naive CD4+ T cell separation efficiency (purity of naive cells) should be >95%, use either FACS or magnetic beads based cell separation kit to get higher separation efficiency.
2. To set up the naive T cell differentiation, use of tissue-culture treated dishes is recommended such as T/C treated Corning 24-well plate.
3. Cell density along with T cell and APCs ratio are important for Th9 cell differentiation. For a 24-well plate, cell density should be 0.5×10^6 naive T cells and 2.5×10^6 irradiated splenocytes (1:5 ratio). Similar cell density and T cells and APCs ratio should be maintained in case of using culture dishes of higher surface area.
4. IL-9 production in Th9 cells is variable on different days; therefore on day 4, cells should be harvested and injected into the mice.
5. B16F10-OVA cells should be maintained in selective culture media to maintain the surface ovalbumin expression.
6. For metastatic melanoma, B16F10-OVA cells should be injected in tail vein. Mice should be sacrificed on day 10 (or day 15) for counting the colonies in the lung.
7. After subcutaneous injection of B16F10-OVA cells, mice should be monitored every second day. Any mice showing necrotic tumor should be excluded from the analysis.
8. Adoptive cell therapy with Th9 cells has been described to regress melanoma tumor growth in subcutaneous and metastatic murine models of melanoma [1, 10].
9. Importantly, blockade of IL-9 by intraperitoneal injection of anti-IL-9 monoclonal antibody (100 μg per mouse) on every alternate day in mice bearing melanoma tumor resulted in increase in tumor growth via mast cell mediated anti-tumor activity [1, 10].

Acknowledgment

This work is supported by grants from Wadhwani foundation, Bristol Myers and Squib (BMS) and intramural start-up grant of IIT Bombay to R.P.

References

1. Purwar R, Schlapbach C, Xiao S, Kang HS, Elyaman W, Jiang X, Jetten AM, Khoury SJ, Fuhlbrigge RC, Kuchroo VK, Clark RA, Kupper TS (2012) Robust tumor immunity to melanoma mediated by interleukin-9-producing T cells. Nat Med 18(8):1248–1253. doi:10.1038/nm.2856
2. Quezada SA, Simpson TR, Peggs KS, Merghoub T, Vider J, Fan X, Blasberg R, Yagita H, Muranski P, Antony PA, Restifo NP, Allison JP (2010) Tumor-reactive CD4(+) T cells develop cytotoxic activity and eradicate large established melanoma after transfer into lymphopenic hosts. J Exp Med 207(3):637–650. doi:10.1084/jem.20091918
3. Ramos CA, Savoldo B, Dotti G (2014) CD19-CAR trials. Cancer J 20(2):112–118. doi:10.1097/PPO.0000000000000031
4. Nishikawa H, Sakaguchi S (2010) Regulatory T cells in tumor immunity. Int J Cancer 127(4):759–767. doi:10.1002/ijc.25429
5. Muranski P, Boni A, Antony PA, Cassard L, Irvine KR, Kaiser A, Paulos CM, Palmer DC, Touloukian CE, Ptak K, Gattinoni L, Wrzesinski C, Hinrichs CS, Kerstann KW, Feigenbaum L, Chan CC, Restifo NP (2008) Tumor-specific Th17-polarized cells eradicate large established melanoma. Blood 112(2):362–373. doi:10.1182/blood-2007-11-120998
6. Overwijk WW, Restifo NP (2001) B16 as a mouse model for human melanoma. Curr Protoc Immunol Chapter 20:Unit 20 21. doi:10.1002/0471142735.im2001s39
7. Beriou G, Bradshaw EM, Lozano E, Costantino CM, Hastings WD, Orban T, Elyaman W, Khoury SJ, Kuchroo VK, Baecher-Allan C, Hafler DA (2010) TGF-beta induces IL-9 production from human Th17 cells. J Immunol 185(1):46–54. doi:10.4049/jimmunol.1000356
8. Purwar R, Campbell J, Murphy G, Richards WG, Clark RA, Kupper TS (2011) Resident memory T cells (T(RM)) are abundant in human lung: diversity, function, and antigen specificity. PLoS One 6(1):e16245. doi:10.1371/journal.pone.0016245
9. Schmitt E, Germann T, Goedert S, Hoehn P, Huels C, Koelsch S, Kuhn R, Muller W, Palm N, Rude E (1994) IL-9 production of naive CD4+ T cells depends on IL-2, is synergistically enhanced by a combination of TGF-beta and IL-4, and is inhibited by IFN-gamma. J Immunol 153(9):3989–3996
10. Lu Y, Hong S, Li H, Park J, Hong B, Wang L, Zheng Y, Liu Z, Xu J, He J, Yang J, Qian J, Yi Q (2012) Th9 cells promote antitumor immune responses in vivo. J Clin Invest 122(11):4160–4171. doi:10.1172/JCI65459

Chapter 18

Th9 Cells and Parasitic Inflammation: Use of *Nippostrongylus* and *Schistosoma* Models

Miguel Enrique Serrano Pinto and Paula Licona-Limón

Abstract

Th9 cells are a new subpopulation of CD4+ T helper cells, characterized by the expression of IL-9 that have been involved in type 2 immune responses, antitumor responses and autoimmune diseases. Here, we describe two different parasitic models frequently maintained in the laboratory where Th9 cells or IL-9 (the cytokine produced by Th9 cells) has been shown to play critical roles in pathogen clearance and immune response activation: the nematode *Nippostrongylus brasiliensis* and the trematode *Schistosoma mansoni*.

Key words IL-9, Type 2 immune response, Helminth, Th9, Parasite infection

1 Introduction

Type 2 immune responses are those classically induced during allergic reactions and parasitic infections that are characterized by IL-4, IL-5, IL-9, IL-10, and IL-13 secretion by T cells, eosinophilia, mucus hypersecretion, increased IgE production, and mast cell hyperplasia [1, 2]. IL-9 is mainly produced by T cells but can also be produced by mast cells and other type of innate sources (Innate Lymphoid Cells type 2: ILC2) according to recent reports [3, 4]. IL-9 was originally described as a Th2 cytokine promoting airway inflammation and induced in response to nematode infections [5–7]. IL-9 stimulates the proliferation of T cells and mast cells, promotes the differentiation of hematopoietic progenitors, and drives IgE secretion from B cells [8–10]. In animal models, IL-9 has been identified as a candidate gene for asthma, and increased IL-9 expression has been observed in asthmatic patients compared to healthy individuals [11]. IL-9 secretion in T cells was initially shown to be dependent on the presence of IL-2 and was greatly increased by the addition of IL-4 and TGF-β in a dose-dependent manner [12]. Induction of IL-9 expression in T cells has been recently reexamined by two groups, who suggested that

Ritobrata Goswami (ed.), *Th9 Cells: Methods and Protocols*, Methods in Molecular Biology, vol. 1585,
DOI 10.1007/978-1-4939-6877-0_18, © Springer Science+Business Media LLC 2017

the T effector cells differentiated in the presence of TGF-β and IL-4 could represent a unique subset of T cells. Th9 cells express very little or nothing of the known transcription factors for T cell differentiation, including those required for Th1, Th2, Th17, or Treg lineages (T-bet, GATA-3, RORγt, and FoxP3, respectively) [13, 14]. By generating an IL-9 reporter mice, we recently identified Th9 cells and ILC2 cells as the major sources of *Il9* expression in vivo in a murine model of parasitic infection [4]. Functionally, we demonstrated that transfer of Th9 cells into *N. brasilienisis*-infected Rag-deficient recipients leads to mast cell accumulation, basophilia and decreased worm loads. Furthermore, IL-9-deficient mice show decreased levels of IL-5, IL-13, mucin 5b, Mcpt1, as well as decreased frequencies of eosinophils, basophils, and mast cells in the infected organs and draining lymph nodes which all together result in decreased anti-helminth immunity and parasite clearance and centers IL-9 and Th9 cells as critical orchestrators of type 2 immunity in vivo [4]. Other authors have also described the role of Th9 cells in antitumor immunity [15], and autoimmunity [16–18]; however, in this chapter we focus on the parasitic infections driving Th9 differentiation and activation in vivo.

1. Nippostrongylus brasiliensis is a nematode commonly used in the laboratory as a model for hookworm infection in humans caused by *Necator americanus* and *Ancylostoma duodenale* affecting approximately two billion people worldwide [19, 20]. It promotes an acute gastrointestinal infection that activates systemic and mucosal type 2 immune responses, which resolves in 7–10 days depending on the mouse strain. *N. brasiliensis* is a parasite whose natural host is rat and periodic passages are recommended to maintain strain virulence but can be successfully maintained in mouse for laboratory purposes. Immunodeficient mice strains lacking Th2 related cytokines (STAT6, IL-9, IL-4Rα, IL-13) or adaptive immunity (Rag) are particularly susceptible and could give better egg yields when necessary [4, 21–23]. *N. brasiliensis* life cycle adapted to the laboratory maintenance is illustrated in Fig. 1 and is complete within 2 weeks. Briefly, L3 infecting larvae harvested from eggs hatched in charcoal plates are subcutaneously injected in the lower back of the mice (close to the base of the tail), they move deep to the epidermis and reach the blood vessels that carry them to the lungs in the first 11 h post-infection. In the lung they mature to the larval stage L4 or migratory stage in 19–50 h post-infection, move up to the trachea and get coughed up and swallowed to reach the gut within 3–4 days. In the gut, specifically within the small intestine, L4 larvae mature to adult or larval stage L5 within 90–108 h, by day 6 worms lay eggs that get expelled in the feces and upon collection and culture, eggs hatch and mature to the larval stages L1, L2 and finally infective stage L3 in around 7 days to restart the cycle.

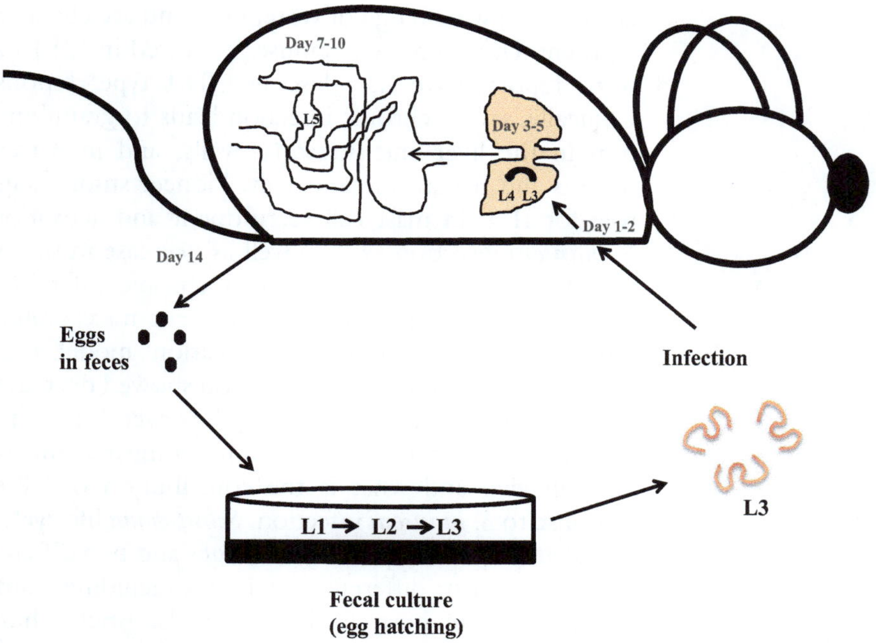

Fig. 1 *Nippostrogylus brasiliensis* life cycle in the laboratory. Larvae stages L1–L3 develop from hatched eggs in charcoal plates within 7 days. Infective stage L3 is subcutaneously injected into the mouse, migrates to the lung and mature to the migratory stage L4. Around day 3–4 post-infection, L4 larvae move up to the trachea and get swallowed to reach the gut. In the gut L4 larvea becomes L5 adult worm and from day 6 post infection start laying eggs that are released in the feces to restart the cycle in appropiate humidity, temperature, and light conditions

N. brasiliensis L3 can penetrate the skin of humans generating a limited infection; personnel, therefore, should wear gloves and laboratory coats when working with this parasite and decontaminate contaminating surfaces with 70% ethanol, 1% hypochlorite, or soap solutions. Given that the infective L3 stage requires several days for free-living development under optimal temperature and humidity conditions, parasites are not transmitted between mice in a normal facility under cleaning and managing procedures, even in a co-house infected with non-infected mice cage. All the different stages of the parasite are sensitive to hot water and detergent cleaning of the cages as well as the low-humidity HEPA-filtered air in the facility room.

2. Schistosoma mansoni is a flatworm from the trematode genus responsible for the second most common parasite infection in humans according to WHO estimates [24]. There are five different species of *Schistosoma* affecting different geographical areas worldwide, *mansoni*, *japonicum*, *mekongi*, *guineensis*, and *haematobium*; however, *S. mansoni* is the most frequently strain maintained in the laboratory and this chapter focuses on this particular species. *Schistosoma* infections can be acute or

chronic in a wide range of vertebrates and are characterized by a potent Th2 immune response (reviewed in [25]), although some reports have also described Th1 type responses [26]. Clinically, severe chronic infection leads to granuloma formation in the liver and intestinal walls, and in extreme cases fibrosis and hepatosplenomegaly. Pioneer studies suggested a role for IL-9 in mast cell recruitment and activation during *S. mansoni* infection [27] as well as decrease in survival when transgenically expressed during a chronic infection murine model [28]. In addition, IL-9-deficient mice evaluated in a lung pulmonary granuloma formation model induced by *Schistosoma* egg intravenous injection showed decreased goblet cell and mast cell hyperplasia [29]; however, future studies will have to address what is the cellular compartment expressing IL-9 in vivo and what is the contribution of Th9 cells in response to *S. mansoni* infection. *Schistosoma* life cycle includes sexual and asexual reproductive stages and two different hosts, a definitive host: different vertebrates including humans, and an intermediate host: freshwater snails. Briefly, humans are infected with cercaria developed in the snail through to the skin, they lose their tails and become schistosomula in the portal vein of the liver where they mature, mate, and sexually reproduce; migration of adult worms to the gut (*S. mansoni*) and bladder (*S. haematobium*) blood vessels allows dissemination and release of eggs in feces and urine. Eggs hatch in fresh water to miracidium and infect different tissues of the snail to become sporocyst. Cercarias develop from sporocyst in the snail and are released to fresh water to restart the cycle. In the laboratory mouse models for *S. mansoni* maintenance has been developed and is illustrated in Fig. 2.

S. mansoni cercarial stage is an extremely efficient infecting parasite and extra caution should be taken when working with them. Laboratory members should wear protective latex gloves at all time when handling snails and cercaria. Cercaria can be killed on contact with 70% ethanol, 1% hypochlorite or hot water (>50 °C). The current drug of choice for treating all species of Schistosomes is praziquantel. It is recommended to collect and store sera at −70 °C from anyone planning to work with schistosomes to have a reference to compare after accidental or suspected cercarial exposure.

For alternative protocols and additional details regarding both parasitic models please refer to the protocols described elsewhere [30, 31].

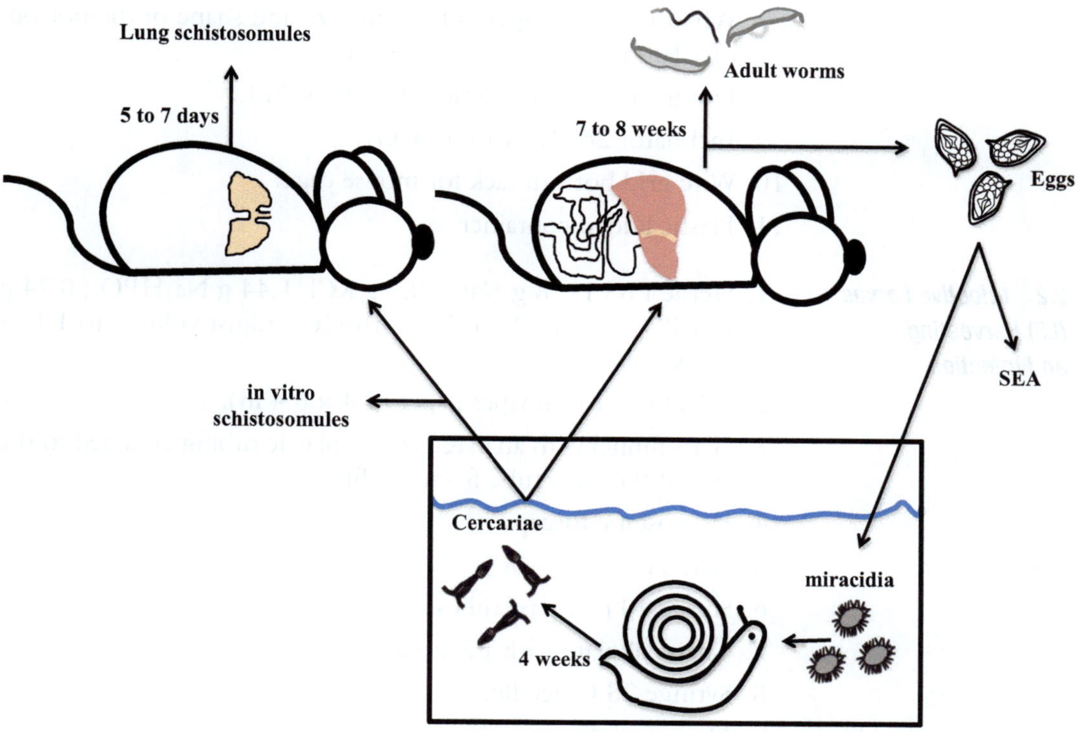

Fig. 2 *Schistosoma mansoni* life cycle in the laboratory. Cercariae are produced 4 weeks post-infection of snails with miracidia. Cercaria can generate schistosomules in vitro, infect mice to harvest lung schistosomules within 5–7 days post-infection or leave chronic infections up to 8 weeks for isolation of adult worms and eggs from infected tissues (liver and intestines). Eggs are used to generate miracidia to restart the cycle, promote lung granuloma formation (not shown in the figure) or for preparation of soluble egg antigen (SEA)

2 Materials

For Nippostrongylus brasiliensis

Prepare all solutions and material sterile, in case of contamination use Fungizone in the worm plates (*see* **Note 1**) and gentamicin in the suspension of L3 larvae for infection (*see* **Note 2**). Store all reagents at room temperature.

2.1 Egg Harvesting and Culture of Infecting Larvae L3

1. Sterile distilled water.
2. Concave metal strainer.
3. Sterile culture petri dish 100 mm × 15 mm.
4. Glass beakers—100 and 250 mL.
5. Spatula.
6. Fecal pellets from *N. brasiliensis* infected C57BL/6 mice (5–10 days post-infection).

7. Whatman filter paper cut by the size and shape of the bottom of a 100 mm × 15 mm petri dish.
8. Fungizone—Amphotericin B 250 μg/mL.
9. Incubator 26 °C (without CO_2).
10. Wire grid bottom rack for mouse cage.
11. Plastic lidded container.

2.2 Infective Larvae (L3) Harvesting and Infection

1. Sterile PBS 1×: 8 g NaCl, 0.2 g KCl, 1.44 g Na_2HPO_4, 0.24 g KH_2PO_4, adjust pH to 7.4 with HCl, adjust volume to 1 L at 37 °C.
2. KIMTECH Kimwipes paper (4.4 × 8.4 in).
3. Glass funnel with an extension of plastic tubing attached to the end of the glass tube from the funnel.
4. Hemostatic forceps.
5. Spatula.
6. Sterile 40 mL glass tubes.
7. Charcoal plate with L3 larvae.
8. Syringe 23 G needle.
9. Mouse restrainer.
10. C57BL/6 mice (young mice and females are more susceptible, do not use mice younger than 2 months).
11. Gentamicin 50 mg/mL.
12. Wahl battery shaver.
13. Retort stand, metal ring, and clamps.
14. Centrifuge.
15. Inverted microscope.

2.3 Isolation and Enumeration of L5 Adult Worms

1. Sterile PBS.
2. Concave metal strainer.
3. Petri dish with 2 mm grid, 60 mm × 15 mm.
4. Glass beaker—100–250 mL.
5. Glass pipettes.
6. Day 6–10 post-infection *N. brasiliensis* infected C57BL/6 mice.
7. Dissecting instruments tools.
8. Ethanol 70%.
9. Inverted microscope.
10. Water bath at 37 °C.

2.4 Determination of Egg Counts

1. Sterile PBS.
2. Wooden stick.

3. 15 mL centrifuge tubes.
4. NaCl saturated solution (5–10 M).
5. McMaster slides.
6. *N. brasiliensis* infected mice.
7. Inverted microscope.

2.5 Collection of L3 and L4 Larvae from the Site of Infection

1. Sterile PBS.
2. 50 mL centrifuge plastic tubes.
3. 40 mm petri dishes.
4. Dissecting instruments.
5. Plastic tissue culture (6-well).
6. *N. brasiliensis* infected mice.
7. Dissecting instruments tools.
8. 70% ethanol.
9. Inverted microscope.
10. Cheesecloth cut into 8 × 8 cm squares.
11. Agarose 1% (w/v).
12. Weigh boat 8 × 8 cm.
13. Clamps.
14. Water bath at 37 °C.
15. Glass pipette.
16. 1 mL syringe.
17. Insyte catheter 18 G.
18. Petri dish 90 mm.

2.6 Generation of L3, L4, and L5 Derived Antigens

1. *N. brasiliensis* infected mice. For optimal antigen collection (*see* **Note 3**).
2. NES wash solution I: 1 L PBS, 4 mL of 50 mg/mL gentamicin, 20 mL of 100× penicillin–streptomycin, filter-sterile through 0.2 mm filter. Can be stored up to 3 weeks at 4 °C.
3. NES wash solution II: 1 L RPMI 1640, 4 mL of 50 mg/mL gentamicin, 20 mL of 100× penicillin–streptomycin, filter-sterile through 0.2 mm filter. Can be stored up to 3 weeks at 4 °C.
4. NES culturing solution: 1 L RPMI 1640, 2 mL of 50 mg/mL gentamicin, 10 mL of 100× penicillin–streptomycin, 10 mL of 100× L-glutamine, 1% (w/v) glucose, filter-sterile through 0.2 mm filter. Can be stored up to 3 weeks at 4 °C.
5. SDS gel 15%.
6. Sterile 40 mL glass tubes.
7. Sterile 50 mL tissue flasks.

8. Incubator 37 °C, 5% CO_2.
9. Filter units 2 μm.
10. Ultrafiltration concentrator unit.

For Schistosoma mansoni

2.7 Percutaneous Exposure of Mice to Cercariae

1. Cercariae (*see* Subheading 2.10).
2. Conditioned water: pass tap water through a charcoal filter and aerate for 2–3 days.
3. C57BL/6 mice.
4. Plastic or glass tubes 12 × 75 mm.
5. Mouse restrainer.
6. Tube rack.
7. Adhesive tape.
8. Plastic gridded 35 mm petri dish.
9. Inverted microscope.
10. Iodine solution: 4 g potassium iodide, 2 g iodine in 100 mL of distilled water.

2.8 Maintenance and Propagation in Snail

1. *Biomphalaria glabrata* snails.
2. Romaine lettuce.
3. Snail food (*see* **Note 4**).
4. Conditioned water (as above).
5. Aquaria (30 gallon) with under-gravel filters and immersible aquarium heaters.
6. Beakers 400 mL.

2.9 Infection of Snails with Miracidia

1. *S. mansoni* infected mice.
2. Sodium pentobarbital with heparin: 50 mL of 65 mg/mL sodium pentobarbital, 2.5 mL of 10, 000 U/mL heparin, 75 mL water, store up to 1 month at room temperature.
3. NaCl in water 1.2% (w/v).
4. Conditioned water (as above).
5. Juvenile *Biomphalaria glabrata* snails (5–8 mm diameter).
6. Waring blender.
7. Centrifuge.
8. Büchner flask with the side arm painted black up to the last mm.
9. Dissecting microscope.

2.10 Collection of Cercariae from Infected Snails

1. Infected *Biomphalaria glabrata* snails.
2. Conditioned water (as above).
3. Iodine solution (as above).

4. Desk lamp.
5. Featherweight forceps or small fish net.
6. Filtration screen apparatus: 300 mL funnel with glass support and 47 mm diameter steel support screen.
7. Plastic pipette 100–1000 mL tips.
8. Plastic gridded 35 mm petri dish.
9. Inverted microscope.

2.11 In Vitro Transformation of Cercariae to Schistosomules

1. Suspension of *S. mansoni* cercariae.
2. DMEM (Dulbelcco's Modified Eagle's Medium, 4.5 g/L D-glucose, 4 mM L-Glutamine, 110 mg/L sodium pyruvate).
3. Percoll gradient suspension: 24 mL Percoll, 4 mL EMEM, 1.5 mL penicillin–streptomycin (10,000 U/mL penicillin, 10,000 mg/mL streptomycin), 1 M 1 mL HEPES, 9.5 mL distilled water, store up to 5 days at 4 °C.
4. Plastic tubes 50 mL.
5. Refrigerated centrifuge.
6. Tissue culture flasks 250 mL.
7. CO_2 incubator at 37 °C.

2.12 Collection of Adult Schistosomal Worms from the Mouse Portal Venous System

1. *S. mansoni* infected mice.
2. Sodium pentobarbital with heparin (as above).
3. Perfusion fluid: 7.5 g trisodium citrate dihydrate, 8.5 g NaCl, in 1 L water, store up to 7 days at 4 °C.
4. Dissecting tools.
5. Peristaltic perfusion pump fitted with 0.16 × 0.48 cm Tygon tube and 20 G needle.
6. Plastic gridded petri dish 60 mm diameter.
7. Fluorescent light box.
8. Dissecting microscope.

2.13 Enumeration of Schistosomal Eggs from Mouse Tissues

1. *S. mansoni* infected mice.
2. Sodium pentobarbital with heparin (*see* above).
3. KOH 4% (w/v).
4. Dissecting tools.
5. Inverted microscope.
6. Repeating pipettor.
7. Glass beakers 400 mL.
8. Scale.
9. Incubator 37 °C (without CO_2).
10. Ethanol 70%.

2.14 Schistosomal Eggs in Pulmonary Granuloma Formation

1. *S. mansoni* infected mice.
2. Sodium pentobarbital (*see* above).
3. NaCl 1.2% (w/v) at 4 °C.
4. Uninfected mice (C57BL/6 are more resistant than BALB/c).
5. Sterile PBS.
6. Dissecting tools.
7. Waring blender with speed control.
8. Stainless steel sieves of decreasing pore sizes (420, 180, 105, and 45 μm).
9. Plastic spray bottle.
10. Glass petri dishes with flat bottom.
11. Pasteur pipettes.
12. Mouse restrainer.
13. Alcohol swabs.
14. Disposable syringes 1 mL with 23 or 25 G needle.

2.15 Preparation of Schistosoma mansoni Soluble Egg Antigen (SEA)

1. Purified *Schistosoma mansoni* eggs.
2. Sterile PBS.
3. Handheld Potter-Elvehjem glass homogenizer (15 mL) with tight pestle, prechilled.
4. Refrigerated centrifuge.
5. Ultracentrifuge.
6. Disposable syringe 10 mL with 0.2 μm pore size filter.

3 Methods

Nippostrongylus brasiliensis.

3.1 Egg Harvest and Culture of Infecting Larvae L3

The following protocol describe the routinely used method for *N. brasiliensis* maintenance in the laboratory, focusing on egg harvest for generation of L3 infective larvae that will complete the life cycle of the parasite. It is recommended to restart the cycle every 15–20 days to ensure parasite propagation and long-term maintenance.

1. One day before harvesting (starting on day 5 post-infection), replace the normal bedding of the *N. brasiliensis* infected mice with a wire grid bottom rack and leave overnight, repeat this step every consecutive day until day 10, post-infection.
2. Fill a 100 mL glass beaker with sterile distilled water and put the concave metal strainer on top to avoid fecal pellets to go to the bottom of the flask.
3. Collect the fecal pellets from the bottom of the wire grid mouse cage with a spatula, avoiding urine and food debris, and

wash them in the pool formed by the concave metal strainer sinking into the glass flask.

4. In a separate dry glass beaker collect the washed pellets and soften them with the spatula to make a paste by adding enough distilled water to dissolve them and make a fecal pellet suspension.
5. Mix equal volumes of charcoal to moist in the fecal pellet suspension, avoid excess water but make sure all the charcoal is completely moist.
6. Place a rounded Whatmann filter paper in the bottom of the 100 mm × 15 mm culture petri dish and dampen it with distilled water, discard excess water and distribute the charcoal–fecal pellet suspension mix on top of the moist filter paper. Avoid contact with the petri dish lid and label with the mouse strain used for infection and date of collection. Several plates can be made per day of collection; however, to increase larvae yield prepare only one plate per day.
7. Place the dishes on a plastic lidded container to ensure maintenance of humidity and incubate in the dark at 26 °C for at least 7 days. Take special attention not to let dry the plates and monitor every 3 days, when necessary add more distilled water on top of the charcoal with a spray bottle. Fecal pellet cultures can get contaminated with fungi (*see* **Note 1** for how to avoid such contamination).

3.2 Infective Larvae (L3) Harvest and Infection

Avoid using any plastic flask, pipette tips, etc., as L3 larvae attach easily to plastic surfaces and could reduce the yield or affect the calculation of L3 quantification.

1. Shave the mice in the lower back (close to the base of the tail) 24 h before infection with L3 larvae.
2. Starting from day 8, charcoal–fecal pellet suspension cultured in petri dish will have L3 infecting larvae; as a reference they will look as white thin fibers growing on the surface of the charcoal by naked eye.
3. Mount the collection set by placing the glass funnel in the metal ring attached with clamps to the retort stand.
4. Fill the funnel with PBS at 37 °C by closing the exit of the plastic tubing attached to the funnel with the hemostatic forceps, avoid any bubbles.
5. Place a Kimwipes paper on top of the funnel forming a concave shape by holding the edges of the paper to the rim of the glass funnel.
6. Carefully add the charcoal with L3 larvae from the plate on the concave structure formed by the Kimwipes paper, avoid adding too much charcoal so the paper do not lose the attachment to

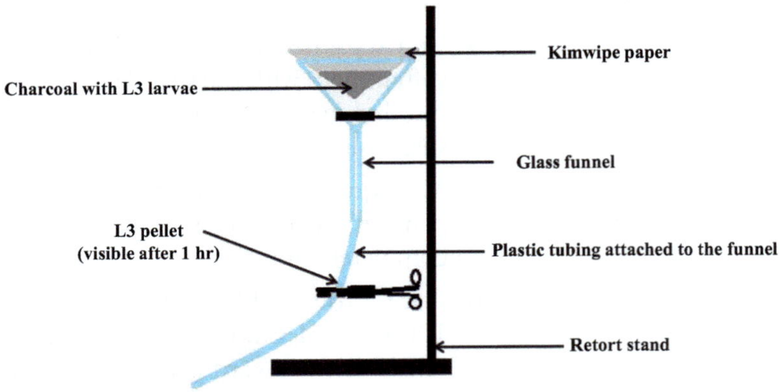

Fig. 3 Diagram of the collection set to harvest infective L3 larvae

the rim of the funnel because of the added charcoal weight, make sure all the charcoal is totally embedded inside the liquid as shown in Fig. 3.

7. Let the larvae sit by gravity for 1 h by ensuring the charcoal is always embedded in the liquid; when necessary add more PBS.
8. Collect the L3 larvae pellet sitting on top of the hemostatic forceps within the plastic tubing attached to the funnel, by opening the forceps with the edge of the plastic tubing inside of a glass tube.
9. Centrifuge at 50 × *g* at room temperature for 5 min.
10. Wash the pellet by taking out the supernatant with a glass pipet and adding PBS at 37 °C, repeat **steps 8** and **9** twice.
11. Resuspend the L3 larvae pellet in 5 mL of PBS and count larvae.
12. Using a glass syringe with a 21–23 G needle take 50 μL of the L3 larvae suspension, mix well before taking the sample as L3 larvae settle quickly.
13. By slowly releasing the 50 μL aliquot, form separated small drops on a petri dish and count the larvae present in an inverted microscope. Make sure you only count viable larvae that move.
14. Repeat counting for an accurate count.
15. Multiply the number of L3 by 20 to give L3 mL^{-1} and adjust the volume for a suspension of 500–700 L3 in 200 μL of PBS. The infection dose per mouse we normally use is 625 L3 but it can change depending on the batch of worms. Titration of the worm dose is recommended in wild-type mice before using in experimental groups of mice to avoid decreased survival of the host. It is also recommended to add gentamicin to avoid any contamination of the L3 suspension (*see* **Note 2**).

16. Restrain the shaved mouse on the holder and subcutaneously inject 200 μL of the L3 suspension with a glass syringe and a 23 G needle in the shaved area. Infect one mouse at a time, mix worm suspension immediately before each drawing of the syringe to ensure consistent dose of larvae. The subcutaneous injection should leave a bubble in the skin of the mouse.
17. To maintain the parasite wait for 6 days before collection of the eggs and repeat Subheadings 3.1 and 3.2.

3.3 Isolation and Enumeration of L5 Adult Worms

Avoid using any plastic flask, pipette tips, etc., as L5 larvae attach easily to plastic surfaces and could reduce the yield or affect the calculation of L5 quantification.

1. On day 7 post-infection euthanize the mice, place on its back and wet the abdominal area with 70% ethanol. If L5 is going to be isolated for antigen preparation (*see* **Note 3**) mice should be starved overnight on day 6.
2. Open the abdominal cavity and harvest the small intestine, from the end of the stomach to the cecum, and place in a petri dish.
3. Prepare a submerging 100–250 mL glass flask with PBS at 37 °C and a concave metal strainer.
4. Open the gut longitudinally with dissecting scissors and submerge in the flask from **step 3** above, shake the gut inside the strainer to help release fecal content and let sit for 1 h at 37 °C in a water bath.
5. Decant excess PBS until reaching a final suspension of the L5 larvae in 10 mL, freshly collected worms become red due to the hemoglobin taken by the parasite when sucking blood from the host.
6. Place the 10 mL final suspension of the L5 larvae in a petri dish with 2 mm grid 60 mm × 15 mm and count in an inverted microscope; for easy counting divide the plate in 4 equal quadrants, mix well the suspension by making slow circular moves, let it sit and count 2 out of the 4 quadrants of the plate.
7. Calculate the total number of L5 larvae per mouse.

3.4 Determination of Egg Counts

The following protocol is adapted for egg counting in one single fecal pellet, it is recommended to isolate and repeat the isolation for at least three different fecal pellets per mouse to assure a more accurate measurement.

1. Collect fecal pellets from *N. brasiliensis* infected mice on 6–10 days, post-infection.
2. Weigh fecal pellet and add 1 mL of PBS, leave at room temperature for 1 h.
3. Using a wooden stick finish destroying the pellet.

4. In a 15 mL plastic tube add 10 mL of saturated NaCl solution (5–10 M), add the 1 mL of PBS and feces and mix.
5. Let it sit for 20–30 min at room temperature.
6. Count eggs taking 750 μL from the top layer of the NaCl/eggs solution in McMaster slides into an inverted microscope.
7. Use the following formula to calculate egg loads per pellet, 150 refers to the McMaster counting chamber volume (150 μL), vol is the final volume of the suspension (11 mL).

$$\frac{1}{\text{Pellet weight}} \times \frac{vol}{150} \times egg\ \text{counts}$$

3.5 Collection of L3 and L4 Larvae from the Site of Infection

Avoid using any plastic flask, pipette tips, etc., as the different larvae stages attach easily to plastic surfaces and could reduce the yield or affect the quantification results.

1. For collection of L3 larvae from mouse skin, use mice infected within the first 2–48 h post-infection. L4 stages are collected on day 2 post-infection.
2. Euthanize the mouse at the recommended time post-infection depending on the larval stage of interest, place the mouse on its back, and wet with 70% ethanol.

   For L3 larvae

3. Excise the skin from the site of infection with dissecting scissors and place it into a 6-well culture dish with 1 mL of PBS at 37 °C.
4. Prepare a submerging glass flask with a concave metal strainer and PBS keep at 37 °C until ready to use.
5. Place an 8 × 8 cm cheesecloth on top of a 8 × 8 cm weigh boat, leave one edge of the cloth hanging over the side of the boat.
6. Cut the skin in small pieces with scissors in a 40 mm petri dish and place it in the cheesecloth prepared in **step 5**.
7. Warm the 1% agarose to 37 °C and pour 3 mL over the skin placed on the cheesecloth, wait for the agarose to set.
8. Use the edge of the cheesecloth to roll and submerge the whole tissue embedded in agarose contained in the cloth inside the concave metal strainer prepared in **step 4**, leave the edge of the cloth over the edge of the flask and trap it with clamps to the strainer.
9. Incubate overnight in a water bath at 37 °C.
10. Discard cheesecloth with agarose, L3 larvae should be at the bottom of the glass flask.
11. Decant excess PBS with a glass pipette to leave 2–3 mL final volume.

12. Place in a petri dish and count using an inverted microscope.
    For L4 larvae
13. As mentioned in **step 1** of this section, use day 2 days post-infection mice for L4 larvae harvest, euthanize and wet with 70% ethanol.
14. Attach the 18 G catheter to the 1 mL syringe and fill with PBS.
15. Expose the throat of the mice by making an incision and pulling back the skin carefully separating the muscle above the trachea, make a horizontal cut across the exposed trachea.
16. Insert catheter and inject content into the lungs, draw liquid in and out at least four times before collection.
17. Place in a 90 mm petri dish and count using an inverted microscope.

3.6 Generation of L3, L4, and L5 Derived Antigens

The following protocol will generate soluble antigens for L3, L4, and L5 larvae including secreted antigens; for generation of a protein extract of somatic tissue, homogenized larvae in cold buffer (Tris, PBS, saline) for protein preparation should suffice and the method is described elsewhere [32].

1. Isolate different larvae stages according to Subheadings 3.2, 3.3, and 3.5.
2. Wash larvae or worms in 25 mL of NES wash solution I, let settle the worms by gravity between washes and aspirate wash solution. For L3 larvae an extensive wash is recommended, at least five times; for L4 and L5 three washes should suffice.
3. Transfer the larvae to a sterile 40 mL glass tube and wash three times with NES wash solution II, let settle the worms by gravity between washes and aspirate wash solution.
4. Transfer worms to a sterile 40 mL glass tube and incubate for 1 h at room temperature in 25 mL of NES wash solution II.
5. Count worms as before and resuspend in NES culturing solution as described below:

    L3: 10,000 larvae/mL.

    L4: 500 larvae/mL.

    L5: 100 larvae/mL.
6. Culture 15 mL of the worm suspension in 50 mL tissue culture flask for 2 days at 37 °C with 5% CO_2.
7. Collect supernatant, filter through 0.2 μm filter and concentrate by ultracentrifugation using an MWCO 5000 membrane (as described in [33]).
8. Determine protein concentration by standard protein quantification methods and analyze the antigen on a 15% SDS-PAGE gel.
    For Schistosoma mansoni

3.7 Percutaneous Exposure of Mice to Cercariae

Mouse is the most common model used for *S. mansoni* experimental infection and no strain preference has been reported. Skin infection resembles the natural route of exposure to cercariae and is the preferred route, although subcutaneous and intraperitoneal injections can also be used [31]. Tail or abdominal skin exposure protocols are available, but in this chapter we focus on the tail infection given the advantages of mouse restrained techniques and quantification of infecting cercariae. Use latex gloves at all times when handling schistosomal suspensions, snails, or any material associated with the snails.

1. Harvest cercariae (*see* Subheading 3.10) and within 5 h after harvesting, pipette them into a 12 × 75 mm glass or plastic tube and add conditioned water up to 10 mm from the top of the tube. Place the tube in a rack taller than the tube, for chronic infections use 25–30 cercariae, for vaccine purposes use 100–150 and for maximum egg and adult worm yields use 200–250.
2. Place the mouse in a plastic restraining tube, the tail should be extended from the bottom of the tube. Use adhesive tape to attach the tail and help to anchor in place. Placing a piece of absorbent paper in the bottom of the restraining tube will help to avoid mouse urine to contaminate and kill cercarial suspension.
3. Moisten a gauze with conditioned water and wipe the mouse tail to clean debris. Insert the tail into the tube by placing the restraining tube on top of the rack.
4. Expose mouse to cercariae for 1 h and move it back to the cage without wiping the tail.
5. Empty the content of the tube into a counting dish, rinse with 2–3 mL of water and estimate the number of cercariae that penetrated the skin by counting on an inverted microscope. Count cercariae and bodies without tails.
6. Wait for 6–7 weeks before collecting adult worms (*see* Subheading 3.12) or eggs (*see* Subheading 3.11).

3.8 Maintenance and Propagation of S. mansoni in Snail

The most commonly used snail species for maintenance of *S. mansoni* is *Biomphalaria glabrata*. Use latex gloves at all times when handling schistosomal suspensions, snails, or any material associated with the snails.

1. The most recommended container for snails is aquaria of 30 gallon, always keep a separate tank of uninfected snails.
2. Avoid snail crowding as it affects their growth; maintain strict temperature control between 26 and 28 °C.
3. Cover the snail tank to limit light exposure until cercarial harvest.

4. Monitor water quality based on snail movement and mortality and avoid contamination with other living organisms (*see* **Note 5**).
5. For feeding the snails romaine lettuce is used along with a prepared supplement (*see* **Note 4**).

3.9 Infection of Snails with Miracidia

Use fresh miracidia to infect (within 3 h from hatching) for better results, eggs will hatch in minutes. Latex gloves should be used at all times when handling schistosomal suspensions, snails, or any material associated with the snails.

1. Euthanize *S. mansoni* infected mice by intraperitoneal injection of 0.3 mL of sodium pentobarbital with heparin.
2. Isolate liver and/or small and large intestine, rinse it in 1.2% NaCl. When using intestines, cut open and rinse out the intestinal content with several washes of 1.2% NaCl.
3. Blend tissues in conditioned water using a Waring blender for 30 s. Centrifuge for 5 min at 300 × *g* room temperature.
4. Decant, add 5 mL of conditioned water and shake the tube vigorously for a few seconds.
5. Dilute the suspension 100 fold in 25 °C conditioned water for better hatching.
6. Place the suspension in a darkened Büchner flask making sure the water covers the unpainted section.
7. Place the flask under a direct light source exposing the unpainted part of the side arm, miracidia will swim into this area after hatching within 10–20 min.
8. Withdraw the miracidia from the side arm with a Pasteur pipette under a dissecting microscope and place in a petri dish.
9. Infect juvenile *B. glabrata* by adding miracidia in a beaker, calculate five miracidia per snail and leave for at least 2 h before returning the snails to the aquaria.

3.10 Collection of Cercariae from Infected Snails

Before infection into mice, expose the infected snails liberating cercariae in a glass beaker under strong light. Use latex gloves at all times when handling schistosomal suspensions, snails, or any material associated with the snails. Take special attention at containment procedures when handling cercariae.

1. Place the snails shedding cercariae (between 3.5 and 5 weeks from miracidia exposure) in a 100 mL beaker with conditioned water at a density of 1 snail/2 mL water to collect cercariae.
2. Expose the flask to a strong light for 1–2 h. Do not allow the water to overheat.
3. Remove the snails from the beaker with a fish net or featherweight forceps and return them to the aquarium.

4. Pour the content of the beaker through the filtration screen apparatus, it will trap snail debris allowing cercariae to pass through.
5. Swirl the cercarial suspension with a 1 mL repeating pipettor and place 200 μL aliquots in separate counting dish. Add 1–2 mL of water and a couple of drops of iodine solution to kill and stain the cercariae.
6. Count the intact cercariae using a dissecting microscope, calculate numbers in several aliquots to ensure accurate quantification. Keep a record of cercariae per snail as an indicator of the efficiency of the life cycle.
7. Use cercariae to infect mice (*see* Subheading 3.7).

3.11 In Vitro Transformation of Cercariae to Schistosomules

Place the cercariae in a 50 mL plastic centrifuge tube and leave on ice for 30 min. To detach the tails from the body of the cercariae, agitation of the organisms is a commonly used method in vitro. Here we describe vortex as a means to generate early schistosomule stage.

1. Centrifuge for 2 min at 100 × *g* at 4 °C.
2. Discard supernatant up to the last 2–3 mL on top of the pellet.
3. Add 3 mL of 4 °C DMEM.
4. Cap the tube and vortex 45 s at maximum speed.
5. Place tube on ice for 3 min and repeat **step 5**.
6. Add 40 mL of 4 °C Percoll gradient suspension to a 50 mL centrifuge tube. Layer on top with a Pasteur pipette up to 10 mL of the cercarial suspension.
7. Centrifuge for 15 min at 500 × *g* at 4 °C.
8. Discard all the liquid except the lowest 10 mL of the tube.
9. Resuspend the pellet and add DMEM (4 °C) to a final 50 mL volume.
10. Centrifuge for 5 min at 100 × *g* at 4 °C.
11. Repeat **steps 10** and **11** twice and remove supernatant.
12. Culture resulting organisms in a 250 mL tissue culture flask with 100 mL DMEM at a density of 500 organisms/mL, incubate at 37 °C in a CO_2 incubator. For cultures longer than 8 h, add antibiotics.

3.12 Collection of Adult Schistosomal Worms from the Mouse Portal Venous System

Use 6–7 weeks post-infection mice to collect adult worms. Personnel should use latex gloves at all times when handling schistosomal suspensions, snails, or any material associated with the snails.

1. Euthanize *S. mansoni* infected mouse by intraperitoneal injection of 0.3 mL of sodium pentobarbital with heparin.

2. Place the mouse on its back and wet the abdominal area with 70% ethanol.
3. Remove the skin from the abdominal and thoracic cavity.
4. Open the abdominal cavity with dissecting scissors, cut the diaphragm and expose the left half of the thoracic cavity. Cut the ribs near the spinal cord.
5. Cut the hepatic portal vein, insert the 20 G needle attached to the peristaltic perfusion pump into the descending aorta and pump perfusion fluid through the needle.
6. Collect the worms in the perfusate in a petri dish.
7. Place the worm suspension over a fluorescent light box and withdraw individual adult worms with a Pasteur pipette. Add perfusion fluid to the worms in a 60 mm gridded petri dish and count under a dissecting microscope.

3.13 Enumeration of Schistosomal Eggs from Mouse Tissues

Counting eggs from tissues (mostly liver and intestines) after perfusion provides an estimate of the level of infection. Fecundity of the worms is reflected on the number of eggs and adult worm burdens. For accurate fecundity estimates it is important to distinguish between mature and immature females characterized as darkly pigmented or smaller and lightly pigmented, respectively.

1. Isolate the tissues to be digested from euthanized mice and determine its wet weight. Intestines should be cut open and washed to clean contents before weighting.
2. Place tissues in an appropriate volume of 4% KOH considering that the tissues should not take more than 10% of the final volume and incubate at 37 °C for 16–24 h.
3. Mix the suspension and count the eggs in several aliquots using an inverted microscope with a 4× objective.

3.14 Schistosomal Eggs Infection for Pulmonary Granuloma Formation

Eggs can be isolated for infection into other animals, for better egg yields 7–8.5 weeks post-infection mice should be used as donors. *Schistosoma* eggs have been used to stimulate egg-induced pathology in a pulmonary granuloma formation model described below.

1. Isolate infected tissues (liver and/or intestines) from euthanized mice and rinse in 1.2% (4 °C) NaCl. Cut open the intestines to clean contents with rinses of NaCl.
2. Cut the tissues in small pieces with scissors, suspend in 1.2% NaCl and place in a Waring blender. Blend for 30 s at low speed and pass the blended material through a crude sieve (420 μm). Blend again for 30 s the remaining tissue on top of the sieve at medium speed. Pass through the sieve as before and blend at the highest speed the remaining tissue on top of the sieve. Pass one last time through the crude sieve.

3. Take the filtrate and pass through the series of stainless steel sieves with decreasing pore sizes, use a plastic spray bottle filled with 1.2% NaCl to rinse between successive screens. Eggs should be retained on the final sieve (45 μm).
4. Collect the eggs from the final sieve into a glass petri dish, by gently swirling concentrate the eggs in the middle of the dish and withdraw with a Pasteur pipette to place in a tube on ice. Repeat harvest of the eggs by adding 1.2% NaCl to the petri dish until no eggs concentrate in the middle.
5. Place the collected eggs back into the dish with fresh 1.2% NaCl and repeat **step 5** to eliminate debris.
6. Count eggs and resuspend at a density of 2000–5000 eggs in 0.25 mL of PBS.
7. Restrain mouse with the tail exposed. Use a sterile alcohol swab to clean the tail and inject 0.25 mL of egg suspension using a 1 mL disposable syringe with a 23 or 25 G needle into one of the lateral tail veins. Upon withdrawing the needle press on the injection site with a sterile gauze to ensure hemostasis before returning the mouse to its cage. Inject one mouse at a time and mix well the egg suspension between loads of the syringe since eggs settle quickly.
8. At 14 days post-infection (peak of lung granuloma formation in naïve mouse), euthanize infected mice by intraperitoneal injection of 0.3 mL of sodium pentobarbital with heparin and remove lungs for histology.

3.15 Preparation of Schistosoma mansoni Soluble Egg Antigen (SEA)

Schistosoma antigen preparation from eggs is typically used to obtain a complex crude homogenate of soluble egg antigen (SEA). Use latex gloves at all times when handling schistosomal suspensions, snails, or any material associated with the snails.

1. Resuspend purified eggs in 5–7 mL of PBS (4 °C) at a density of 100,000 eggs/mL. To decrease protein degradation, a protease inhibitor (leupeptin at 10 μg/mL) can be included in the extraction buffer.
2. Homogenize the eggs with a prechilled glass Potter-Elvehjem hand held homogenizer using a tight pestle on ice. Check homogenization by placing a drop on a microscope slide and examining at 40×; intact eggs are easily distinguished from empty shells. Repeat grinding until breaking most of the eggs (500 cycles of grinding).
3. Centrifuge the crude mixture 20 min at 2000 × *g* at 4 °C, take the supernatant and keep at 4 °C. Re-homogenize the pellet as described in **step 2**.

4. Withdraw the crude supernatant and ultracentrifuge 90 min at 100,000 × *g* at 4 °C.
5. Pass supernatant through a 0.2 μm filter to sterilize, determine protein concentration, and store at −70 °C.

4 Notes

1. To avoid contamination with fungi of charcoal–fecal pellet cultures and generation of L3 larvae, use amphotericin B diluted 1:500 in the distilled water to moist the charcoal plates.
2. Add gentamicin at a 1:500 dilution to the L3 larvae suspension before infecting to avoid contamination with bacteria.
3. For isolation of L5 larvae for antigen preparation is better to starve the mice overnight 1 day before isolation to decrease fecal content in the gut and avoid debris.
4. Prepare snail food by mixing: 2.5 g wheat germ, 1.25 g dried milk, 2.5 g cerophyl, 2.5 g fish food, and 2.5 g sodium alginate (medium viscosity) in 250 mL of heated (near to boiling) distilled water. Dissolve by continue heating during 15–20 min. Pour the content into a pan and cool at 4 °C overnight. Pour 500 mL of 2% $CaCl_2$ over solidified mixture, 24 h later replace the $CaCl_2$ with water and keep it at 4 °C until use. Typically 3–4 g of the food is enough for 40 snails.
5. To avoid contamination of the snail container pay special attention to the quality of the water. Cloudy conditions and odor are indicative of bacterial overgrowth. Fungi can also grow on the shell of the snails and can be removed mechanically with a Q-tip or water spray. Combat typical contamination by metazoans (especially rotifers) by incubating snail eggs in 1% of Chlorox solution in conditioned water for 10 min at room temperature and wash extensively.

Acknowledgment

We apologize to the researchers whose work could not be cited due to space limitations. This work was supported by the following grants from CONACYT (CB-2015-01-255287, S008-2015-2-261227) and DGAPA (IA202116-PAPIIT).

References

1. Lukacs NW, Strieter RM, Kunkel SL (1995) Leukocyte infiltration in allergic airway inflammation. Am J Respir Cell Mol Biol 13(1):1–6. doi:10.1165/ajrcmb.13.1.7598934
2. Robinson DS, Hamid Q, Ying S, Tsicopoulos A, Barkans J, Bentley AM, Corrigan C, Durham SR, Kay AB (1992) Predominant TH2-like bronchoalveolar T-lymphocyte population in atopic asthma. N Engl J Med 326(5):298–304. doi:10.1056/NEJM199201303260504
3. Wilhelm C, Hirota K, Stieglitz B, Van Snick J, Tolaini M, Lahl K, Sparwasser T, Helmby H, Stockinger B (2011) An IL-9 fate reporter demonstrates the induction of an innate IL-9 response in lung inflammation. Nat Immunol 12(11):1071–1077. doi:10.1038/ni.2133
4. Licona-Limon P, Henao-Mejia J, Temann AU, Gagliani N, Licona-Limon I, Ishigame H, Hao L, Herbert DR, Flavell RA (2013) Th9 Cells Drive Host Immunity against Gastrointestinal Worm Infection. Immunity 39(4):744–757. doi:10.1016/j.immuni.2013.07.020
5. Nicolaides NC, Holroyd KJ, Ewart SL, Eleff SM, Kiser MB, Dragwa CR, Sullivan CD, Grasso L, Zhang LY, Messler CJ, Zhou T, Kleeberger SR, Buetow KH, Levitt RC (1997) Interleukin 9: a candidate gene for asthma. Proc Natl Acad Sci U S A 94(24):13175–13180
6. Levitt RC, McLane MP, MacDonald D, Ferrante V, Weiss C, Zhou T, Holroyd KJ, Nicolaides NC (1999) IL-9 pathway in asthma: new therapeutic targets for allergic inflammatory disorders. J Allergy Clin Immunol 103(5 Pt 2):S485–S491
7. Faulkner H, Renauld JC, Van Snick J, Grencis RK (1998) Interleukin-9 enhances resistance to the intestinal nematode Trichuris muris. Infect Immun 66(8):3832–3840
8. Uyttenhove C, Simpson RJ, Van Snick J (1988) Functional and structural characterization of P40, a mouse glycoprotein with T-cell growth factor activity. Proc Natl Acad Sci U S A 85(18):6934–6938
9. Petit-Frere C, Dugas B, Braquet P, Mencia-Huerta JM (1993) Interleukin-9 potentiates the interleukin-4-induced IgE and IgG1 release from murine B lymphocytes. Immunology 79(1):146–151
10. Dugas B, Renauld JC, Pene J, Bonnefoy JY, Peti-Frere C, Braquet P, Bousquet J, Van Snick J, Mencia-Huerta JM (1993) Interleukin-9 potentiates the interleukin-4-induced immunoglobulin (IgG, IgM and IgE) production by normal human B lymphocytes. Eur J Immunol 23(7):1687–1692. doi:10.1002/eji.1830230743
11. Shimbara A, Christodoulopoulos P, Soussi-Gounni A, Olivenstein R, Nakamura Y, Levitt RC, Nicolaides NC, Holroyd KJ, Tsicopoulos A, Lafitte JJ, Wallaert B, Hamid QA (2000) IL-9 and its receptor in allergic and nonallergic lung disease: increased expression in asthma. J Allergy Clin Immunol 105(1 Pt 1):108–115
12. Schmitt E, Germann T, Goedert S, Hoehn P, Huels C, Koelsch S, Kuhn R, Muller W, Palm N, Rude E (1994) IL-9 production of naive CD4+ T cells depends on IL-2, is synergistically enhanced by a combination of TGF-beta and IL-4, and is inhibited by IFN-gamma. J Immunol 153(9):3989–3996
13. Veldhoen M, Uyttenhove C, van Snick J, Helmby H, Westendorf A, Buer J, Martin B, Wilhelm C, Stockinger B (2008) Transforming growth factor-beta 'reprograms' the differentiation of T helper 2 cells and promotes an interleukin 9-producing subset. Nat Immunol 9(12):1341–1346. doi:10.1038/ni.1659
14. Dardalhon V, Awasthi A, Kwon H, Galileos G, Gao W, Sobel RA, Mitsdoerffer M, Strom TB, Elyaman W, Ho IC, Khoury S, Oukka M, Kuchroo VK (2008) IL-4 inhibits TGF-beta-induced Foxp3+ T cells and, together with TGF-beta, generates IL-9+ IL-10+ Foxp3(−) effector T cells. Nat Immunol (12):1347–1355. doi:10.1038/ni.1677
15. Purwar R, Schlapbach C, Xiao S, Kang HS, Elyaman W, Jiang X, Jetten AM, Khoury SJ, Fuhlbrigge RC, Kuchroo VK, Clark RA, Kupper TS (2012) Robust tumor immunity to melanoma mediated by interleukin-9-producing T cells. Nat Med 18(8):1248–1253. doi:10.1038/nm.2856
16. Jager A, Dardalhon V, Sobel RA, Bettelli E, Kuchroo VK (2009) Th1, Th17, and Th9 effector cells induce experimental autoimmune encephalomyelitis with different pathological phenotypes. J Immunol 183(11):7169–7177. doi:10.4049/jimmunol.0901906
17. Gerlach K, Hwang Y, Nikolaev A, Atreya R, Dornhoff H, Steiner S, Lehr HA, Wirtz S, Vieth M, Waisman A, Rosenbauer F, McKenzie AN, Weigmann B, Neurath MF (2014) TH9 cells that express the transcription factor PU.1 drive T cell-mediated colitis via IL-9 receptor signaling in intestinal epithelial cells. Nat Immunol 15(7):676–686. doi:10.1038/ni.2920
18. Ciccia F, Guggino G, Rizzo A, Manzo A, Vitolo B, La Manna MP, Giardina G, Sireci G, Dieli F, Montecucco CM, Alessandro R, Triolo G (2015) Potential involvement of IL-9 and Th9 cells in the pathogenesis of rheumatoid

arthritis. Rheumatology (Oxford) 54(12): 2264–2272. doi:10.1093/rheumatology/kev252

19. Kennedy MWHW (2013) Parasitic nematodes: molecular biology, biochemistry and immunology, 2nd edn. CAB International, Wallingford
20. Organization WH (2016) Soil transmitted helminth infections. WHO Media Center, Geneva
21. Urban JF Jr, Maliszewski CR, Madden KB, Katona IM, Finkelman FD (1995) IL-4 treatment can cure established gastrointestinal nematode infections in immunocompetent and immunodeficient mice. J Immunol 154(9): 4675–4684
22. Barner M, Mohrs M, Brombacher F, Kopf M (1998) Differences between IL-4R alpha-deficient and IL-4-deficient mice reveal a role for IL-13 in the regulation of Th2 responses. Curr Biol 8(11):669–672
23. McKenzie GJ, Emson CL, Bell SE, Anderson S, Fallon P, Zurawski G, Murray R, Grencis R, McKenzie AN (1998) Impaired development of Th2 cells in IL-13-deficient mice. Immunity 9(3):423–432
24. WHO (2016) Schistosomiasis Media center fact sheet. WHO, Geneva
25. Schramm G, Haas H (2010) Th2 immune response against *Schistosoma mansoni* infection. Microbes Infect 12(12–13):881–888. doi:10.1016/j.micinf.2010.06.001
26. Fallon PG, Smith P, Dunne DW (1998) Type 1 and type 2 cytokine-producing mouse CD4+ and CD8+ T cells in acute *Schistosoma mansoni* infection. Eur J Immunol 28(4):1408–1416
27. Khalil RM, Luz A, Mailhammer R, Moeller J, Mohamed AA, Omran S, Dormer P, Hultner L (1996) *Schistosoma mansoni* infection in mice augments the capacity for interleukin 3 (IL-3) and IL-9 production and concurrently enlarges progenitor pools for mast cells and granulocytes-macrophages. Infect Immun 64(12):4960–4966
28. Fallon PG, Smith P, Richardson EJ, Jones FJ, Faulkner HC, Van Snick J, Renauld JC, Grencis RK, Dunne DW (2000) Expression of interleukin-9 leads to Th2 cytokine-dominated responses and fatal enteropathy in mice with chronic *Schistosoma mansoni* infections. Infect Immun 68(10):6005–6011
29. Townsend JM, Fallon GP, Matthews JD, Smith P, Jolin EH, McKenzie NA (2000) IL-9-deficient mice establish fundamental roles for IL-9 in pulmonary mastocytosis and goblet cell hyperplasia but not T cell development. Immunity 13(4):573–583
30. Camberis M, Le Gros G, Urban J Jr (2003) Animal model of Nippostrongylus brasiliensis and Heligmosomoides polygyrus. Curr Protoc Immunol Chapter 19:Unit 19.12. doi:10.1002/0471142735.im1912s55
31. Lewis FA (1998) Schistosomiasis. Current protocols in immunology. Jonh Wiley & Sons inc, New York, NY
32. Nawa Y, Miller HR, Hall E, Jarrett EE (1981) Adoptive transfer of total and parasite-specific IgE responses in rats infected with Nippostrongylus brasiliensis. Immunology 44(1): 119–123
33. SM A (1997) Dialisis and concentration of protein solutions. Current protocols in immunology. Jonh Wiley & Sons inc, New York, NY

Chapter 19

Isolation and Purification of Th9 Cells for the Study of Inflammatory Diseases in Research and Clinical Settings

Patricia Keating and James X. Hartmann

Abstract

Th9 cells are associated with atopic and inflammatory diseases, and their increased levels and function correlate with the severity of symptoms in various inflammatory disorders including asthma, food allergy, atopic dermatitis, ulcerative colitis, and psoriatic arthritis. Thus, clinical trials are warranted to evaluate the role of Th9 cells in allergic diseases with the goal of controlling these ailments.

Circulating T cells (naïve or memory CD4+ T cells) purified from human blood and expanded using anti-CD3 and anti-CD28 antibodies can be treated with appropriate cytokines in order to polarize them to the Th9 phenotype as evidenced by their production of IL-9. When treated in vitro with cholecalciferol or 1,25$(OH)_2$ vitamin D3, cells polarized under Th9 conditions significantly downregulate production of IL-9. The percentage of polarized Th9 memory cells from patients treated with steroids or other modalities can be monitored during clinical trials and compared to control populations.

Key words Th9, IL-9, Inflammation, Allergy, Atopy, T cell polarization, Vitamin D

1 Introduction

When T cells become activated by antigen-presenting cells, their function and their cytokine profile is subsequently influenced by the microenvironment. Th1, Th2, Tfh, Th17, Th22, Th9, Treg, and other T helper cell subsets have been described and are associated with specific immune responses. Th9 cells are CD4+ T cells that produce IL-9, low levels of IL-17 and, in mouse models, produce IL-10 [1]. They express the transcription factors PU.1 [2], IRF4 [3], BATF [4] and low levels of GATA3. They do not produce IL-4, IL-5 or IL-13 and do not express CD294 (CRTH2), the prostaglandin receptor, which defines Th2 cells [5]. IL-9-producing T cells induce differentiation and antibody production by B cells, [6] therefore this cytokine is associated with many autoimmune inflammatory conditions.

Ritobrata Goswami (ed.), *Th9 Cells: Methods and Protocols*, Methods in Molecular Biology, vol. 1585,
DOI 10.1007/978-1-4939-6877-0_19,

Th9 cells have been implicated in eliciting symptoms in both allergies and asthma [7]. In children with asthma and peanut allergy, Th9 levels correlate with the severity of the disease [8–10]. Th9 cells are found in skin lesions of atopic dermatitis patients and both their percentage in blood and PU.1 transcription factor expression were increased when compared with healthy controls and psoriasis patients [11]. Patients with ulcerative colitis had elevated levels of IL-9+PU.1+ cells, and in an animal model, IL-9 deficiency protected from the symptoms of colitis [12]. IL-9 has also been implicated in rheumatoid arthritis and psoriatic arthritis as abundant IL-9 positive cells have been detected in circulation and in the gut mucosa of psoriatic arthritis patients [13].

Strategies to ameliorate the symptoms of atopic and inflammatory autoimmune diseases have relied on chronic corticosteroid treatment to control immune responses. Vitamin D is an endogenous anti-inflammatory steroid, whose function is to downregulate late phase immune responses. The storage form of the steroid (25(OH) cholecalciferol) is hydroxylated by CYP27B1 to generate the active form, $1{,}25(OH)_2$ cholecalciferol. This enzyme is induced by activated immune cells such as macrophages, T cells and B cells and also by epithelial cells when they are infected or irritated (reviewed by Szymczak et al.) [14].

The therapeutic effects of vitamin D and its analogues are being evaluated in numerous diseases, with varying results. Vitamin D ameliorates the symptoms of atopic dermatitis [15]. Chambers et al. showed vitamin D altered the T cell phenotype of glucocorticoid resistant asthma patients, characterized by high IFN-γ, and IL-17 production, and skewed towards a steroid sensitive phenotype [16]. Vitamin D downregulated IL-9 production in human Th9 cells in vitro [5] by inhibiting BATF binding to the *Il9* promoter and the suppression of the aryl hydrocarbon receptor [17]. Prenatal supplementation with vitamin D resulted in a lower incidence of asthma in children [18], and asthma exacerbations in children tend to be lower when they are supplemented with vitamin D. Therefore, vitamin D is now recommended as adjunctive therapy for children with asthma [19].

A large, long term clinical trial to evaluate vitamin D supplementation, the VITAL study, is underway. In preliminary reports, vitamin D supplementation lowered pneumonia risk, respiratory exacerbation episodes, asthma control, and improved lung function in adults [20].

Enokizumab, (also known as MEDI-528) is a humanized immunoglobulin G1k anti-IL-9 monoclonal antibody that has been developed for the treatment of asthma. In an open-label, phase I dose-escalation study, single doses of MEDI-528 were administered to 53 healthy participants (by either endogenous or subcutaneous route). MEDI-528 had an acceptable safety profile and exhibited linear pharmacokinetics over the dose range studied

(NCT00116168). In subsequent clinical trials (NCT00507130, NCT00590720), patients with mild to moderate asthma, or with exercise-induced bronchoconstriction, were treated with MEDI-528. Adverse events (AEs), pharmacokinetics (PK), immunogenicity, asthma control (including asthma exacerbations), and exercise challenge tests were evaluated. The results were not conclusive, but MEDI-528 had a good safety profile and suggested certain degree of clinical activity.

In a prospective double-blind, multicenter, parallel-group study (NCT00968669), 329 subjects were randomized to subcutaneous placebo or MEDI-528 every 2 weeks for 24 weeks, in addition to their usual asthma medications. There were no significant differences in FEV1% or in asthma exacerbations between the placebo and the treated groups [21–23]. As IL-9 targets multiple cell types and tissues, and its action is paracrine, designing monoclonal antibodies against the IL-9 receptor might prove to be more effective.

The Th9 subset is an important target for therapeutic strategies to treat atopic and mucosal inflammatory diseases. Therefore, clinical trials are further warranted to evaluate if this subset can be used as a reliable biomarker of atopic disease, if this subset correlates with vitamin D deficiency, and if vitamin D supplementation could be recommended to prevent atopic disease and used as an adjuvant in autoimmune therapies.

2 Materials

2.1 Reagents

1. 96-well plates, flat bottom, tissue culture treated, sterile.
2. Polystyrene 15 or 50 mL centrifuge tubes, sterile.
3. Pipettes.
4. Ficoll/Histopaque 1077.
5. Phosphate buffered saline (PBS).
6. Cell culture media: AIM-V (or other serum free media), L-glutamine (2 mM), 2-β mercaptoethanol (50 μM/L), anti-human CD28 (clone CD28.2) (2 μg/mL), recombinant human IL-2 (20 IU/mL).
7. Trypan blue.
8. Negative selection kit for memory T cell (CD4+CD45RO+) purification.
9. Negative selection kit for naïve T cell (CD4+CD45RA+) purification.
10. Activation media: AIM-V, L-glutamine (2 mM), phorbol 12-myristate 13-acetate (PMA)(5–20 ng/mL), ionomycin (1 μg/mL), cytokines according to each polarization condition.

11. Golgi Stop or brefeldin (amount according to manufacturer's instructions).
12. FACS buffer: PBS, 2% fetal bovine serum or 1% albumin (v/v), 2–5 mM EDTA, 0.1% sodium azide (w/v).
13. Fixation buffer (2–4% paraformaldehyde) (*see* **Note 1**).
14. Permeabilization buffer (This buffer is obtained commercially; preparations are proprietary and vary according to the manufacturer).
15. Cholecalciferol (25D3) and 1,25 vitamin D3(1,25D3).
16. Vacutainer tubes.

2.2 Antibodies and Recombinant Human Cytokines

1. Anti-human CD3 antibody (clone UCHT1).
2. Anti-human CD28 (clone CD28.2).
3. Recombinant human (rh) IL-2.
4. rh IL-7.
5. rh IL-4.
6. rh transforming growth factor-beta (TGF-β).

2.3 Flow Cytometry Reagents (Labeled with Fluorescent Dyes)

1. Cell surface markers: anti-human CD4, anti-human CD45RO, anti-human CD45RA.
2. Intracellular markers: anti-human IL-9, IL-17, IL-13, IL-5, IL-4.
3. Intranuclear markers: anti-human PU.1, GATA3, IRF4, BATF.

2.4 Instruments

1. Centrifuge (benchtop).
2. Laminar flow hood (class II).
3. Incubator (37 °C, 5% CO_2).
4. Microscope.
5. Hemocytometer.
6. Flow cytometer.

3 Methods

Work should be performed using sterile technique, and sterile materials and reagents, in a class II biosafety laminar flow hood.

3.1 Preparation of 96-Well Plates for the Clonal Expansion of T Cells

1. Use a tissue culture treated 96-well plate (*see* **Note 2**).
2. Add 50 μL/well of sterile PBS containing 10 μg/mL of anti-human CD3 antibody (clone UCHT1).
3. Incubate preferably overnight at 4 °C (alternatively, at 37 °C for 4 h).

4. After the incubation period, wash out the anti-CD3 three times with sterile PBS (*see* **Note 3**).

3.2 Separate Peripheral Blood Mononuclear Cells (PBMC) from Circulating Blood

1. Obtain blood from patients or healthy donors by phlebotomy using vacutainer tubes containing heparin as an anticoagulant (50–70 mL). Dilute the blood 1:1 in RPMI 1640, or PBS.
2. Dispense Ficoll-Histopaque in 50 mL tubes (20 mL each) and carefully lay over 20 mL of diluted blood in each tube without disturbing the interface.
3. Centrifuge at 800 × *g* for 20 min, at room temperature, without brake.
4. Collect the buffy coat containing PBMC.
5. Wash the PBMCs three times with PBS (with 5% FBS).
6. Take an aliquot of the cells and dilute as necessary with trypan blue (for example 1:1). Using a hemocytometer, count PBMCs, record viability, specifying live (bright yellow cells) versus dead (dull blue cells). Viability should be greater than 90%.

3.3 Purification of Naïve or Memory T Cells from PBMCs

1. Use a Negative Selection kit for purification of Naïve or Memory T cells, following the manufacturer's instructions (*see* **Note 4**).
2. Immediately after purification, separate an aliquot (1–2 × 10^5 cells) and check purity by flow cytometry. Label cells with anti-CD4, anti-CD45RA or anti-CD45RO antibodies, fix with 2% paraformaldehyde, and analyze by flow cytometry. CD45RA is expressed in naïve T cells and represent 40–50% of the CD4+ T cell population. CD45RO is complementarily expressed in activated memory CD4+ T cells.

3.4 Clonal Expansion of Memory or Naïve T Cells Using Nonspecific Stimulation

1. Prepare cell culture media and add rh IL-7 (20 ng/mL) if needed (*see* **Note 5**).
2. Resuspend purified T cells in cell culture media at a concentration of 25 × 10^4 cell/mL. Place 200 μL of the cell suspension in wells of a 96-well plate that has been pre-coated with anti-CD3 (clone UCHT1) at 10 μg/mL.
3. Add rh IL-4 (50 ng/mL) and rh TGF-β (10 ng/mL) to appropriate wells according to the different polarizing conditions, not to non-polarized control wells (*see* **Note 6**).
4. Incubate the plate in a humidified incubator at 37 °C, 5% CO_2, for 3 days.
5. Naïve or memory CD4+ T cells will have undergone clonal expansion after 3 days of incubation, and clusters of T cells will be visible under a microscope.

6. Treatment of T cell cultures with compounds of interest such as vitamin D can be performed to evaluate the effects of these compounds on the polarization process.
7. Add vitamin D (25(OH)D_3) storage form (40 ng/mL or less as a titration experiment), or 1,25$(OH)_2D_3$ (20 ng/mL or less as a titration experiment) or vehicle (ethanol) to the cultures at their inception (*see* **Note 7**).

3.5 Collect Supernatants for the Quantification of Cytokines by ELISA

1. On day 3–4, carefully aspirate 150 μL of supernatants for ELISA assays (IL-9, IL-17).
2. Add fresh complete media (100–200 μL/well) with the appropriate cytokines, to the cells.
3. Add PMA (50 ng/mL) and ionomycin (250 ng/mL) to all wells.
4. Incubate for 5–6 h.
5. Add brefeldin during the last 4 h of incubation.
6. Harvest the cells and count them, assess viability with trypan blue (*see* **Note 8**).

3.6 Flow Cytometry Analysis

1. Surface marker staining: Harvest the expanded T cells (2–5 × 10^5 cells). Place the cells in 5 mL polystyrene tubes, centrifuge at 250 × *g* for 5 min, at room temperature.
2. Resuspend cells in FACS buffer. Cells can be stained in the 5 mL tubes or placed in wells (V-shaped) of a 96-well plate. (Resuspend accordingly in 2 mL or 200 μL of FACS buffer.)
3. Add fluorescently labeled anti-human CD4, and anti-CD45RO (amount previously determined by titration) (optional, *see* **Note 9**).
4. Incubate cells for 30 min, wash twice with FACS buffer, resuspending in 2 mL or 200 μL of FACS buffer according to tube or plate based assay.
5. For intracellular staining add fixation buffer (2% or 4% paraformaldehyde) (*see* **Note 1**) (200 or 500 μL), while vortexing cells, to avoid clusters of cells fixed with the paraformaldehyde.
6. Incubate with fixation buffer for 30 min at 4 °C, protected from light.
7. Wash cells three times with 1 mL or 200 μL of permeabilization buffer.
8. Resuspend cells in 100 μL of permeabilization buffer.
9. Add fluorescently labeled anti human IL-9, (also IL-4, IL-13, IL-5) (amounts previously determined by titration).
10. Incubate cells for 30 min at 4 °C.

11. Resuspend cells in permeabilization buffer (1 mL or 200 μL).
12. Centrifuge at 250 × *g* for 5 min at room temperature and then discard supernatant.
13. Resuspend cells in permeabilization buffer (1 mL or 200 μL).
14. Centrifuge and then discard supernatant.
15. Resuspend cells in cold FACS buffer (4 °C) (200–500 μL).
16. Analyze cells in the flow cytometer.
17. Gating strategy: Establish a FS vs SS gate to encompass lymphocytes, use a dot plot of CD4 vs IL-9 (or other cytokine or protein) and record double positive cells (*see* **Note 10**).

4 Notes

1. Fixation buffer: use 1–2% paraformaldehyde for surface staining alone, 2–4% to detect intracellular cytoplasmic proteins such as IL-9 and IL-13 and 4% paraformaldehyde for the detection of intranuclear proteins such as transcription factors (PU.1, GATA3, IRF4, BATF).
2. Plates with larger wells such as 24-well or 48-well plates do not work as well as a 96-well plate would for clonal expansion of T cells.
3. Soluble anti-CD3 antibody should be thoroughly removed from the wells before adding the T cells. The coated plates can be kept at 4 °C for several days.
4. Memory T cells express PU.1 and are epigenetically "poised" for ready synthesis of IL-9 [24].
5. IL-7 is needed for the memory T cell cultures.
6. A control sample in which no polarizing cytokines (rh TGF-β and rh IL-4) are added, may be included in the protocol for a baseline determination of IL-9.
7. Vitamin D is an inhibitor of cyclins and treatment with this steroid will inhibit proliferation of T cells. Comparing vitamin D treated with untreated CD4+ T cells, about a 10–20% decrease in proliferation is observed. This must be taken into account in interpreting the results of ELISA tests if they are performed. This does not alter the flow cytometry results as these are given on a per-cell, percentage basis.
8. If a robust proliferation, with large clusters of T cells is not observed, the production of IL-9 will be very scarce. Using cells with lower than optimal viability, using very large wells (as in 6-well or 24-well plates) instead of the recommended 96-well plates, or using cytokines that have lost their integrity (such as occurs when they have been repeatedly frozen and

thawed), are some of the instances whereby proliferation will not be robust and IL-9 production will be compromised.

9. CD4 expression is slightly downregulated in CD4+ T cells after they have been subjected to activation signals such as anti-CD3, anti-CD28 antibodies. After antigen stimulation cells will not be naïve, they will express CD45RO, not CD45RA.
10. Th9 cells are positive for IL-9 and negative for CD294, IL-13, IL-4, IL-5, IFN-γ, Foxp3, T-bet, low for GATA-3, high for IRF-4, high for PU.1 in human.

References

1. Palmer MT, Lee YK, Maynard CL, Oliver JR, Bikle DD, Jetten AM, Weaver CT (2011) Lineage-specific effects of 1,25-dihydroxyvitamin D(3) on the development of effector CD4 T cells. J Biol Chem 286(2):997–1004. doi:10.1074/jbc.M110.163790
2. Chang HC, Sehra S, Goswami R, Yao W, Yu Q, Stritesky GL, Jabeen R, McKinley C, Ahyi AN, Han L, Nguyen ET, Robertson MJ, Perumal NB, Tepper RS, Nutt SL, Kaplan MH (2010) The transcription factor PU.1 is required for the development of IL-9-producing T cells and allergic inflammation. Nat Immunol 11(6):527–534. doi:10.1038/ni.1867
3. Goswami R, Jabeen R, Yagi R, Pham D, Zhu J, Goenka S, Kaplan MH (2012) STAT6-dependent regulation of Th9 development. J Immunol 188(3):968–975. doi:10.4049/jimmunol.1102840
4. Jabeen R, Goswami R, Awe O, Kulkarni A, Nguyen ET, Attenasio A, Walsh D, Olson MR, Kim MH, Tepper RS, Sun J, Kim CH, Taparowsky EJ, Zhou B, Kaplan MH (2013) Th9 cell development requires a BATF-regulated transcriptional network. J Clin Invest 123(11):4641–4653. doi:10.1172/JCI69489
5. Keating P, Munim A, Hartmann JX (2014) Effect of vitamin D on T-helper type 9 polarized human memory cells in chronic persistent asthma. Ann Allergy Asthma Immunol 112(2):154–162. doi:10.1016/j.anai.2013.11.015
6. Knoops L, Louahed J, Renauld JC (2004) IL-9-induced expansion of B-1b cells restores numbers but not function of B-1 lymphocytes in xid mice. J Immunol 172(10):6101–6106
7. Yao W, Tepper RS, Kaplan MH (2011) Predisposition to the development of IL-9-secreting T cells in atopic infants. J Allergy Clin Immunol 128(6):1357–1360.e1355. doi:10.1016/j.jaci.2011.06.019
8. Erpenbeck VJ, Hohlfeld JM, Volkmann B, Hagenberg A, Geldmacher H, Braun A, Krug N (2003) Segmental allergen challenge in patients with atopic asthma leads to increased IL-9 expression in bronchoalveolar lavage fluid lymphocytes. J Allergy Clin Immunol 111(6):1319–1327
9. Shimbara A, Christodoulopoulos P, Soussi-Gounni A, Olivenstein R, Nakamura Y, Levitt RC, Nicolaides NC, Holroyd KJ, Tsicopoulos A, Lafitte JJ, Wallaert B, Hamid QA (2000) IL-9 and its receptor in allergic and nonallergic lung disease: increased expression in asthma. J Allergy Clin Immunol 105(1 Pt 1):108–115
10. Brough HA, Cousins DJ, Munteanu A, Wong YF, Sudra A, Makinson K, Stephens AC, Arno M, Ciortuz L, Lack G, Turcanu V (2014) IL-9 is a key component of memory TH cell peanut-specific responses from children with peanut allergy. J Allergy Clin Immunol 134(6):1329–1338.e1310. doi:10.1016/j.jaci.2014.06.032
11. Ma L, Xue HB, Guan XH, Shu CM, Zhang JH, Yu J (2014) Possible pathogenic role of T helper type 9 cells and interleukin (IL)-9 in atopic dermatitis. Clin Exp Immunol 175(1):25–31. doi:10.1111/cei.12198
12. Gerlach K, Hwang Y, Nikolaev A, Atreya R, Dornhoff H, Steiner S, Lehr HA, Wirtz S, Vieth M, Waisman A, Rosenbauer F, McKenzie AN, Weigmann B, Neurath MF (2014) TH9 cells that express the transcription factor PU.1 drive T cell-mediated colitis via IL-9 receptor signaling in intestinal epithelial cells. Nat Immunol 15(7):676–686. doi:10.1038/ni.2920
13. Ciccia F, Guggino G, Ferrante A, Raimondo S, Bignone R, Rodolico V, Peralta S, Van Tok M, Cannizzaro A, Schinocca C, Ruscitti P, Cipriani P, Giacomelli R, Alessandro R, Dieli F, Rizzo A, Baeten D, Triolo G (2016) IL-9 over-expression and Th9 polarization characterize the inflamed gut, the synovial tissue and the peripheral blood of patients with psoriatic arthritis. Arthritis Rheumatol. doi:10.1002/art.39649

14. Szymczak I, Pawliczak R (2016) The active metabolite of vitamin D3 as a potential immunomodulator. Scand J Immunol 83(2):83–91. doi:10.1111/sji.12403
15. Kim G, Bae JH (2016) Vitamin D and atopic dermatitis: a systematic review and meta-analysis. Nutrition. doi:10.1016/j.nut.2016.01.023
16. Chambers ES, Nanzer AM, Pfeffer PE, Richards DF, Timms PM, Martineau AR, Griffiths CJ, Corrigan CJ, Hawrylowicz CM (2015) Distinct endotypes of steroid-resistant asthma characterized by IL-17A(high) and IFN-gamma(high) immunophenotypes: Potential benefits of calcitriol. J Allergy Clin Immunol 136(3):628–637.e624. doi:10.1016/j.jaci.2015.01.026
17. Takami M, Fujimaki K, Nishimura MI, Iwashima M (2015) Cutting edge: AhR is a molecular target of calcitriol in human T cells. J Immunol 195(6):2520–2523. doi:10.4049/jimmunol.1500344
18. Litonjua AA, Carey VJ, Laranjo N, Harshfield BJ, McElrath TF, O'Connor GT, Sandel M, Iverson RE Jr, Lee-Paritz A, Strunk RC, Bacharier LB, Macones GA, Zeiger RS, Schatz M, Hollis BW, Hornsby E, Hawrylowicz C, Wu AC, Weiss ST (2016) Effect of prenatal supplementation with vitamin D on asthma or recurrent wheezing in offspring by age 3 years: The VDAART Randomized Clinical Trial. JAMA 315(4):362–370. doi:10.1001/jama.2015.18589
19. Jiao J, Castro M (2015) Vitamin D and asthma: current perspectives. Curr Opin Allergy Clin Immunol 15(4):375–382. doi:10.1097/ACI.0000000000000187
20. Gold DR, Litonjua AA, Carey VJ, Manson JE, Buring JE, Lee IM, Gordon D, Walter J, Friedenberg G, Hankinson JL, Copeland T, Luttmann-Gibson H (2016) Lung VITAL: rationale, design, and baseline characteristics of an ancillary study evaluating the effects of vitamin D and/or marine omega-3 fatty acid supplements on acute exacerbations of chronic respiratory disease, asthma control, pneumonia and lung function in adults. Contemp Clin Trials 47:185–195. doi:10.1016/j.cct.2016.01.003
21. Parker JM, Oh CK, LaForce C, Miller SD, Pearlman DS, Le C, Robbie GJ, White WI, White B, Molfino NA, M-CT G (2011) Safety profile and clinical activity of multiple subcutaneous doses of MEDI-528, a humanized anti-interleukin-9 monoclonal antibody, in two randomized phase 2a studies in subjects with asthma. BMC Pulm Med 11:14. doi:10.1186/1471-2466-11-14
22. White B, Leon F, White W, Robbie G (2009) Two first-in-human, open-label, phase I dose-escalation safety trials of MEDI-528, a monoclonal antibody against interleukin-9, in healthy adult volunteers. Clin Ther 31(4):728–740. doi:10.1016/j.clinthera.2009.04.019
23. Oh CK, Leigh R, McLaurin KK, Kim K, Hultquist M, Molfino NA (2013) A randomized, controlled trial to evaluate the effect of an anti-interleukin-9 monoclonal antibody in adults with uncontrolled asthma. Respir Res 14:93. doi:10.1186/1465-9921-14-93
24. Ramming A, Druzd D, Leipe J, Schulze-Koops H, Skapenko A (2012) Maturation-related histone modifications in the PU.1 promoter regulate Th9-cell development. Blood 119(20):4665–4674. doi:10.1182/blood-2011-11-392589

Index

A

Adoptive transfer 9, 12, 24, 161, 190, 203, 210–214
Airway hyperresponsiveness 23, 26, 39, 40, 189–191
Allergic inflammation 8, 12, 13, 23, 24, 41, 42, 51, 60, 189–197, 202
Alum 191, 192
Antigen presenting cells (APCs) 1, 60, 64, 65, 68, 79, 94, 141, 197, 204–207, 213, 221, 247
Antitumor responses 9
Asthma 11–13, 23, 24, 31, 40, 42, 43, 59, 61, 84, 94, 95, 127, 141, 180, 189–191, 202, 223, 248, 249
Autoimmune diseases 6, 24–25, 46, 95, 128, 180, 203, 248

B

B cells 2, 8, 38–40, 42, 73, 127, 142, 180, 223, 247, 248
BIOBASE 118, 123
Biomphalaria glabrata 230, 238, 239
B16-murine melanoma 9, 217–221
Bronchoalveolar lavage (BAL) 59, 189, 191, 193, 197
Brugia malayi Antigen (BmA) 84, 85, 88, 90

C

Catheter 191, 193, 229, 237
Cell lines 9, 22, 23, 28, 38, 41, 73, 156–158, 217, 219
Cercaria 39, 226, 227, 230–231, 238–240
Chemokines 9, 10, 42, 43
Chloroquine 159, 165
Chromatin 7, 9, 113–116, 119–122, 167–176
Chromatin immunoprecipitation (ChIP) 112–117, 119–121, 167–176, 181, 182, 184, 186
Clinical trials 217, 248, 249
Concanavalin A 22, 93
Conserved non-coding sequences (CNS) 7, 10, 113, 203, 210
Co-stimulation 44
Crohn's disease (CD) 11, 203
Cross-linking 26
Cytokine bead array 207
Cytokines 1, 11, 21, 37, 51, 59, 73, 83, 94, 129, 141, 155, 170, 180, 189, 201, 219, 223, 247
Cytospin 192, 193

D

Dextran sulfate sodium-induced colitis 94
Digoxigenin-labeled DNA 169
DNA purification 117, 121
DNA shearing 119, 183–184, 186

E

ELISA plate reader 84
Enzyme linked immunosorbent assay (ELISA) 27, 28, 31, 32, 56, 68, 81, 84, 87–88, 91, 95, 133, 148, 181, 192, 193, 195–197, 207, 208, 214, 252
Eosinophils 8, 10, 12, 40–43, 73, 83, 127, 190, 224
Epigenetics 3, 7, 113, 179–186
Epithelial cells 2, 8, 11, 38, 41–46, 51, 83, 93, 127, 248
Experimental autoimmune encephalomyelitis (EAE) 8, 10–11, 60, 61, 128, 180, 201–215

F

False negative 117, 122
False positive 81, 117, 118, 122
Ficoll-Hypaque 182
Flow cytometer 62, 68, 78, 84, 87, 102, 131, 135, 145, 148, 149, 162, 209, 250, 253
Fluorescence-activated cell-sorting (FACS) 131, 132, 206
Fluorescence minus one (FMO) 86, 89
Fluorochrome 53, 55–57, 74, 79, 106, 158, 181, 192, 195, 197
Frosted slides 52, 53, 145

G

Gamma-irradiator 220
Glutathione S-transferase (GST)-fused DNA 169
Glutathione-Sepharose beads 169
GST fusion protein 169, 173, 174

H

HEK293T cells 162
Helminth 45
Hemocytometer 63, 64, 96, 99, 104, 131, 250, 251
Histone modification 7, 181, 182
Homogenizer 232, 242
Human peripheral blood 41, 74, 97

Ritobrata Goswami (ed.), *Th9 Cells: Methods and Protocols*, Methods in Molecular Biology, vol. 1585, DOI 10.1007/978-1-4939-6877-0, © Springer Science+Business Media LLC 2017

I

Inflammatory bowel disease (IBD)...................... 6, 8, 11, 43, 44, 61, 95, 215
Innate lymphoid cells (ILCs) 27, 45–46, 94
Intracellular cytokine staining (ICS) 86–89, 96, 161–162, 165, 193, 194, 197, 204, 207, 209–210, 212
Intranasal...193
Intraperitoneal.. 193, 238, 239, 242
Invariant natural killer T (iNKT) cells 25, 94, 96, 97, 99–104
In vitro differentiation 60, 128, 203, 213
Ionomycin............... 26, 52, 55, 56, 63, 67, 68, 84, 85, 88, 91, 98, 101, 105, 130, 133, 137, 143, 147, 152, 158, 161, 163, 164, 191, 194, 197, 204, 207, 209, 213, 249, 252

J

Jak-STAT 37, 38

L

Larvae............................... 224, 225, 227–229, 232–237, 243
Liquid luminescent DNA precipitation assay...........167–176
Live/dead staining 96, 101
Luciferase reporter..158
Luminometer...158

M

Magnetic beads......................... 100, 101, 104, 113, 116, 119, 172, 182, 206, 221
Mast cells......................... 2, 8–12, 21–26, 31, 37–40, 46, 51, 60, 83, 93, 95, 180, 189, 190, 221, 223, 224, 226
McMaster slides 229, 236
MEDI-528... 13, 248, 249
Mice
  2D2... 207, 210, 211, 214
  2D2 X Foxp3-GFP 203, 214
  BALB/c 142, 157, 158, 191, 232
  C57BL/6 .... 22, 52, 61, 130, 157, 158, 170, 191, 203, 204, 210–212, 220, 227, 228, 230, 232
  Foxp3-GFP reporter...203
  OTII..190
  Rag-1-deficient 203, 211, 212
  transgenic 5, 23, 24, 39–41, 127, 141, 210, 217
Microfilariae antigen (Mf) 84
Miracidia 227, 230, 239
Mitomycin C... 62, 68, 213
Monoclonal antibodies (MAbs) 24, 75, 86, 89, 96, 98, 221, 248, 249
Mouse restrainer... 228, 230, 232
Mucus hyperproduction 189, 190
Mycobacterial Purified Protein Derivative (PPD) antigen...84

N

Nematode..223, 224
Neoplasia..24, 31
Nippostrongylus brasiliensis 10, 24, 27, 45, 60, 224, 225, 227–229, 232, 235
Nucleofector..158, 164

O

Ovalbumin (OVA)...................................12, 24, 42, 59, 127, 190–193, 195, 197, 217, 221

P

Parasite infections...167, 225
Peristaltic perfusion pump...231, 241
Permeabilization...52, 55, 57, 62, 89, 97, 105, 130, 138, 143, 149, 158, 165, 191, 195, 204, 250, 252, 253
Pertussis toxin...210
Phorbol myristate acetate (PMA).........................52, 55, 56, 63, 67, 68, 85, 88, 91, 98, 101, 105, 130, 133, 137, 143, 147, 152, 158, 161, 163, 164, 191, 194, 197, 204, 207, 209, 213, 249, 252
Polybrene..157, 161
Praziquantel...226
Proliferation assay...27–32
Promoters 3, 4, 6, 7, 21, 26, 40, 94, 112, 113, 117, 118, 156, 162, 173, 185, 202, 248

R

Real time PCR 27, 63, 65–67, 133, 136–137, 142, 147, 182
Red blood cell (RBC) lysis 205, 218
Reporter assay...165
Retrovirus 129, 158–161, 163

S

Schistosoma mansoni................................39, 40, 60, 225–227, 230–232, 238–239, 241–243
siRNA 4, 167–176
Soluble egg antigen (SEA) 227, 232, 242–243
Sonication.. 123, 174, 184
Sonicator 115, 145, 182, 186
Surface staining 78, 79, 89, 101–102, 105, 131, 133, 138, 161, 197, 253

T

T cell
  cytotoxic 9
  helper................................ 1–13, 22, 51, 56, 83, 93–106, 155, 156, 162, 181, 182, 190, 195, 247
  memory 7, 74–77, 249, 251
  polarization...207

receptor.. 45, 59–69, 94, 148
stimulation...160, 204
Tetramers...97, 100
Transactivation 112
TRANScription FACtor database
(TRANSFAC)..................................... 113, 115, 117
Transcription factors
BATF 4, 5, 7, 12, 51, 60, 74, 94, 95, 112, 129, 142, 156, 163, 202, 207, 247, 248, 250, 253
BCL6..6, 113
GATA-3...73, 94, 129, 181, 224
IRF4.................4–7, 11, 12, 44, 51, 60, 94, 95, 111, 112, 129, 156, 168, 173, 190, 202, 207, 208, 247, 250, 253
Maf...5
NF-κB.. 5, 37, 44, 142
PU.1... 3, 45, 112, 129
RBPJ...112
RORγt...26, 73
SMAD2/3 112
T-bet... 26, 73, 112
Transduction.....................................2, 37, 83, 129, 155–165
Transfection.............. 156, 157, 159–160, 164, 165, 167–176
Trematode 225

U

Ulcerative colitis (UC).............. 11, 44, 45, 95, 202, 203, 248

V

Vitamin D 248, 249, 252, 253

W

Western blotting (WB) 141–152, 171